ANALYSES FOR... PHYSICIANS

By the
Analytic Laboratories
of Merck & Co.
New York

Examinations of Water, Milk, Blood, Urine, Sputum, Pus, Food Products, Beverages, Drugs, Minerals, Coloring Matters, etc., for diagnostic, prophylactic, or other scientific purposes.

All analyses at these Laboratories are so conducted as to assure the best service attainable on the basis of the latest scientific developments. The laboratories are amply supplied with a perfect quality of reagent materials, and with the most efficient constructions of modern apparatus and instruments. The probable cost for some of the most frequently needed researches is approximately indicated below :

Sputum, for tuberculosis bacilli,	$3.00
Urine, for tuberculosis bacilli,	3.00
Milk, for tuberculosis bacilli,	3.00
Urine, qualitative, for one constituent,	1.50
Urine, qualitative, for each additional constituent,	1.00
Urine, quantitative, for each constituent,	3.00
Urine, sediment, microscopical,	1.50
Blood, for ratio of white to red corpuscles,	2.00
Blood, for Widal's typhoid reaction,	2.00
Water, for general fitness to drink,	10.00
Water, for typhoid germs,	25.00
Water, quantitative determination of any one constituent,	10.00
Pus, for gonococci,	3.00

The cost for other analyses—more variable in scope—can only be given upon closer knowledge of the requirements of individual cases.

All pharmacists in every part of the United States will receive and transmit orders for the MERCK ANALYTIC LABORATORIES.

*Physicians are earnestly requested to com-
municate to Merck & Co., University Place,
New York, any suggestions that may tend to
improve this book for its Second Edition,
which will soon be in course of preparation.*

*Whatever the Publishers can do to make
Merck's Manual of still greater service to the
Medical Profession will be gladly undertaken
and promptly performed for all subsequent
editions.*

*Therefore, any Physician who will propose
improvements in the subject-matter (especially
as regards the Newer Materia Medica), or
in the arrangement, style, and form of this
work, for future editions, will thus be rendering valuable service, not only to its Publishers,
but to the entire Profession as well!*

MERCK'S 1899 MANUAL

OF THE

MATERIA MEDICA

TOGETHER WITH A SUMMARY OF THERAPEUTIC INDICATIONS AND A
CLASSIFICATION OF MEDICAMENTS

A READY-REFERENCE POCKET BOOK

FOR THE

PRACTICING PHYSICIAN

CONTAINING

NAMES AND CHIEF SYNONYMS, PHYSICAL FORM AND APPEARANCE, SOLUBILI-
TIES, PERCENTAGE STRENGTHS AND PHYSIOLOGICAL EFFECTS, THERA-
PEUTIC USES, MODES OF ADMINISTRATION AND APPLICATION,
REGULAR AND MAXIMUM DOSAGE, INCOMPATIBLES,
ANTIDOTES, PRECAUTIONARY REQUIREMENTS,
ETC., ETC., — OF THE

CHEMICALS AND DRUGS USUAL IN MODERN MEDICAL PRACTICE

Compiled from the Most Recent Authoritative Sources and Published by

MERCK & CO., NEW YORK

This facsimile of the first edition of *The Merck Manual*, published as *Merck's 1899 Manual*, has been produced in celebration of the 100th anniversary of this not-for-profit publication. *The Merck Manual* has been continuously published longer than any other general textbook of medicine in the English language.

We hope that you enjoy revisiting medical history through this facsimile edition. It illustrates the enormous advances in medical knowledge and practice that have taken place over the past 100 years.

MERCK'S MANUAL is designed to meet a need which every general practitioner has often experienced. Memory is treacherous. It is particularly so with those who have much to do and more to think of. When the best remedy is wanted, to meet indications in cases that are a little out of the usual run, it is difficult, and sometimes impossible, to recall the whole array of available remedies so as to pick out the best. Strange to say, too, it is the most thoroughly informed man that is likely to suffer to the greatest extent in this way ; because of the very fact that his mind is overburdened. But a mere reminder is all he needs, to make him at once master of the situation and enable him to prescribe exactly what his judgment tells him is needed for the occasion.

In MERCK'S MANUAL the physician will find a complete Ready-Reference Book covering the entire eligible Materia Medica. A glance over it just before or just after seeing a patient will refresh his memory in a way that will facilitate his coming to a decision. In this book, small as it is, he will find the essential data found in the ponderous Dispensatories, together with the facts of newest record, which can appear only in future editions of those works.

Part I affords at a glance a descriptive survey, in one alphabetic series, of the entire Materia Medica to-day in general use by the American profession. Part II contains a summary of Therapeutic Indications for the employment of remedies, arranged according to the Pathologic Conditions to be combated. Part III presents a Classification of Medicaments in accordance with their Physiologic Actions.

The publishers may be allowed to state that they have labored long and earnestly, so to shape this little volume that it shall prove a firm and faithful help to the practitioner in his daily round of duty. They now send it forth in the confident hope that, the more it is put to the test of actual use, the more it will grow in the esteem of its possessor.

CONTENTS.

———

Part First.—THE MATERIA MEDICA, as in actual
use to-day by American Physicians. (Alpha-
betically arranged.)

THIS PART EMBRACES all those Simple Medicinal Substances
(that is, drugs and chemicals) which are in current and
well-established use in the medical practice of this country;
or which, if too recently introduced to be as yet in general
use, are vouched for by eminent authorities in medical
science;—also, the medicinally employed Pharmaceutic Preparations recognized by the United States Pharmacopœia.

(Added thereto, for the convenience of those practitioners who
prescribe them, are Medicamentous Mixtures advertised only to
the Profession, but whose composition or mode of manufacture
has not been made known with sufficient completeness or exactness
to satisfy all members of the Profession. In the selection the publishers have been guided solely by the recognition accorded the various preparations by the Profession, according to the best information obtained.)

There has also been included, under the title of "Foods
and Dietetic Preparations," a list of such preparations as are
frequently prescribed for infants' diet, or for the sick or con-
valescent.

OMITTED from the Materia Medica chapter are: Medicaments that have become obsolete, or that are too rarely used
to be of general interest; and such new remedies as are not yet
safely accredited on reliable authority; also those galenic
preparations (syrups, extracts, pills, essences, elixirs, wines,
emulsions, etc.) which are not standardized according to the
U. S. Pharmacopœia; likewise all articles that are put up
and advertised for self-medication by the lay public.

SEPARATE TITLES in the alphabetic series are accorded,
as a rule, to the botanical drugs and other pharmaceutical

mother-substances, to proximate principles (alkaloids, glucos-
ides, organic acids, etc.), and to chemical compounds (salts,
"synthetics," etc.); while the official galenic preparations,
solutions and dilutions, derived from them, are mostly men-
tioned under the titles of their respective mother-substances.
(Thus, for instance, "Dover's Powder" will be found under
"Opium," while "Morphine" is described under its own
title.)

> (*Smaller type* has been employed—in order to economize space—
> for botanic drugs, gums, and some others of the older drugs and
> preparations which are so long and well known that but little
> reference will need be made to them.)
>
> (Those substances of the Materia Medica which can be had of the
> MERCK brand are—for the convenience of prescribers—so designated).

Pages
83 to 184.

Part Second.—THERAPEUTIC INDICATIONS for the use of the Materia Medica and other agents. (Arranged alphabetically under the titles of the various Pathologic Conditions.)

THIS PART SUMMARIZES in brief form, the principal means
of treatment for each form of disease, as reported to be
in good use with practitioners at the present time. The
statements hereon are drawn from the standard works of the
leading modern writers on Therapeutics, and supplemented —
in the case of definite chemicals of more recent introduction
—by the reports of reputable clinical investigators.

Pages
185 to 192.

Part Third.—CLASSIFICATION OF MEDICA-MENTS according to their Physiologic Actions. (Arranged alphabetically under the titles of the Actions.)

THIS PART RECAPITULATES, for ready survey, such statements as are already given in "PART I," as to the modes of
action of the various medicaments.

INDEX.

---

*For Details, see Descriptive Table of Contents, on
pages 6 and 7.*

ABBREVIATIONS.

alm. = almost
amorph. = amorphous
arom. = aromatic
comp. = compound
cryst. = crystals or crystalline
D. = dose
decoct = decoction
dil. = dilute or diluted
emuls. = emulsion
ext. = extract
extern. = externally
F. E. or fl. ext. = fluid extract

fl. dr. = fluid dram
grn. = grain or grains
infus. = infusion
inject. = injection
insol. = insoluble
intern. = internally
lin. = liniment
liq. = liquid or liquor
Max. D. = maximum dose
min. or ℳ =minim or minims
odorl. = odorless
oint. = ointment

oz. = ounce or ounces
powd. = powder
q. v. = which see (*quod vide*)
sl. = slightly
sol. = soluble or solubility
solut. = solution
spt. = spirit
syr. = syrup
tastel. = tasteless
tr. = tincture
wh. = white
3 t., 4 t. = 3 times, 4 times

MERCK'S 1899 MANUAL.

PART FIRST.

THE MATERIA MEDICA,

As in Actual Use To-day by American Physicians.

Reader please note:—

The **GALENIC PREPARATIONS** of the United States Pharmacopœia, when not listed under their own titles, will be found under the titles of the drugs from which they are derived.

FOODS AND DIETETIC PRODUCTS proper will be found under the title : "Foods"; while Digestants, Hematinics, etc., are listed under their own titles.

SMALL TYPE is employed for botanic drugs, gums, and some others of the older drugs and preparations which are so well known as to require but little description.

Those articles of which the **MERCK** brand is on the market, are — for convenience in prescribing — designated accordingly.

Absinthin Merck.

ABSINTHIIN—Yellow-brown, amorph. or cryst. powd.; very bitter.—SOL. in alcohol, chloroform; slightly in ether; insol. in water.—Bitter Tonic (in anorexia, constipation, chlorosis, etc.).—**Dose :** 1½—4 grn.

Absinthium—U. S. P.

WORMWOOD.—Dose: 20–40 grn.—Infus. (1–2:64) and oil (D., 1–3 min.) used.

Acacia—U. S. P.

GUM ARABIC.—SOL. in water, insol. in alcohol.—*Preparations:* Mucilage; Syr.—both vehicles.

Acetanilid Merck.—U. S. P.—Cryst. or Powd.

ANTIFEBRIN.—Wh. scales or powd.; odorl.; burning taste.—SOL. in 194 parts water, 5 alcohol, 18 ether; very sol. in chloroform.—Antipyretic, Analgesic, Antirheumatic, Antiseptic.—USES : *Intern.*, fever, rheumatism, headache, alcoholism, delirium, neuralgia, sleeplessness in children, etc.; *extern.*, like iodoform, and as a preservative of hypodermic solutions (1:500).—**Dose :** 3—10 grn., in powd., alcoholic solut., or hot water cooled down and sweetened to taste.—MAX. D.: 15 grn. single, 60 grn. daily.—CAUTION: Avoid large doses in fever!

Acetyl-phenyl-hydrazine Merck.

HYDRACETIN; PYRODIN.—Prisms, or tablets; silky luster; odorl.; tastel. —SOL. in 50 parts water; in alcohol, chloroform.—Antipyretic, Antiparasitic.—USES : *Intern.*, to reduce fever generally, in rheumatism, etc.; *extern.*, psoriasis and other skin diseases.—**Dose :** ½—3 grn.— EXTERN. in 10% oint.

Acid, Acetic, Merck.—Glacial.—U. S. P.—99.5%.—C. P.

Caustic (in warts or corns) and Vesicant. Not used internally.—ANTIDOTES : Emetics, magnesia, chalk, soap, oil, etc.

Acid, Acetic, Merck.—U. S. P.—36%.

Dose : 15—40 ℳ, well diluted.

Acid, Acetic, Diluted—U. S. P.

6 per cent.—Dose: 2–4 drams.

Acid, Agaric, Merck.

AGARIC, LARICIC or AGARICINIC, ACID.—Powd.; odorl.; almost tastel.—SOL. in ether or chloroform; in 130 parts cold and 10 parts boiling alcohol.—Antihidrotic.—USES: Night-sweat of phthisis, and to check the sudorific effects of antipyretics.—**Dose:** $\frac{1}{8}$–$\frac{1}{2}$ grn., at night, in pills.

Acid, Arsenous, Merck.—U. S. P.—Pure, Powder.

White powd.; odorl.; tastel.—SOL. very slightly in water or alcohol.—Antiperiodic, Antiseptic, Alterative.—USES: *Intern.*, malarial fever, skin diseases, chorea, neuralgia, gastralgia, uterine disorders, diabetes, bronchitis; *extern.*, to remove warts, cancers, etc.—**Dose:** $\frac{1}{60}$–$\frac{1}{30}$ grn. 4 t. daily.—MAX. D.: About $\frac{1}{12}$ grn. single; about $\frac{1}{8}$ grn. daily.—*Preparation :* Solut. (1%).—EXTERN. on neoplasms in large amounts to get *quick results ;* otherwise it is dangerous. Keep from healthy tissues, lest dangerous absorption may occur.—ANTIDOTES: Emetics; stomach pump or siphon if seen immediately; hot milk and water with zinc sulphate or mustard. After vomiting, give milk or eggs, and magnesia in milk. If saccharated oxide of iron or dialyzed iron is handy, use it. If tincture of iron and ammonia water are within reach, precipitate former with latter, collect precipitate on a strainer, and give it wet. Always give antidotes, be the case ever so hopeless.—INCOMPATIBLES: Tannic acid, infusion cinchona, salts of iron, magnesium, etc.

Acid, Benzoic, from Benzoin,—Merck.—U. S. P.—Sublimed.

Pearly plates, or needles; aromatic odor and taste.—SOL. in 2 parts alcohol; 3 parts ether; 7 parts chloroform; 10 parts glycerin; 500 parts water. (Borax, or sod. phosphate, increases sol. in water.)—Antiseptic, Antipyretic, Expectorant.—USES: *Intern.*, to acidify phosphatic urine, reduce acidity of uric-acid urine, control urinary incontinence, also in chronic bronchitis and jaundice; *extern.*, wound dressing (1:100), in urticaria, etc.—**Dose:** 10–40 grn. 6 t. daily.—INCOMPATIBLES: Corrosive sublimate, lead acetate, etc.

Acid, Boric, Merck.—U. S. P.—C. P., Cryst. or Impalpable Powder.

Dose: 5–15 grn.

Acid, Camphoric, Merck.—C. P., Cryst.

Colorl. needles or scales; odorl.; feebly acid taste.—SOL. in alcohol, ether; very slightly in water; 50 parts fats or oils.—Antiseptic, Antiseptic, Astringent, Anticatarrhal.—USES: *Extern.*, 2–6% aqueous solut., with 11% of alcohol to each 1% of acid, in acute skin diseases, as gargle or spray in acute and chronic affections of respiratory tract; *intern.*, night-sweats, chronic bronchitis, pneumonia, gonorrhea, angina, chronic cystitis, etc.—**Dose:** 8–30 grn., in powd.—MAX. D.: 60 grn.

Acid, Carbolic, Merck.—Absolute, C. P., Loose Crystals or Fused.—U. S. P.

PHENOL.—**Dose:** $\frac{1}{2}$–2 grn., well diluted or in pills.—*Preparations :* Glycerite (20%); Oint. (5%).—ANTIDOTES: Soluble alkaline sulphates after emesis with zinc sulphate; raw white of egg; calcium saccharate; stimulants hypodermically.—INCOMPATIBLES : Chloral hydrate, ferrous sulphate.

MERCK'S "Silver Label" Carbolic Acid is guaranteed not to redden under the proper precautions of keeping.

Acid, Carbolic, Iodized, Merck.—N. F.

IODIZED PHENOL.—Solut. of iodine in carbolic acid.—Antiseptic, Escharotic.—USES: Uterine dilatation.—APPLIED pure, by injection.

Acid, Caryophyllic,—see EUGENOL.

Acid, Cetraric,—see CETRARIN.

Acid, Chloracetic, caustic, Merck.

Mixture of chlorinated acetic acids.—Colorl. liq.—USES: Escharotic.

Acid, Chromic, Merck.—Highly Pure, Cryst.; also Fused, in Pencils.

INCOMPATIBLES: Alcohol, ether, glycerin, spirit of nitrous ether, arsenous acid, and nearly every organic substance.—CAUTION: Dangerous accidents may occur by contact with organic substances. Avoid cork stoppers!

Acid, Chrysophanic, medicinal,—so-called,—see CHRYSAROBIN.

Acid, Cinnamic, Merck.—C. P.

CINNAMYLIC ACID.—White scales; odorl.—SOL. in alcohol, ether; very slightly in water.—Antitubercular, Antiseptic.—USES: Tuberculosis and lupus, parenchymatously and intravenously.—APPLIED in 5% emulsion or alcoholic solut.—INJECTION (intravenously): ¼—¾ grn., in 5% oily emulsion, with 0.7% solut. sodium chloride, twice a week.

Acid, Citric, Merck.—C. P., Cryst. or Powd.

SOL. in water, alcohol.—Antiseptic, Antiscorbutic, and Refrigerant.—USES: *Extern.*, post-partum hemorrhage; pruritus; agreeable application in diphtheria, angina or gangrenous sore mouth; *intern.*, cooling beverage to assuage fever, and remedy in scurvy.—**Dose**: 10—30 grn.—*Preparation:* Syr. (1%).—EXTERN., for painting throat, 5—10% solut. in glycerin; gargle, 1—2%; cooling drink, 80 grn. to 1 quart.

Acid, Cresotic, Para-, Merck.—Pure.

White needles.—SOL. in alcohol, ether, chloroform.—Children's Antipyretic, Intestinal Antiseptic.—USES: Febrile affections, gastro-intestinal catarrh. Mostly used as Sodium paracresotate (which see).—**Dose** (acid): *Antipyretic*, 2—20 grn., according to age; *intestinal antiseptic* (children's diseases), ¼—1 grn., in mixture.—MAX. D.: 60 grn.

Acid, Dichlor-acetic, Merck.—Pure.

Colorl. liq.—SOL. in water, alcohol.—Caustic.—USES: Venereal and skin diseases.

Acid, Filicic, Merck.—Amorph.

FILICINIC ACID.—Amorph., sticky powd.; odorl.; tastel.—Anthelmintic.—**Dose**: 8—15 grn.

Acid, Gallic, Merck.—U. S. P.—Pure, White Cryst.

Dose: 5—20 grn.—INCOMPATIBLES: Ferric salts.

Acid, Hydrobromic, Merck.—Diluted.—U. S. P.—10%.

Dose: 30—90 ℳ, in sweet water.

Acid, Hydrochloric—U. S. P.

31.9 per cent. HCl.—**Dose**: 5–10 minims, well diluted.—ANTIDOTES : Chalk, whiting, magnesia, alkali carbonates, and albumen.— INCOMPATIBLES : Alkalies, silver salts, chlorates, salts of lead, etc.

Acid, Hydrochloric, Diluted—U. S. P.

10 per cent. HCl.—**Dose**: 10–30 minims, in sweet water.

Acid, Hydrocyanic, Diluted—U. S. P.

2 per cent. HCN—**Dose**: 2–5 min.—MAX. DOSE: 10 min.—EXTERN. 1: 8–16 as lotion. only on unbroken skin.—CAUTION: Very liable to decomposition. When brown in color it is unfit for use.

Acid, Hydro-iodic, Merck.—Sp. Gr. 1.5.—47%.

Deep-brown, fuming liq.—Antirheumatic, Alterative.—USES:Rheumatism, bronchitis (acute or chronic), asthma. syphilis, obesity, psoriasis, to eliminate mercury or arsenic from the system, etc.—**Dose:** 5—10 ℔, in much sweet water.

Acid, Hypophosphorous, Merck.—Diluted.—10%.

Dose: 10—60 ℔.

Acid, Lactic, Merck.—U. S. P.—C. P.

Caustic.—APPLIED as 50—80% paint.
MERCK'S Lactic Acid is perfectly colorless and odorless.

Acid, Laricic,—see ACID, AGARICIC.

Acid, Monochlor-acetic, Merck.

Very deliquescent cryst.—SOL. in water.—Caustic.—USES: Warts, corns, etc.—APPLIED in concentrated solut.

Acid Nitric.—U. S. P.

68 per cent. HNO₃.—APPLIED (as an Escharotic) pure.—ANTIDOTES : Same as of hydrochloric acid.

Acid Nitric, Diluted—U. S. P.

10 per cent. HNO₃.—Dose: 5-30 minims, well diluted.

Acid, Nitro-hydrochloric, Diluted—U. S. P.

One-fifth strength of concentrated, which is not used therapeutically.—USES: *Intern.,* jaundice, biliary calculi, dyspepsia, chronic rheumatism. etc.; *extern.,* diluted, as sponge- or foot-bath, 2 or 3 t. a week.—**Dose:** 5-20 minims, well diluted.—ANTIDOTES and INCOM- PATIBLES: Same as of hydrochloric acid.

Acid, Osmic,—see ACID, PEROSMIC.

Acid, Oxalic, Merck.—C. P., Cryst.

Transparent cryst.; very acid taste.—SOL. in water, alcohol; slightly in ether.—Emmenagogue, Sedative.—USES: Functional amenorrhea, acute cystitis.—**Dose:** ⅙—1 grn. every 4 hours, in sweet water.—ANTIDOTES: Calcium saccharate, chalk, lime-water, magnesia.—INCOMPATIBLES: Iron and its salts, calcium salts, alkalies.

Acid, Oxy-naphtoic, Alpha-, Merck.—Pure.

White or yellowish powd.; odorl.; sternutatory.—SOL. in alcohol, chloro- form, fixed oils, aqueous solut's of alkalies and alkali carbonates; insol. in water.—Antiparasitic, Antizymotic.—USES: *Intern.,* disin- fectant intestinal tract (reported 5 times as powerful as salicylic acid); *extern.,* in parasitic skin diseases (in 10% oint.), coryza, etc.—**Dose:** 1¼–3 grn.

Acid, Perosmic, Merck.

OSMIC ACID.—Yellowish needles; very pungent, disagreeable odor.—SOL. in water, alcohol, ether.—Antineuralgic, Discutient, Anti-epileptic.— USES: *Intern.,* muscular rheumatism, neuralgia; *extern.,* remove tumors, and in sciatica (by injection).—**Dose:** ₆₄⁄ grn., several t. daily.—INJEC- TION: ₇₀⁄—⅛ grn. as 1% solut. in aqueous glycerin (40%) —ANTIDOTE: Sulphu- retted hydrogen.—INCOMPATIBLES: Organic substances. phosphorus, fer- rous sulphate, etc.—CAUTION: Vapor exceedingly irritating to the air- passages.

Acid, Phosphoric, (Ortho-), Merck.—Syrupy.—85%.

Dose: 2–6 ℔, well diluted.

do. Merck.—Diluted.—10%.

Dose: 20—60 ℔.—INCOMPATIBLES; Ferric chloride, lead acetate, etc.

Acid, Picric, Merck.—C. P., Cryst.

PICRONITRIC, PICRINIC, or CARBAZOTIC, ACID.—Yellow cryst.; odorl.; intensely bitter.—SOL. in alcohol, ether, chloroform; sl. in water.—Antiperiodic, Antiseptic, Astringent.—USES: *Intern.*, in malaria, trichiniasis, etc.; *extern.*, in erysipelas, eczema, burns, fissured nipples, etc.; $\frac{1}{10}$% solut. for cracked nipples, $\frac{1}{6}$—1% hydro-alcoholic solut. on compress renewed only every 3 to 7 days in burns.—**Dose:** $\frac{1}{2}$—2 grn., in alcoholic solut.—MAX. D.: 5 grn.—ANTIDOTE: Albumen.—INCOMPATIBLES: All oxidizable substances. Dangerously explosive with sulphur, phosphorus, etc.—CAUTION: Do not apply in substance or in oint., lest toxic symptoms appear!

Acid, Pyrogallic, Merck, (Pyrogallol, U. S. P.)—Resublimed.

Used only *extern.*, in 5—10% oint.

Acid, Salicylic, Merck.—U. S. P.—C. P., Cryst.; and Natural (from Oil Wintergreen).

Dose: 10—40 grn.

Acid, Sozolic,—see ASEPTOL.

Acid, Sulpho-anilic, Merck.—Cryst.

White efflorescent needles.—SOL. slightly in alcohol; 112 parts water.—Anticatarrhal, Analgesic.—USES: *Intern.*, coryza, catarrhal laryngitis, etc.—**Dose:** 10—20 grn. 1 to 2 t. daily, in aqueous sol. with sodium bicarb.

Acid, Sulpho-salicylic, Merck.

White cryst.—SOL. in water, alcohol.—USES: Delicate urine-albumin test.

Acid, Sulphuric, Aromatic—U. S. P.

20 per cent. H_2SO_4.—Best form for administration.—**Dose:** 10-20 min.

Acid, Sulphuric, Diluted—U. S. P.

10 per cent. H_2SO_4.—(Concentrated Sulphuric acid is not used medicinally.)—USES: *Intern.*, gastro-intest. disorders, phthisical sweats, exophthalmic goiter, etc.; also as solvent for quin. sulph., etc.—**Dose:** 15-30 min., well diluted.—ANTIDOTES: Same as of hydrochloric acid.

Acid, Sulphurous—U. S. P.

6.4 per cent. SO_2.—Antiseptic, Antizymotic.—**Dose:** 15-60 min., well diluted. EXTERN. 10-25 per cent. solut.

Acid, Tannic, Merck.—U. S. P.—C. P., Light.

Dose: 2—20 grn.—*Preparations:* Styptic Collodion (20%); Glycerite (20%); Oint. (20%); Troches (1 grn.).—INCOMPATIBLES: Ferrous and ferric salts, antimony and potassium tartrate, lime water, alkaloids, albumen, gelatin, starch.

Acid, Tartaric, Merck.—U. S. P.—C. P., Cryst. or Powd.

Dose: 10—30 grn.

Acid, Trichlor-acetic, Merck.—Pure, Cryst.

Deliquescent cryst.; pungent, suffocating odor; caustic. SOL. freely in water, alcohol, ether.— Escharotic, Astringent, Hemostatic. — USES: Venereal and cutaneous warts, papillomata, vascular nævi, pigment patches, corns, nosebleed, obstinate gleet, gonorrhea, nasopharyngeal affections and indolent ulcers.—APPLIED: As *escharotic*, pure, or in conconcentrated solut.; *astringent* and *hemostatic*, 1—3% solut.—CAUTION: Keep in glass-stoppered bottle.

Acid, Valerianic, Merck.

Oily liq., strong valerian odor; bitter, burning taste.—SOL. in water, alcohol, ether, chloroform.—Antispasmodic, Sedative.—USES: Nervous affections, hysteria, mania, etc.—**Dose:** 2—10 drops, in sweetened water.

Aconite Root—U. S. P.

Preparations: Ext. (D., ¼–½ grn.), F. E. (D., ¼–1 min.); Tr. (q.v.).—See also, Aconitine.

Aconitine, Potent, Merck.—Cryst.—(*Do not confound with the " Mild"!*)

Alkaloid from Aconite, prepared according to process of Duquesnel.—White cryst.; feebly bitter taste.—Sol. in alcohol, ether, chloroform; insol. in water.—Anti-neuralgic, Diuretic, Sudorific, Anodyne.—Uses: *Intern.*, neuralgia, acute or chronic rheumatism, gout, toothache, etc.; *extern.*, rheumatism, other pains.—**Dose:** $1/640$—$1/200$ grn. several t. daily, in pill or solut., with caution.—Max. D.: $1/64$ grn. single; $1/20$ grn. daily.—Extern.: 1 : 2000–500 parts lard.—Antidotes: Small repeated doses of stimulants; artificial respiration, atropine, digitalis, ammonia. —Caution: Never use on abraded surfaces. Danger of absorption! 10 times as toxic as the mild amorph. aconitine (below)!

Aconitine, Mild, Merck.—Amorph.—(*Do not confound with the "Potent"!*)

Uses: As aconitine, potent, cryst.; but only $1/10$ as powerful.—**Dose:** $1/64$—$1/20$ grn., very carefully increased.—Extern. ½–2% oint. or solut.

Aconitine Nitrate Merck.—Cryst.

Dose: Same as of aconitine, potent, cryst.

Adeps Lanæ Hydrosus Merck.—U. S. P.

Hydrous Wool-fat.—Yellowish-white, unctuous mass. Contains about 25% water. Freely takes up water and aqueous solut's.—Non-irritant, permanent emollient, and base for ointments and creams; succedaneum for lanolin in all its uses.

Adeps Lanæ Anhydricus.

(Anhydrous Wool-fat).—Contains less than 1% of water.

Adonidin Merck.

Adonin.—Yellowish-brown, very hygroscopic, odorl. powd.; intensely bitter.—Sol. in water, alcohol; insol. in ether, chloroform.—Cardiac Stimulant, mild Diuretic.—Uses: Heart diseases, especially mitral and aortic regurgitation, and relieving precordial pain and dyspnea.—**Dose:** $1/16$—¼ grn. 4 t. daily, in pill, or solut. in chloroform water with ammonium carbonate. —Max. D.: 1 grn.

Agaricin Merck.

White powd.; sweet, with bitter after-taste.—Sol. in alcohol; slightly in water, ether, or chloroform.—Antihidrotic.—Uses : Phthisical night-sweats, sweating from drugs.—**Dose:** ¼–1 grn.

Airol.

Bismuth Oxyiodogallate, *Roche.*—Grayish-green, odorl., tastel. powd. —Insol. in water, alcohol, etc.—Surgical Antiseptic, like iodoform; also Antigonorrhoic and Intestinal Astringent.—**Dose:** 2–5 grn. 3 t. daily.— Extern. pure, 10% emuls. in equal parts glycerin and water, or 10—20% oint.

Alantol Merck.

Amber liq.; odor and taste like peppermint.—Sol. in alcohol, chloroform, ether.—Internal Antiseptic, Anticatarrhal.—Uses: Instead of turpentine, in pulmonary affections.—**Dose:** ⅛ ℳ, 10 t. daily, in pill, powd., or alcoholic solut.

Alcohol—U. S. P.

91 per cent.—Sp. Gr. 0.820.

Aletris Cordial.

Not completely defined.—(Stated: "Prepared from Aletris farinosa [or True Unicorn], combined with aromatics.—Uterine Tonic and Restorative.—**Dose:** 1 fl. dr. 3 or 4 t. daily.")

Allyl Tribromide Merck.

Yellow liq.; cryst. mass in cold.—SOL. in alcohol, ether.—Sedative, Antispasmodic.—USES: Hysteria, asthma, whooping cough, etc.—**Dose:** 5—10 ♏, 2 or 3 t. daily, in capsules.—INJECTION: 2 or 3 drops, in 20 drops ether.

Almond, Bitter—U. S. P.

Preparations: Oil (D., one-sixth to ½ min.); Spt. (1 per cent. oil); Water (q. v.).

Almond, Sweet—U. S. P.

Preparations: Emuls. (as vehicle); Oil (D., 2–8 drams); Syr. (as vehicle).

Aloes, Barbadoes—U. S. P.

Dose: 2–20 grn.

Aloes, Purified—U. S. P.

From Socotrine Aloes.—**Dose:** 1–10 grn.—*Preparations:* Pills (2 grn.); Pills Aloes and Asafetida; Pills Aloes and Iron; Pills Aloes and Mastic; Pills Aloes and Myrrh; Tr. (1:10); Tr. Aloes and Myrrh.

Aloes, Socotrine—U. S. P.

Dose: 2–10 grn.—*Preparation:* Ext. (D., 1–5 grn.).

Aloin Merck.—U. S. P.—C. P.

BARBALOIN.—**Dose:** ⅙—2 grn.—MAX. D.: 4 grn. single, 10 grn. daily.—INJECTION: ¾ grn. dissolved in formamide.

MERCK'S Aloin, C. P., is *clearly soluble*, and meets all other requirements of U. S. P.

Althea—U. S. P.

MARSHMALLOW.—*Preparation:* Syr. (1:20), as vehicle.

Alums:—Ammonium; Ammonio-ferric; Potassium,—see ALUM-

INIUM AND AMMONIUM SULPHATE; IRON AND AMMONIUM SULPHATE, FERRIC; AND ALUMINIUM AND POTASSIUM SULPHATE.

Aluminium Acetate Merck.—Basic.

Gummy mass or granular powd.—Insol. in water.—USES: *Intern.*, diarrhea and dysentery; *extern.*, washing foul wounds.—**Dose:** 5—10 grn. 3 t. daily.

Aluminium Aceto-tartrate Merck.

Lustrous, yellowish granules; sour-astringent taste.—SOL. freely but very slowly in water; insol. in alcohol, ether, glycerin.—Energetic Disinfectant and Astringent.—USES: Chiefly in diseases of the air-passages.—APPLIED in ½ to 2% solutions: or as snuff, with ½ its weight of powdered boric acid; 50% solut. for chilblains.

Aluminium Sulphate Merck.—U. S. P.—Pure.

White lumps or powd.; odorl.; sweet-astringent taste.—SOL. in 1.2 parts water.—External Antiseptic, Caustic, Astringent.—USES: Fetid ulcers, fetid discharges; enlarged tonsils, scrofulous and cancerous ulcers; endometritis; nasal polypi, etc.—APPLIED in 1: 20 to 1: 100 solut., or concentrated solut.

Aluminium & Potassium Sulphate Merck.—(*Alum, U. S. P.*)—

C. P. Cryst. or Powd.; Pure, Burnt; and in Pencils (Plain or Mounted).

Dose: 5—15 grn.; *emetic*, 1—2 teaspoonfuls.

Aluminum, etc.,—see ALUMINIUM, ETC.

Ammonia Water—U. S. P.

10 per cent. NH₃.—Dose: 10–30 min.—*Preparations:* Lin. (3 per cent. NH3); Arom. Spt. (0.9 per cent. NH₃).

Ammonia Water, Stronger—U. S. P.

28 per cent. NH₃.—Dose: 4–10 min., well diluted.—ANTIDOTES: Acetic, tartaric, dil. hydrochloric acids, after vomiting.—INCOMPATIBLES: Strong mineral acids, iodine, chlorine water, alkaloids.—*Preparation:* Spt. (10 per cent. NH₃).

Ammoniac—U. S. P.

GUM or RESIN AMMONIAC.—Dose: 5–15 grn.—*Preparations:* Emuls. (4 per cent.); Plaster (with mercury).

Ammonium Arsenate Merck.

White, efflorescent cryst.—SOL. in water.—Alterative.—USES: Chiefly in skin diseases.—**Dose :** ½ grn., gradually increased, 3 t. daily in water.

Ammonium Benzoate Merck.—U. S. P.

Dose: 10—30 grn. 3 or 4 t. daily, in syrup or water.

Ammonium Bicarbonate Merck.—Pure, Cryst.

SOL. in water, alcohol.—Antacid, Stimulant.—USES: Acid fermentation of stomach; stimulant depressed condition.—**Dose :** 5—15 grn.

Ammonium Bromide.—U. S. P.

Dose: 15—30 grn.

Ammonium Carbonate Merck.—U. S. P.—C. P.

Dose: 5—20 grn.

Ammonium Chloride Merck.—U. S. P.—Pure, Granul.

Dose: 5—20 grn.—*Preparation:* Troches (1½ grn.).

Ammonium Embelate Merck.

Red, tastel. powd.—SOL. in diluted alcohol.—USES: Tape-worm.—**Dose:** Children, 3 grn.; adults, 6 grn., in syrup or honey, or in wafers, on empty stomach, and followed by castor oil.

Ammonium Fluoride Merck.—C. P.

Very deliquescent, colorl. cryst.; strong saline taste.—SOL. in water.; slightly in alcohol.—Antiperiodic, Alterative.—USES: Hypertrophy of spleen and in goitre.—**Dose :** 5—20 ℳ of a solut. containing 4 grn. to 1 ounce water.—CAUTION: Keep in gutta-percha bottles!

Ammonium Hypophosphite Merck.

White cryst.—SOL. in water.—USES: Phthisis, and diseases with loss of nerve power.—**Dose:** 10—30 grn., 3 t. daily.

Ammonium Ichthyol-sulphonate,—see ICHTHYOL.

Ammonium Iodide—U. S. P.

Deliquescent, unstable powd.—Alterative, Resolvent.—Dose: 3–10 grn.

Ammonium Phosphate, Dibasic, Merck.—C. P.

Colorl. prisms; odorl.; cooling, saline taste.—SOL. in 4 parts water.— USES: Rheumatism, gout.—**Dose:** 5—20 grn., 3 or 4 t. daily, in water.

Ammonium Picrate Merck.

AMMONIUM PICRONITRATE or CARBAZOTATE.—Bright-yellow scales or prisms.—SOL. in water.—Antipyretic, Antiperiodic.—USES: Malarial neuralgia, periodic fevers, and headache.—**Dose:** ¼—1½ grn., 3 t. daily, in pills.

Ammonium Salicylate Merck.

Colorl. prisms.—Sol. in water.—Antirheumatic, Antipyretic, Germicide, Expectorant.—Uses : In febrile conditions, bronchitis, etc.—**Dose :** 2—10 grn., in wafers.

Ammonium Sulpho-ichthyolate,—see Ichthyol.

Ammonium Tartrate Merck.—Neutral, Cryst.

Colorl.—Sol. in water.—Expectorant.—**Dose :** 5—30 grn.

Ammonium Valerianate Merck.—White, Cryst.

Dose : 2—8 grn.

Ammonium & Iron Tartrate Merck.—U. S. P

Dose : 10—30 grn.

Ammonium Double-Salts,—see under Bismuth, Iron, Potass-

ium, Sodium, etc.

Ammonol.

Not completely defined.—(Stated to be "Ammoniated Phenylaceta-mide.—Yellowish alkaline powd.; ammoniacal taste and odor.—Anti-pyretic, Analgesic.—**Dose :** 5—20 grn., 3—6 t. daily, in caps., tabl., or wafers.")

Amyl Nitrite Merck.—U. S. P.—Pure, or in Pearls (1-3 drops).

Caution: Amyl Nitrite is so very volatile that it is practically impossible to so stopper bottles that they will carry it without loss, especially in warm weather. Shipped in cool weather and kept in a cool place, the loss is not material, but if kept in a warm place, or if agitated much, so as to keep up any pressure of the vapor within the bottle, the loss will be considerable, proportionately to the pressure.—**Dose :** 2—5 drops, in brandy.

Amylene Hydrate Merck.

Colorl., oily liq.: ethereal, camphoric taste.—Sol. in 8 parts water; all pro-portions of alcohol, ether, chloroform, benzene, glycerin.—Hypnotic, Seda-tive.—Uses : Insomnia, alcoholic excitement, epilepsy, whooping-cough, etc.—**Dose :** Hypnotic, 45—90 ℳ; sedative, 15—30 ℳ; in beer, wine, brandy, syrup, etc., or in capsules.

Anemonin Merck.

Colorl., odorl., neutral needles.—Sol. in hot alcohol, chloroform; insol. in water.—Antispasmodic, Sedative, Anodyne.—Uses : Asthma, bronchitis, whooping-cough, dysmenorrhea, orchitis and oöphoritis and other painful affections of female pelvis.—**Dose :** ¼—¾ grn., 2 t. daily.—Max. D.: 1½ grn. single, 3 grn. daily.

Anise—U. S. P.

Preparations: Oil (D., 5–10 min.); Spt. (10 per cent. oil); Water (one-fifth per cent. oil).

Anthrarobin Merck.

Yellowish-white powd.—Sol. in weak alkaline solut.; slightly in chloro-form and ether; in 10 parts alcohol.—Deoxidizer, Antiseptic.—Uses : *Ex-tern.*, instead of chrysarobin in skin diseases, especially psoriasis, tinea tonsurans, pityriasis versicolor, and herpes.—Applied in 10 to 20% oint. or alcoholic solut.

Antifebrin,—see Acetanilid.

Antikamnia.

Not completely defined.—(Stated : "Coal-tar derivative. — Wh., odorl. powd.—Antipyretic, Analgesic.—**Dose**: 5—15 grn., in powd. or tabl.")

Antimony Oxide, Antimonous, Merck.

Expectorant.—**Dose**: 1—3 grn.—*Preparation :* Antimonial Powder (33%).

Antimony Sulphide, Black, Merck.—(*Purified Antimony Sulphide, U. S. P.*).

Diaphoretic, Alterative.—**Dose**: 10—30 grn.

Antimony Sulphide, Golden, Merck.—C. P.

Alterative, Diaphoretic, Emetic, Expectorant.—**Dose**: $\frac{1}{8}$—1½ grn.— INCOMPATIBLES: Sour food, acid syrups, metallic salts.

Antimony, Sulphurated, Merck.

KERMES MINERAL.—Alterative, Diaphoretic, Emetic.—USES: Cutaneous diseases and syphilis; alterative generally.—**Dose**: 1—2 grn. in pill; as emetic, 5—20 grn.—*Preparation :* Pills Antimony Compound (0.6 grn.).

Antimony & Potassium Tartrate Merck.—U. S. P.—Pure, Cryst. or Powd.

TARTAR EMETIC.—**Dose**: *alter.,* $\frac{1}{32}$—$\frac{1}{8}$ grn.; *diaphor.* and *expect.,* $\frac{1}{12}$ —$\frac{1}{3}$ grn.; *emetic,* ⅓ grn. every 20 minutes.—*Preparation:* Wine Antimony (0.4%).—ANTIDOTES (as for antimonial compounds in general): Tannic acid in solut., freely; stimulants and demulcents.

Antinosine.

Sodium salt of nosophen.—Greenish-blue powd., of faint iodine odor.— SOL. in water.—Antiseptic.—USES: Chiefly in vesical catarrh.—EXTERN. in $\frac{1}{10}$—½ per cent. solut.

Antipyrine.

PHENYL-DIMETHYL-PYRAZOLONE.—SOL. in 1 part of water, 2 alcohol.— **Dose**: 10—20 grn.—APPLIED (as Styptic) in 20 per cent. solut. or pure.— INCOMPATIBLES: Acids, alkalies, cinchona preparations, copper sulphate, spirit nitrous ether, syrup ferrous iodide; also tinctures of catechu, ferric chloride, iodine, kino, and rhubarb.

Antispasmin.

NARCEINE-SODIUM and SODIUM SALICYLATE, *Merck.*—Reddish, slightly hygroscopic powd.; 50% narceine.—SOL. in water.—Antispasmodic, Sedative, and Hypnotic.—USES : Whooping-cough, laryngitis stridula, irritating coughs, etc.—**Dose** (5% solut., 3—4 t. daily): under ½ year 3—5 drops, ½ year 5—8 drops, 1 year 8—10 drops, 2 years 10—12 drops, 3 years 15—20 drops, older children 20—40 drops.—CAUTION: Keep from air!

Antitoxin, Diphtheria.

From serum of blood that has been subjected to poison of diphtheria.— Limpid liq., generally preserved with ½% carbolic acid or other preservative.—**Dose** (children): *Prophylactic,* 200—250 antitoxic units: *ordinary* cases, 600—1000 units ; *severe* cases (or those seen late, or of nasal or laryngeal type), 1500—3000 units; given hypodermically, and repeated in about 8 hours if necessary. Adults receive twice as much. CAUTION: The various brands differ in strength.

Apiol, Green, Merck.—Fluid.

Greenish, oily liq.—SOL. in alcohol, ether.—Emmenagogue, Antiperiodic. —USES: Dysmenorrhea, malaria.—**Dose**: 5—10 ℳ, 2 or 3 t. daily, in capsules; in malaria 15—30 ℳ.

Apioline.

Not completely defined.—(Stated: "True active principle of parsley, in 4-min. capsules.—Emmenagogue.—**Dose:** 2 or 3 caps., with meals.")

Apocodeine Hydrochlorate Merck.

Yellow-gray, very hygroscopic powd.—SOL. in water.—Expectorant, Sedative, Hypnotic.—USES: Chronic bronchitis, and other bronchial affections. Acts like codeine, but weaker; induces large secretion of saliva, and accelerates peristalsis.—**Dose:** 3–4 grn. daily, in pills.—INJECTION: $\frac{1}{8}$–$\frac{1}{2}$ grn., in 2% aqueous solut.

Apocynum—U. S. P.

CANADIAN HEMP.—Diuretic.—Dose: 5–20 grn.—*Preparation:* F. E. (1:1).

Apomorphine Hydrochlorate Merck.—U. S. P.—Cryst. or Amorphous.

Dose: *Expect.*, $\frac{1}{60}$–$\frac{1}{20}$ grn ; *emetic,* $\frac{1}{15}$–$\frac{1}{8}$ grn.—INJECT. (emetic): $\frac{1}{10}$–$\frac{1}{8}$ grn.—ANTIDOTES: Strychnine, chloral, chloroform.—INCOMPATIBLES: Alkalies, potassium iodide, ferric chloride.—CAUTION: Keep dark and well-stoppered!

Aqua Levico, Fortis and Mitis.

NATURAL ARSENO-FERRO-CUPRIC WATERS, from springs at Levico, Tyrol. —ALTERANT TONIC.—USES: Anemic, chlorotic, neurasthenic, and neurotic conditions; in scrofulous, malarial, and other cachexias; and in various chronic dermatoses.—**Dose:** Tablespoonful of Aqua Levico Mitis, diluted, after meals, morning and night. After a few days, increase dose gradually, up to 3 tablespoonfuls. After one or two weeks, substitute for the two doses a single daily dose of one tablespoonful of Aqua Levico Fortis, best with principal meal. Some days later, augment this dose gradually as before. Constitutional effects and idiosyncrasies are to be watched, and dosage modified accordingly. Decreasing dosage at conclusion of treatment, with a return to the "Mitis," is usual.

Arbutin Merck.

White needles; bitter.—SOL. in alcohol; slightly in water.—Diuretic.—USES: Instead of uva-ursi.—**Dose:** 5–15 grn. 4 t. daily.

Arecoline Hydrobromate Merck.

White cryst.—SOL. in water, alcohol.—Myotic.—APPLIED in 1% solut.

Argentamine.

8% solut. silver phosphate in 15% solut. ethylene-diamine.—Alkaline liq., turning yellow on exposure.—Antiseptic and Astringent, like silver nitrate.—USES: Chiefly gonorrhea.—INJECT. in 1:4000 solut.

Argonin.

Silver-casein compound; 4.25 per cent. silver.—Wh. powd.—SOL. in hot water; ammonia increases solubility.—Antiseptic.—USES: Chiefly in gonorrhea, in 1—2 per cent. solut.

Aristol.

DITHYMOL DI-IODIDE.—Reddish-brown, tastel. powd.; 46% iodine.—SOL. in chloroform, ether, fatty oils; sparingly in alcohol; insoluble in water or glycerin.—Succedaneum for iodoform externally.—APPLIED like the latter.—INCOMPATIBLES: Ammonia, corrosive sublim., metallic oxides, starch, alkalies or their carbonates; also heat.—CAUTION: Keep from light !

Arnica Flowers—U. S. P.

Preparation: Tr. (D., 10–30 min.).

Arnica Root—U. S. P.

Preparation: Ext. (D. 1–2 grn.); F. E. (D., 5–10 min.); Tr. (D., 20–40 min.).

Arsenauro.

Not completely defined.—(Stated: "10 min. contain $\frac{1}{32}$ grn. each gold and arsenic bromides.—Alterative Tonic.—**Dose:** 5–15 min., in water, after meals.")

Arsen-hemol Merck.

Hemol with 1% arsenous acid.—Brown powd.—Alterative and Hematinic; substitute for arsenic, without untoward action on stomach.—**Dose:** 1½ grn., in pill, 2 to 5 t. daily, adding one pill to the daily dose every fourth day until 10 pills are taken per day.

Arsenic Bromide Merck.

Colorless, deliquescent prisms; strong arsenic odor.—SOL. in water.—USES: Diabetes.—**Dose:** $\frac{1}{60}$—$\frac{1}{15}$ grn.—MAX. D.: $\frac{1}{8}$ grn.—ANTIDOTES: Same as arsenous acid.—INCOMPATIBLE: Water.—CAUTION: Keep well-stoppered!

Arsenic Chloride Merck.

Colorless, oily liq.—Decomposes with water.—SOL. in alcohol, ether, oils. —**Dose:** $\frac{1}{60}$—$\frac{1}{15}$ grn.

Arsenic Iodide Merck.—U. S. P.—Pure, Cryst.

Dose: $\frac{1}{60}$—$\frac{1}{15}$ grn., in pills.—MAX. D.: ⅛ grn. –INCOMPATIBLE: Water. —CAUTION: Keep from air and light!

Asafetida—U. S. P.

Dose: 5–15 grn.—*Preparations:* Emuls. (4 per cent.); Pills (3 grn.); Tr. (1:5).

Asaprol Merck.

CALCIUM BETA-NAPHTOL-ALPHA-MONO-SULPHONATE.—Whitish to reddish-gray powd.; slightly bitter, then sweet, taste.—SOL. in water; 3 parts alcohol.—Analgesic, Antiseptic, Antirheumatic, Antipyretic.—USES: Tuberculosis, rheumatism, pharyngitis, gout, typhoid fever, sciatica, diphtheria, etc.—**Dose:** 8–15 grn.—EXTERN. in 2–5% solut.—INCOMPATIBLES: Antipyrine and quinine.—CAUTION: Keep from heat and moisture!

Asclepias—U. S. P.

PLEURISY ROOT.—*Preparation:* F. E. (D. 20–60 min.).

Aseptol Merck.

SOZOLIC ACID.—33⅓% solut. ortho-phenol-sulphonic acid.—Yellow-brown liq.; odor carbolic acid.—SOL. in alcohol, glycerin; all proportions water. —Antiseptic, Disinfectant.—USES: *Extern.*, in diseases of bladder, eye, skin, and in diphtheria, laryngitis, gingivitis, etc.—APPLIED in 1 to 10% solut.—CAUTION: Keep from light!

Aspidium—U. S. P.

MALE FERN.—Dose: 30–90 grn.—*Preparation:* Oleoresin (q. v.).

Aspidosperma—U. S. P.

QUEBRACHO.—*Preparation:* F. E. (D., 30–60 min.).

Aspidospermine Merck.—Amorph., Pure.

Brown-yellow plates; bitter taste.—SOL. in alcohol, ether, chloroform, benzene.—Respiratory Stimulant, Antispasmodic.—USES: Dyspnea, asthma, spasmodic croup, etc.—**Dose:** 1–2 grn., in pills.

Atropine (Alkaloid) Merck.—U. S. P.—C. P., Cryst.

Dose: $\frac{1}{120}$—$\frac{1}{60}$ grn.—ANTIDOTES: Emetics; pilocarpine, muscarine nitrate, or morphine, hypodermically; tannin, or charcoal before absorption.—IN-

COMPATIBLES: *Chemical*, alkalies, tannin, salts of mercury; *physiological*, morphine, pilocarpine, muscarine, aconitine, and eserine.

Atropine Sulphate Merck.—U. S. P.—C. P., Cryst.

USES AND DOSE: Same as of alkaloid.

(Other salts of Atropine are not described because used substantially like the above.)

Balsam Peru—U. S. P.

SOL. in absol. alcohol, chloroform; insol. in water.—**Dose**: 10–30 min.

Balsam Tolu—U. S. P.

SOL. in alcohol, ether, chloroform; insol. in water.—**Dose**: 5–15 grn.—*Preparations*: Syr. (1:100); Tr. (1:10.).

Baptisin Merck.—Pure.

Brownish powd.—SOL. in alcohol.—Purgative in large doses; Tonic, Astringent in small doses.—USES: Scarlet fever, chronic dysentery, etc.—**Dose**: ½–5 grn., in pills.

Barium Chloride Merck.—C. P., Cryst.

Colorl.; bitter, salty taste.—SOL. in 2½ parts water; almost insol. in alcohol.—Cardiac Tonic and Alterative.—USES: *Intern.*, arterial sclerosis and atheromatous degeneration, syphilis, scrofula, etc.; *extern.*, eyewash.—**Dose**: $1/_{10}$–½ grn., 3 t. daily, in 1% sweetened, aromatic solut.—ANTIDOTES: Sodium or magnesium sulphate; emetic; stomach pump.

Barium Iodide Merck.

Deliquescent cryst.—Decomposes and reddens on exposure.—SOL. in water, alcohol.—Alterative.—USES: Scrofulous affections, morbid growths.—**Dose**: $\frac{1}{10}$–⅙ grn., 3 t. daily.—EXTERN. as oint. 4 grn. in 1 ounce lard.—CAUTION: Keep well stoppered!

Barium Sulphide Merck.—Pure.

Amorph., light-yellow powd.—SOL. in water.—Alterative.—USES: Syphilitic and scrofulous affections; depilatory (with flour).—**Dose**: ½–1 grn. in keratin-coated pills.

Bebeerine Merck.—Pure.

BEBIRINE; BIBIRINE; supposed identical with BUXINE and PELOSINE.—Yellowish-brown, amorph. powd.; odorl.; bitter.—SOL. in alcohol, ether; insol. in water.—Antipyretic, Tonic, similar to quinine.—**Dose**: *Febrifuge*, 6–12 grn.; *tonic*, ½–1½ grn. 3 or 4 t. daily.

Bebeerine Sulphate Merck.

Reddish-brown scales.—SOL. in water, alcohol.—USES AND DOSES: As of bebeerine.

Belladonna Leaves—U. S. P.

Preparations: Ext. (D., ⅛–½ grn.); Tr. (D., 5–20 min.); Plaster (20 per cent. **ext.**); Oint. (10 per cent. ext.)

Belladonna Root—U. S. P.

Preparations: F. E. (D., ½–2 min.); Lin. (95 per cent. F. E., 5 per cent. camphor).

Benzanilide Merck.

White powd., or colorl. scales.—SOL. in 58 parts alcohol; slightly in ether; almost insol. in water.—Antipyretic, especially for children.—**Dose**: *Children*, 1⅙–6 grn., according to age, several t. daily; *adults*, 10–15 grn.

Benzene, from Coal Tar, Merck.—Highly Purified, Crystallizable.

MISCIBLE with alcohol, ether, chloroform, oils.—Antispasmodic and Anticatarrhal.—USES: Whooping-cough, influenza, etc.—**Dose:** 2—10 ℳ every 3 hours, in emulsion, or on sugar or in capsules.—MAX. D.: 45 ℳ.

Benzoin—U. S. P.

Preparations: Tr. (D., 20–40 min.), Comp. Tr. (D., 30–60 min.).

Benzolyptus.

Not completely defined.—(Stated: "Alkaline solution of various highly approved antiseptics of recognized value in catarrhal affections ; Dental and Surgical Disinfectant ; Antifermentative.—Liq.—SOL. in water. —**Dose:** 1 fl. dr., diluted.—EXTERN. in 10—30% solut.")

Benzosol.

BENZOYL-GUAIACOL ; GUAIACOL BENZOATE.—Wh., odorl., alm. tastel., cryst. powd.—SOL. in alcohol ; insol. in water.—Antitubercular, Intest. Antiseptic.—**Dose:** 3—15 grn., in pill, or powd. with peppermint-oil sugar.

Benzoyl-pseudotropeine Hydrochlorate Merck,—see TROPACOCAINE, ETC.

Berberine Carbonate Merck.

Yellowish-brown cryst. powd.; bitter taste.—SOL. in diluted acids.—Anti periodic, Stomachic, Tonic.—USES: Malarial affections, amenorrhea, enlargement of spleen, anorexia, chronic intestinal catarrh, vomiting of pregnancy, etc.—**Dose:** *Antiperiodic,* 8—15 grn.; *stomachic and tonic,* ½—1½ grn. 3 t. daily; in pills or capsules.

Berberine Hydrochlorate Merck.—Cryst.

Yellow, microcrystalline needles.—SOL. in water.—USES and DOSE: Same as berberine carbonate.

Berberine Phosphate Merck.—Cryst.

Yellow powd.—SOL. in water.—Most sol. salt of berberine, and easiest to administer, in pills, hydro-alcoholic solut., or aromatic syrup.—USES and DOSE: Same as berberine carbonate.

Berberine Sulphate Merck.—Cryst.

Yellow needles.—SOL. with difficulty in water; almost insol. in alcohol.— USES and DOSE: Same as berberine carbonate.

Betol Merck.

NAPHTALOL ; NAPHTO-SALOL ; SALI-NAPHTOL ; BETA-NAPHTOL SALICYLATE.—White powd.; odorl.; tastel.—SOL. in boiling alcohol, in ether, benzene ; insol. in water, glycerin.—Internal Antiseptic, Antizymotic, Antirheumatic.—USES : Putrid processes of intestinal tract, cystic catarrh, rheumatism, etc.—**Dose :** 4—8 grn., 4 t. daily, in wafers, milk or emulsion.

Bismal.

BISMUTH METHYLENE-DIGALLATE, *Merck.*—Gray-blue powd.—SOL. in alkalies ; insol. in water or gastric juice.—Intestinal Astringent (especially in diarrheas not benefited by opiates).—**Dose:** 2—5 grn. every 3 hours, in wafers or powd.

Bismuth Benzoate Merck.—C. P.

White, tastel. powd.—27% of benzoic acid.—SOL. in mineral acids; insol. in water.—Antiseptic.—USES : *Intern.*, gastro-intestinal diseases; *extern.*, like iodoform on wounds, etc.—**Dose :** 5—15 grn.

Bismuth Beta-naphtolate.

ORPHOL.—Brown, insol. powd.; 23% beta-naphtol.—Intestinal Antiseptic. —**Dose:** 8—15 grn., in pills or wafers ; children half as much.

Bismuth Citrate Merck.—U. S. P.

White powd.; odorl.; tastel.—Sol., very slightly in water.—Stomachic and Astringent.—Uses : Diarrhea, dyspepsia, etc.—**Dose** : 1–3 grn.

Bismuth Nitrate Merck.—Cryst.

Bismuth Ter-nitrate or Trinitrate.—Colorl. hygroscopic cryst.; acid taste.—Changed to sub-nitrate by water.—Sol. in acids, glycerin.—Astringent, Antiseptic.—Uses : Phthisical diarrhea, etc.—**Dose** : 5–10 grn., dissolved in glycerin and then diluted with water.

Bismuth Oxyiodide Merck.

Bismuth Subiodide.—Brownish-red, amorph., insol. powd.; odorl., tastel.—Antiseptic.—Uses : *Extern.*, on suppurating wounds, ulcers, in skin diseases, gonorrhea, etc.; *intern.*, gastric ulcers, typhoid fever, and diseases of mucous membranes.—**Dose** : 3–10 grn., 3 t. daily, in mixture, powd., or capsule.—Extern. like iodoform; in gonorrhea in 1% injection.

Bismuth Phosphate, Soluble, Merck.

White powd.—Sol. in 3 parts water.—Intestinal Antiseptic and Astringent. —Uses : Acute gastric or intestinal catarrh.—**Dose** : 3–8 grn.

Bismuth Salicylate Merck.—Basic.—64% Bi$_2$O$_3$.

White, odorl., tastel. powd.; insol. in water.—External and Intestinal Antiseptic and Astringent.—Uses : *Intern.*, phthisical diarrhea, summer complaint, typhoid, etc.; *extern.*, like iodoform.—**Dose** : 5–15 grn.

Bismuth Sub-benzoate Merck.

White powd.—Antiseptic, like iodoform.—Uses : As dusting-powd. for syphilitic ulcers, etc.

Bismuth Subcarbonate Merck.—U. S. P.—C. P.

Dose: 5–30 grn.

Bismuth Subgallate Merck.

Odorl., yellow, insol. powd.; 55% Bi$_2$O$_3$.—Siccative Antiseptic, and substitute for bismuth subnitrate internally.—Uses : *Extern.*. on wounds, ulcers, eczemas, etc.; *intern.*, in gastro-intestinal affections.—**Dose** : 4–8 grn., several t. daily.—Extern. like iodoform.

Bismuth Subiodide,—see Bismuth Oxyiodide.

Bismuth Subnitrate Merck.—U. S. P.—C. P.

Dose: 5–40 grn.
    Merck's Bismuth Subnitrate is a very light powder and fully conforms to the pharmacopœial requirements.

Bismuth Valerianate Merck.

White powd., valerian odor.—Insol. in water, alcohol.—Sedative, Antispasmodic.--Uses : Nervous headache, cardialgia, chorea, etc.—**Dose** : 1–3 grn.

Bismuth and Ammonium Citrate Merck.—U. S. P.

Sol. in water ; slightly in alcohol.—**Dose** : 2–5 grn.

Black Haw—U. S. P.
    Viburnum Prunifolium.—Nervine, Oxytocic, Astringent.—*Preparation:* F. E. (D., 30–60 min.)

Borax,—see Sodium Borate.

Boro-fluorine.

Not completely defined.—(Stated: " Contains 19¼% boric acid, 5¾% sodium fluoride, 3% benzoic acid, 42% gum vehicle, ½% formaldehyde, 29½% water.—Colorl. liq.; miscible with water in all proport.—Surgical Antiseptic, Internal Disinfectant.—**Dose**: ½—1 fl. dr., in water.—EXTERN. mostly in 5—20% solut.")

Borolyptol.

Not completely defined.—(Stated: " 5% aceto-boro-glyceride, 0.1% formaldehyde, with the antiseptic constituents of pinus pumilio, eucalyptus, myrrh, storax, and benzoin."—Arom., slightly astring., non-staining liq. —Antiseptic, Disinfectant.—**Dose**: ½—1 fl. dr., diluted.—EXTERN. in 5—50% solut.")

Brayerin,—see KOUSSEIN.

Bromalin.

HEXAMETHYLENE-TETRAMINE BROMETHYLATE, *Merck*.—Colorl. laminæ, or white powd.—SOL. in water.—Nerve-sedative, Anti-epileptic; free from untoward effects of inorganic bromides.—USES : As substitute for potassium bromide.—**Dose**: 30—60 grn., several t. daily, in wafers or sweetened water.

Bromides (Peacock's).

Not completely defined.—(Stated: " Each fl. dr. represents 15 grn. combined bromides of potass., sod., calc., ammon., lithium.—Sedative, Antiepileptic.—**Dose**: 1—2 fl. drs., in water, 3 or 4 t. daily.")

Bromidia.

Not completely defined.—(Stated: " Each fl. dr. contains 15 grn. each chloral hydrate and potass. bromide, ⅛ grn. each ext. cannab. ind. and ext. hyoscyam.—Hypnotic, Sedative.—**Dose**: 1—2 fl. drs.")

Bromine—U. S. P.

SOL. in alcohol, ether, chloroform, solut. bromides; also 30 parts water.—Dose: 1–3 min., well diluted. EXTERN. ¼–1 per cent. washes or oily paints; as caustic, pure or 1:1 alcohol.—ANTIDOTES: Stomach irrigation, croton oil in alkaline solut., inhalation of ammonia.

Bromipin Merck.

Bromine addition-product of sesame oil.—Yellow oily fluid, of purely oleaginous taste ; contains 10% bromine.—Nervine and Sedative.—**Dose** : Tea- to tablespoonful, 3 or 4 t. daily, in emulsion with peppermint water and syrup.

Bromoform Merck.—C. P.

Heavy liq., odor and taste similar to chloroform ; darkens on exposure. —SOL. in alcohol, ether; almost insol. in water.—Antispasmodic, Sedative. —USES : Chiefly whooping-cough.—**Dose** (3 or 4 t. daily) : Under 1 year, 1—3 drops ; 1—4 years, 4—5 drops ; 5-7 years, 6—7 drops, in hydro-alcoholic solut. or in emulsion.—CAUTION : Keep well-stoppered !

Bromo-hemol Merck.

Hemol with 2.7% bromine.—Brown powd.—Organic, easily assimilable Nerve-tonic and Sedative ; without the deleterious effect on the blood common to the inorganic bromides.—USES : Hysteria, neurasthenia, epilepsy.—**Dose**: 15—30 grn., 3 t. daily.

Brucine Merck.—Pure.

White powd.—SOL. in alcohol, chloroform.—Nerve-tonic, like strychnine, but much milder.—**Dose** : 1/12—½ grn., in pills or solut.—MAX D.: ¾ grn. —ANTIDOTES : Chloral, chloroform, tannic acid.

Bryonia—U. S. P.

Preparation: Tr. (D., 1-4 drams).

Buchu—U. S. P.
  Preparation: F. E. (D., 15–60 min.).

Butyl-Chloral Hydrate Merck.

" CROTON "-CHLORAL HYDRATE.—Light, white, cryst. scales ; pungent odor.—SOL. in water, alcohol, glycerin.—Analgesic, Hypnotic.—USES : Trigeminal neuralgia, toothache, etc., insomnia of heart disease.—**Dose:** *Hypnotic,* 15—30 grn.; *analgesic,* 2—6 grn.; in solut. water, alcohol, or glycerin.—MAX. D.: 45 grn.—EXTERN. with equal part phenol.—ANTIDOTES : Atropine, strychnine, caffeine, artificial respiration.

Cadmium Iodide Merck.

Lustrous tables. - SOL. in water, alcohol.—Resolvent, Antiseptic.—USES : Scrofulous glands, chronic inflammation of joints, chilblains, and skin diseases.—APPLIED in oint. 1 in 8 lard.

Cadmium Sulphate Merck.—Pure.

White cryst.—SOL. in water, alcohol.—Antiseptic, Astringent.—USES : Instead of zinc sulphate in eye washes (½—1% solut.).

Caesium and Ammonium Bromide Merck.

White, cryst. powd.—SOL. in water.—Nerve Sedative.—USES : Epilepsy, etc.—**Dose :** 15—45 grn., 1 or 2 t. daily.

Caffeine Merck.—U. S. P.—Pure.

THEINE; GUARANINE.—**Dose:** 1—5 grn.—MAX. D.: 10 grn. single, 30 grn. daily.

Caffeine, Citrated, Merck.—U. S. P.

(Improperly called " Citrate of Caffeine ").—50% caffeine.—White powd.; acid taste.—**Dose:** 2—10 grn.

Caffeine Hydrobromate Merck.—True salt.

Glass-like cryst.; reddish or greenish on exposure.—SOL. in water, with decomposition.—USES: Chiefly as diuretic, hypodermically.—INJECTION : 4—10 ℔ of solut. caffeine hydrobromate 10 parts, hydrobromic-acid 1 part, distilled water 3 parts.—CAUTION: Keep well stoppered, in brown bottles!

Caffeine and Sodium Benzoate Merck.

45.8% caffeine.—White powd.—SOL. in 2 parts water.—USES: By injection, 2—10 grn.

Caffeine and Sodium Salicylate Merck.

62.5% caffeine.—White powd.—SOL. in 2 parts water.—USES: By injection; in rheumatism with heart disease, and in threatened collapse of pneumonia.—**Dose :** 1½—6 grn.

Calamus—U. S. P.
  SWEET FLAG.—Dose: 15-60 grn.—*Preparation:* F. E. (1:1).

Calcium Bromide Merck.—U. S. P.

White granules; very deliquescent; sharp, saline taste.—SOL. in water, alcohol.—Nerve Sedative, like potassium bromide.—USES : Epilepsy, hysteria, etc.—**Dose :** 10—30 grn., 2 t. daily.

Calcium Carbonate, Precipitated, Merck.—U. S. P.

PRECIPITATED CHALK.—**Dose:** 10—40 grn.

Calcium Carbonate, Prepared—U. S. P.
  DROP CHALK.—*Preparations :* Comp. Powd. (D., 10-30 grn.); Mercury with Chalk (D., 3-10 grn.), Chalk Mixt. (D., 1-4 fl. drs.); Troches (4 grn.).

Calcium Chloride Merck.—U. S. P.—Pure.

Dose: 5—20 grn.

Calcium Glycerino-phosphate Merck.

White cryst. powd.—SOL. in water; almost insol. in boiling water.—
Directly assimilable Nerve-tonic and Reconstructive.—USES: In rachitis,
wasting diseases, and convalescence.—**Dose:** 2—5 grn., 3 t. daily, in syrup
or solut.

Calcium Hippurate Merck.

White powd.—SOL. slightly in hot water.—Alterative and Antilithic.—
USES: Cystitis, lithiasis, scrophulosis, phthisis, difficult dentition, etc.—
Dose : 5—15 grn.

Calcium Hypophosphite Merck.—Purified.

Dose: 10–30 grn.

Calcium Lactophosphate Merck.—Cryst., Soluble.

White, hard crusts; 1% phosphorus.—SOL. in water.—Stimulant and
Nutrient.—USES: Rachitis, and conditions of malnutrition.—**Dose :** 3—
10 grn., 3 t. daily.—*Preparation:* Syr. (3%).

Calcium Permanganate Merck.—C. P., Cryst.

Deliquescent, brown cryst.—SOL. in water.—USES: *Intern.*, gastro-enter-
itis and diarrhea of children; *extern.*, as other permanganates for mouth
lotions and for sterilizing water; and vastly more powerful than potassium permanganate.—**Dose:** ¾—2 grn., well diluted.

Calcium Phosphate, Tribasic, Merck.—*(Precipitated Calcium Phosphate, U. S. P.).*—Pure, Dry.

Dose: 10—20 grn.

Calcium Sulphite Merck.—Pure.

White powd.—SOL. in 20 parts glycerin, 800 parts water.—Antizymotic.—
USES : Flatulence, diarrhea, and some dyspepsias.—**Dose :** 2—5 grn., in
pastilles.

Calomel,—see MERCURY CHLORIDE, MILD.

Calumba—U. S. P.

COLUMBO.—**Dose:** 5-20 grn.—*Preparations:* F. E. (1.1); Tr. (1:10).

Camphor—U. S. P.

Dose: 2-5 grn.—*Preparations:* Cerate (1:50); Lin. (1:5); Spt. (1:10); Water (1:125).

Camphor, Monobromated, Merck.

Dose : 2—5 grn., in pill or emulsion.

Cannabine Tannate Merck.

Yellow or brownish powd.; slightly bitter and strong astringent taste.
—SOL. in alkaline water or alkaline alcohol, very slightly in water or
alcohol.—Hypnotic, Sedative.—USES: Hysteria, delirium, nervous insom
nia, etc.—**Dose:** 8—16 grn., at bedtime, in powd. with sugar.—MAX. D.:
24 grn.

Cannabis Indica—U. S. P.

INDIAN HEMP.—*Preparations:* Ext. (D., ¼-1 grn.); F. E. (D., 2–5 min.); Tr. (D., 5–20
min.).—See also, Cannabine Tannate.

Cantharides—U. S. P.

Preparations: Cerate (32 per cent.); Collodion (q. v.); Tr. (D., 3–10 min.).—See also,
Cantharidin.—ANTIDOTES: Emetics, flaxseed tea; opium per rectum; morphine subcut.;
hot bath. Avoid oils!

Cantharidin Merck.—C. P., Cryst.

Colorl., cryst. scales; blister the skin.—Sol. in alcohol, ether, chloroform.—Stimulant, Vesicant, Antitubercular.—Uses: In lupus and tuberculosis; also cystitis.—**Dose:** Teaspoonful of 1:100,000 solut. in 1% alcohol (with still more water added before taking), 3 or 4 t. daily.—Injection is given in form of potassium cantharidate, which see.

Capsicum—U. S. P.

CAYENNE PEPPER; AFRICAN PEPPER.—Dose: 1-5 grn.—*Preparations:* F. E. (1:1); Oleores. (D., ¼-1 grn.); Plaster; Tr. (1:20).

Cardamom—U. S. P.

Dose: 5-15 grn.—*Preparations:* Tr. (1:10); Comp. Tr. (vehicle).

Carnogen.

Not completely defined.—(Stated: "Combination of medullary glyceride and unalterable fibrin of ox-blood.—Hematinic.—Uses: Chiefly grave or pernicious anemia, and neurasthenia.—**Dose:** 2—4 fl. drs., in cold water or sherry. 3—4 t. daily; avoid hot fluids !")

Cascara Sagrada—U. S. P.

Preparation: F. E. (D., 15-60 min.).

Castanea—U. S. P.

CHESTNUT.—*Preparation:* F. E. (D., 1-2 drams).

Catechu—U. S. P.

Dose: 5-20 grn.—*Preparations:* Comp. Tr. (1:10); Troches (1 grn.).

Celerina.

Not completely defined.—(Stated: "Each fl. dr. represents 5 grn. each celery, coca, kola, viburnum, and aromatics.—Nerve Tonic, Sedative.—**Dose:** 1—2 fl. drs.")

Cerium Oxalate, Cerous, Merck.—Pure.

White granular powd.; odorl.; tastel.—Sol. in diluted sulphuric and hydro-chloric acids.—Sedative, Nerve-tonic.—Uses: Vomiting of pregnancy, sea-sickness, epilepsy, migraine, chronic diarrhea.—**Dose:** 1—5 grn.

Cetraria—U. S. P.

ICELAND MOSS.—*Preparation:* Decoct. (D., 1-4 oz.).—See also, Cetrarin.

Cetrarin Merck.—C. P., Cryst.

CETRARIC ACID.—White needles, conglomerated into lumps; bitter.—Sol. in alkalies and their carbonates; slightly in water, alcohol, ether.—Hematinic. Stomachic, Expectorant.—Uses: Chlorosis, incipient phthisis, bronchitis, digestive disturbances with anemia, etc.—**Dose:** 1½—3 grn.

Chalk,—see CALCIUM CARBONATE.

Chamomilla Compound (Fraser's).

Not completely defined.—(Stated: "Mixture of mother tinctures of cin-chona, chamomilla, ignatia, and phosphorus, with aromatics and nux vomica.—Nerve Tonic, Stomachic.—**Dose:** 1 fl. dr. before meals and at bedtime, with tablespoonful hot water.")

Charcoal—U. S. P.

WOOD CHARCOAL.—Dose: 10-30 grn.

Chelidonium—U. S. P.

CELANDINE.—Dose: 10-40 grn.

Chenopodium—U. S. P.

AMERICAN WORMSEED.—Dose: 10-40 grn.—*Preparation:* Oil (D., 10 min. 3 t. daily; castor oil next day).

Chimaphila—U. S. P.

PIPSISSEWA; PRINCE'S PINE.—*Preparation:* F. E. (D., 30-60 min.).

Chirata—U. S. P.

Dose: 10-30 grn.—*Preparations:* F. E. (1:1); Tr. (1:10).

Chloralamide.

CHLORAL-FORMAMIDE.—Colorl., bitter cryst.—SOL. in abt. 20 parts water (slowly); in 2 alcohol; decomp. by hot solvents.—Hypnotic, Sedative, Analgesic.—**Dose:** 15—45 grn.

Chloral Hydrate Merck.—U. S. P.—Loose Cryst.; also Flakes.

Dose: 10—30 grn.—MAX. D.: 60 grn.—CONTRA-INDICATED in gastritis; large doses must not be given in heart disease; in children and the aged, use with caution.—ANTIDOTES: Emetics, stomach siphon; cocaine, strychnine, or atropine, hypodermically; stimulants, oxygen, mucilage acacia.—INCOMPATIBLES: Carbolic acid, camphor, alcohol, potassium icdide, potassium cyanide, borax; alkaline hydrates and carbonates.

Chloral-ammonia Merck.

White, cryst. powd.; chloral odor and taste.—SOL. in alcohol, ether; insol. in cold water; decomposed by hot water.—Hypnotic, Analgesic.—USES: Nervous insomnia, neuralgia, etc.—**Dose:** 15—30 grn.

Chloralimide Merck.—(*Not: Chloralamide.*)

Colorl. needles; odorl.; tastel.—SOL. in alcohol, ether, chloroform, oils; insol. in water.—Hypnotic, Analgesic.—USES: Insomnia, headache, etc. —**Dose:** 15—30 grn., 2 or 3 t. daily.—MAX. D.: 45 grn. single; 90 grn. daily.

Chloralose Merck.

Small, colorl. cryst.; bitter, disagreeable taste.—SOL. in alcohol; slightly in water.—Hypnotic.—USES: Insomnia. Free from disagreeable cardiac after-effects and cumulative tendency.of chloral hydrate. Acts principally by reducing excitability of gray matter of brain.—**Dose:** 3—12 grn.

Chlorine Water—U. S. P.

0.4 per cent. Cl.—Dose: 1-4 drams.—ANTIDOTES: Milk and albumen.

Chloroform Merck.—Recryst. and Redistilled, for Anesthesia.

Dose: 10—20 ℳ.—MAX. D.: 30 ℳ.—*Preparations:* Emuls. (4%); Lin. (3%); Spt. (6%); Water (½%).—ANTIDOTES: Vomiting, stomach siphon; cold douche, fresh air, artificial respiration, etc.—CAUTION: Keep in dark amber. Never administer as anesthetic near a flame, as the vapor then decomposes, evolving very irritating and perhaps poisonous gases!

MERCK'S Chloroform is prepared by a new process insuring the highest attainable purity. It is absolutely free from all by-products that are liable to cause untoward effects.

Chrysarobin Merck.—U. S. P.

So-called "CHRYSOPHANIC ACID"; Purified Goa-Powder.—Antiparasitic, Reducing Dermic, etc. Not used internally.—EXTERN. 2-10% oint. or paint.—*Preparation:* Oint. (5%).—CAUTION: Very dangerous to the eyes!

Cimicifuga—U. S. P.

BLACK SNAKEROOT; BLACK COHOSH.—Dose: 15-45 grn. *Preparations:* Ext. (D., 2–6 grn.); F. E. (1:1); Tr. (1:5).—See also, Cimicifugin.

Cimicifugin Merck.

MACROTIN.—Yellowish-brown, hygroscopic powd.—SOL. in alcohol.— Antispasmodic, Nervine, Oxytocic.—USES: Rheumatism, dropsy, hysteria, dysmenorrhea, etc.—**Dose:** 1—2 grn.

Cinchona—U. S. P.

Dose: *Tonic*, 5-15 grn.; *antiperiodic*, 40–120 grn.—*Preparations:* Ext. (D., 1–10 grn.); F. E. (1:1); Infus. (6:100); Tr. (1:5); Comp. Tr. (vehicle).—See also, its var. alkaloids.

Cinchonidine Merck.—Pure, Cryst.
SOL. in dil. acids; insol. in water.—**Dose:** *Tonic*, 1—2 grn., in pills or syrup ; *antiperiodic*, 15—30 grn., between paroxysms.

Cinchonidine Sulphate.—U. S. P.
SOL. in alcohol; sl. in water.—**Dose:** Same as Cinchonidine.

Cinchonine Merck.—U. S. P.—Pure, Cryst.
SOL. in dil. acids; insol. in water.—**Dose:** Same as Cinchonidine.

Cinchonine Sulphate Merck.—U. S. P.
SOL. in 10 parts alcohol, 70 water.—**Dose:** Same as Cinchonine.

Cinnamon, Cassia—U. S. P.

CASSIA BARK.—Dose: 10-30 grn.—*Preparation:* Oil (D., 1-3 min.).

Cinnamon, Ceylon—U. S. P.

Dose: 10-30 grn.—*Preparations:* Oil (D., 1-3 min.); Spt. (10 per cent. oil); Tr. (1:10); Water (one-fifth per cent. oil).

Coca—U. S. P.

ERYTHROXYLON.—*Preparation:* F. E. (D., 20-60 min.).—See also, Cocaine.

Cocaine Hydrochlorate Merck.—U. S. P.—C. P., Cryst. or Powder.
Dose: ½—1½ grn.—MAX. D.: 2 grn. single; 6 grn. daily.—ANTIDOTES: Chloral, amyl nitrite, caffeine, morphine, digitalis, alcohol, ammonia.

MERCK's Cocaine Hydrochlorate strictly conforms to the U. S. P. and all other known tests for its purity.

(Other salts of Cocaine are not described because used substantially as the above.)

Codeine Merck.—U. S. P.—Pure, Cryst. or Powd.
Dose: ½—2 grn.—INJECTION: ¼—1 grn.

Codeine Phosphate Merck.
White powd.—SOL. in 4 parts water; slightly in alcohol.—Best codeine salt for hypodermic use; most sol., least irritating.—INJECTION: ½—1 grn.

(Other salts of Codeine are not described because used substantially as the above.)

Colchicine Merck.—Cryst.
Yellow cryst. powd.; very bitter taste.—SOL. in water, alcohol, ether, chloroform.—Alterative, Analgesic.—USES: Rheumatism, gout, uremia, chronic sciatica, asthma, cerebral congestion, and rheumatic sciatica.—**Dose:** $\frac{1}{120}$—$\frac{1}{30}$ grn., 2 or 3 t. daily.—ANTIDOTES: Stimulants.

Colchicum Root—U. S. P.

Preparations: Ext. (D., ½-2 grn.); F. E. (D., 2-8 min.); Wine (5-20 min.).

Colchicum Seed—U. S. P.

Preparations: F. E. (D., 3-10 min.); Tr. (D., 20-60 min.); Wine (D., 20-60 min.).

Colchi-sal.
Not completely defined.—(Stated: " Caps. each containing $\frac{1}{250}$ grn. colchicine dissolved in 3 min. methyl salicylate.—Antirheumatic, Antipodagric.—**Dose:** 2—4 caps. with meals and at bedtime.")

Collodion, Cantharidal, Merck.—U. S. P.
(Blistering, or Vesicating, Collodion).—Olive-green, syrupy liq.—Represents 60% cantharides.—USES: Blister instead of cantharides.

Collodion, Styptic.—U. S. P.

20 per cent. tannic acid.—USES: Bleeding wounds.

Colocynth—U. S. P.

Dose: 3-10 grn.—*Preparations:* Ext. (D., 1-3 grn.); Comp. Ext. (D., 3-10 grn.).

Colocynthin (Glucoside) Merck.—C. P.

Yellow powd.—SOL. in water, alcohol.—Cathartic (not drastic and toxic, as the extract).—**Dose:** $\frac{1}{6}-\frac{2}{3}$ grn.—INJECTION: $\frac{1}{6}$ grn.; rectal 4–16 ℳ of 4% solut. in equal parts glycerin and alcohol.

Coniine Hydrobromate Merck.

White needles.—SOL. in 2 parts water, 2 parts alcohol; chloroform, ether. —Antispasmodic, Antineuralgic, etc.—USES: Tetanus, cardiac asthma, sciatica and whooping-cough; large doses have been given in traumatic tetanus.—**Dose:** $\frac{1}{30}-\frac{1}{12}$ grn., 3–5 t. daily: children, $\frac{1}{240}-\frac{1}{40}$ grn., 2–4 t. daily.—INJECTION: $\frac{1}{20}-\frac{1}{15}$ grn.—ANTIDOTES: Emetics, stomach siphon; atropine, strychnine; picrotoxin with castor oil; caffeine, and other stimulants.

Conium—U. S. P.

HEMLOCK.—*Preparations:* Ext. (D., ½-2 grn.); F. E. (D., 2-5 min.).—See also, Coniine Hydrobromate.

Convallaria—U. S. P.

LILY OF THE VALLEY.—*Preparation:* F. E. (D., 15-30 min.).—See also, Convallamarin.

Convallamarin Merck.

Yellowish-white, amorph. powd.—SOL. in water, alcohol.—Cardiac Stimulant, Diuretic.—USES: Heart disease, œdema, etc.—**Dose:** ¾–1 grn., 6 to 8 t. daily.—INJECTION: ⅛ grn. every 4 hours, in sweet solut., gradually increasing to 5 grn. daily.—MAX. D.: 1 grn. single; 5 grn. daily.

Copaiba—U. S. P.

Dose: 20-60 grn.—*Preparations:* Mass (94 per cent.); Oil (D., 5-15 min.); Resin (D., 5-15 grn.).

Copper Acetate, Normal, Merck.—Pure, Cryst.

Dose: ⅛–¼ grn.—ANTIDOTES (*for all copper salts*): Encourage vomiting, stomach pump, then milk and sugar or white of egg freely; pure potassium ferrocyanide (10 or 15 grn.).

Copper Arsenite Merck.

Yellowish-green powd.—SOL. in alkalies; slightly in water.—Intestinal Antiseptic, Antispasmodic, Sedative.—USES: Cholera infantum, dysentery, whooping-cough, dysmenorrhea, etc.—**Dose:** $\frac{1}{120}$ grn. every ½ hour until relieved, then every hour.—MAX. D.: 1 grn. single and daily.

Copper Sulphate Merck.—U. S. P.—Pure, Cryst.

Dose: *Nervine* and *alterative*, ⅛–⅓ grn.; *emetic*, 2–5 grn.

Cornutine Citrate Merck.

Brown, very hygroscopic scales or mass.—SOL. in water (incompletely). —USES: Hemorrhage from genito-urinary organs, paralytic spermatorrhea, etc.—**Dose:** *Hemostatic*, $\frac{1}{12}-\frac{1}{6}$ grn.; *spermatorrhea*, $\frac{1}{20}-\frac{1}{10}$ grn. daily.

Cotarnine Hydrochlorate,—see STYPTICIN.

Cotton-Root Bark—U. S. P.

Emmenagogue, Oxytocic.—*Preparation:* F. E. (D., 30-60 min.).

Creolin.

SAPONIFIED DEPHENOLATED COAL-TAR CREOSOTE, *Pearson.*—Dark syrupy liq.; tar odor.—SOL. in alcohol, ether, chloroform; milky emulsion with water; sol. in water to 2½%.—Disinfectant, Deodorizer,

Styptic, Anticholeraic, etc.—USES: Non-poisonous substitute for carbolic acid, etc. Removes odor of iodoform. *Intern.*, dysentery, diarrhea, meteorism, gastric catarrh, worms, thrush, diphtheria, etc.; enema ½% solut. in dysenteric troubles; *extern.*, ½ to 2% solut. in surgical operations, ₁⁰₀–¼% injection for gonorrhea, 2 5% ointment in scabies and pediculi, erysipelas, cystitis, burns, ulcers, etc.—**Dose:** 1–5 ℳ 3 t. daily, in pills. In cholera 16 ℳ every ½—1 hour for 5 doses, then at longer intervals.—CAUTION: Aqueous solut. should be freshly made when wanted.

Creosote Carbonate.

CREOSOTAL.—Light-brown, odorl., sl. bitter liq.—SOL. in oils (5 parts cod-liver oil), alcohol, ether; insol. in water.—Antitubercular.—**Dose:** 20 min., grad. increased to 80 min., 3 t. per day.

Creosote from Beechwood, Merck.—U. S. P.

Dose: 1–3 ℳ, gradually increased to limit of tolerance, in pills, capsules, or with wine or brandy.—MAX. INITIAL D.: 5 ℳ single; 15 ℳ daily.—ANTIDOTES: Emetics, stomach pump, soluble sulphates (such as Glauber or Epsom salt).—CAUTION: Wherever Creosote is indicated for internal medication, Creosote from Beechwood should be dispensed; and under no circumstances should "Creosote from Coal Tar" be given, unless explicitly so directed. Wood Creosote and Coal-Tar Creosote differ very widely in their action on the human body: Wood Creosote is comparatively harmless; Coal-Tar Creosote decidedly poisonous.—*Preparation:* Water (1%).
MERCK'S Beechwood Creosote is *absolutely free from the poisonous cærulignol* found in some of the wood creosote on the market.

Creosote Phosphite.

PHOSPHOTAL.—Oily liq.: 90% creosote.—SOL. in alcohol, glycerin, oils.—Antitubercular, Anticachectic.—**Dose:** Same as of creosote; in pills, wine, or elixir.

Cubebs—U. S. P.

Dose: 15–60 grn.—*Preparations:* F. E. (1:1); Oil (D., 5–15 min.); Oleores. (D., 10–30 min.); Tr. (1:5); Troches (½ min. oleores.).

Cupro-hemol Merck.

Hemol with 2% copper.—Dark-brown powd.—USES: Substitute for usual copper compounds in tuberculosis, scrofula, nervous diseases, etc.—**Dose:** 1–3 grn., 3 t. daily, in pills.

Curare Merck.—Tested.

Dose: ₁⁄₁₂–⅛ grn., hypodermically, 1 or 2 t. daily, or until effect is noticed.—CAUTION: Avoid getting it into a wound, as this may prove fatal!

Curarine Merck.—C. P.

Deliquescent brown powd.—SOL. in water, alcohol, chloroform.—Antitetanic, Nervine, etc.—USES: Rectal tetanus, hydrophobia, and severe convulsive affections.—INJECTION: ₆⁰₀–₁⁄₁₂ grn.—ANTIDOTES: Strychnine, atropine, artificial respiration and stimulants.

Cypripedium—U. S. P.

LADIES' SLIPPER.—Dose: 15–30 grn.—*Preparation:* F. E. (1:1).

Dermatol,—see BISMUTH SUBGALLATE.

Diabetin.

LEVULOSE.—Wh. powd.—SOL. in water.—Substitute for sugar in diabetes.

Diastase (of Malt) Merck.—Medicinal.

Yellowish-white to brownish-yellow, amorph. powd.; tastel.—USES: Aid to digestion of starchy food.—**Dose:** 1–3 grn. pure or with pepsin.

Dietetic Products,—see FOODS AND DIETETIC PRODUCTS.

Digitalin, "German," Merck.

Yellowish-white powd.—Sol. in water, alcohol; almost insol. in ether, chloroform.—Non-cumulative, reliable Heart-tonic, Diuretic; well adapted to injection.—**Dose:** $\frac{1}{16}$–¼ grn., 3 or 4 t. daily, in pills or subcutaneously. - Antidotes: Emetics, stomach pump, tannic acid, nitroglycerin, morphine early, strophanthin later; alcoholic stimulants, etc.

Digitalis—U. S. P.

Dose: 1–3 grn.—*Preparations:* Ext. (D., ¼–½ grn.); F. E. (1:1); Infus. (15:1000); Tr. (15:100).—See also, Digitalin and Digitoxin.

Digitoxin Merck.—Cryst.

Most active glucoside from digitalis.—White cryst. powd.—Sol. in alcohol, chloroform; slightly in ether; insol. in water.—Prompt, reliable, powerful Heart-tonic; of uniform chemical composition and therapeutic activity. –Uses: Valvular lesions, myocarditis, etc.—**Dose:** $\frac{1}{240}$–$\frac{1}{60}$ grn., 3 t. daily, with 3 ℳ chloroform, 60 ℳ alcohol, 1½ fl. oz. water. Enema: $\frac{1}{60}$ grn. with 10 ℳ alcohol, 4 fl. oz. water, 1 to 3 t. daily.—Max. D.: Daily, $\frac{1}{12}$ grn.

Dioviburnia.

Not completely defined.—(Stated: "1 fl. oz. represents 45 ℳ each fl. extracts viburn. prunifol., viburn. opulus, dioscorea villosa, aletris farinosa, helonias dioica, mitchella repens, caulophyllum, scutellaria.—Antispasmodic, Anodyne.—Uses : Dysmenorrhea, amenorrhea, etc.—**Dose:** 10–30 ℳ.")

Diuretin,—see Theobromine and Sodium Salicylate.

Duboisine Sulphate Merck.

Yellowish, very deliquescent powd.—Sol. in water, alcohol.—Hypnotic, Sedative, Mydriatic.—Uses: Principally as mydriatic, much stronger than atropine; also in mental diseases, usually hypodermically.—**Dose:** $\frac{1}{80}$–$\frac{1}{20}$ grn.—Extern. in 0.2 to 0.8% solut.

Dulcamara—U. S. P.

Preparation: F. E. (D., 30–120 min.).

Duotal,—see Guaiacol Carbonate.

Elaterin Merck.—U. S. P.—Cryst.

Cryst. powd.: very bitter taste.—Sol. in alcohol, chloroform; slightly in ether.—Drastic Purgative.—Uses: Ascites, uremia, pulmonary œdema, poisoning by narcotics, etc.—**Dose:** $\frac{1}{20}$–$\frac{1}{12}$ grn.

Elaterium Merck.—(According to Clutterbuck).

Dose: ⅛–½ grn.

Emetin (Resinoid) Merck.—(*Do not confound with the Alkaloid !*)

Yellowish-brown lumps.—Emetic, Diaphoretic, Expectorant.—**Dose:** *Emetic,* ⅛–¼ grn.; *expectorant,* $\frac{1}{80}$–$\frac{1}{30}$ grn.

Emetine (Alkaloid) Merck.—Pure.—(*Do not confound with the Resinoid !*)

Brownish powd.; bitter taste; darkens on exposure.—Sol. in alcohol, chloroform; slightly in ether; very slightly in water.—Emetic, expectorant.—**Dose:** *Emetic,* $\frac{1}{16}$–⅛ grn.; *expectorant,* $\frac{1}{120}$–$\frac{1}{60}$ grn.

Ergot Aseptic.

Standardized, sterilized preparation of ergot for hypodermatic use; free from extractive matter and ergotinic acid. Each 1Cc. bulb represents 2 Gm. (30 grn.) ergot.

Ergot—U. S. P.

Dose: 20–90 grn.—*Preparations:* Ext. (D., 5–15 grn.); F. E. (1:1); Wine (15:100).

Ergotin (Bonjean) Merck.

Dose: 3—10 grn.—CAUTION: Decomposes in solut.; should be sterilized and kept with great care.

Ergotole.

Liq. prepar. of ergot, 2½ times strength of U. S. P. fl. ext.; stated to be permanent.—INJECT.: 5—20 min.

Eriodictyon—U. S. P.

    YERBA SANTA.—*Preparation:* F. E. (D., 20–60 min.).

Erythrol Tetranitrate Merck.

Cryst. mass, exploding on percussion; therefore on the market only in *tablets* with chocolate, each containing ½ grn. of the salt.—Vasomotor Dilator and Antispasmodic, like nitroglycerin.—USES: Angina pectoris, asthma, etc.—**Dose:** 1—2 tablets.

Eserine Salicylate Merck.

PHYSOSTIGMINE SALICYLATE.—Slightly yellowish cryst.—SOL. in 150 parts water; solut. reddens on keeping.—Spinal Depressant, Antitetanic, Myotic.—USES: Tetanus, tonic convulsions, strychnine poisoning, etc.; in 5% solut. to contract pupil.—**Dose :** $\frac{1}{120}$—$\frac{1}{30}$ grn.—MAX. D.: $\frac{1}{20}$ grn.

Eserine Sulphate Merck.

PHYSOSTIGMINE SULPHATE.—Yellowish, very deliquescent powd.; bitter taste; rapidly reddens.—SOL. easily in water, alcohol.—USES, DOSES, ETC As Eserine Salicylate.

Ether—U. S. P.

    SULPHURIC ETHER.—Dose: 10-40 min.—ANTIDOTES: Emetics, fresh air, ammonia.—CAUTION: Vapor inflammable!—*Preparations* Spt. (32.5 per cent.); Comp. Spt. (32.5 per cent.).

Ethyl Bromide Merck.—C. P.

HYDROBROMIC ETHER.—Colorl., inflammable, volatile liq.; burning taste, chloroform odor.—SOL. in alcohol, ether, chloroform.—Inhalant and Local Anesthetic, Nerve Sedative.—USES: Minor surgery, spray in neuralgia, etc.; epilepsy, hysteria, etc. [It is of great importance to have a pure article for *internal* use, since with an impure one, alarming after-effects may occur; MERCK's is strictly pure.]—**Dose:** 150—300 ♏ for inhalation; by mouth, 5—10 drops on sugar, or in capsules.—CAUTION: Keep from light and air!—*N. B.* This is *not* Ethylene Bromide, which is poisonous!

Ethyl Chloride Merck.

Gas at ordinary temperatures and pressures; when compressed, colorl. liq. —SOL. in alcohol.—Local Anesthetic.—USES: Minor and dental surgery, and neuralgia, as spray; heat of hand forcing the stream from the tubes. Hold 6—10 inches away from part.—CAUTION: Highly inflammable!

Ethyl Iodide Merck.

HYDRIODIC ETHER.—Clear, neutral liq.; rapidly turns brown on keeping.—SOL. in alcohol, ether; insol. in water.—Alterative, Antispasmodic, Stimulant.—USES: *Intern.,* chronic rheumatism, scrofula, secondary syphilis, chronic bronchitis, asthma, chronic laryngitis, and by inhalation in bronchial troubles; *extern.,* in 10—20% oint.—**Dose:** 5—16 ♏, several t. daily, in capsules or on sugar; *inhal.,* 10—20 drops. —CAUTION: Even in diffused daylight Ethyl Iodide decomposes quite rapidly, the light liberating iodine which colors the ether. When not exposed to light at all the decomposition is very slow; and with the least practicable exposure, by care in using it, it is not rapid. The decomposition is rendered still slower by the presence in each vial of about 10 drops of a very dilute solution of soda. When deeper than a pale wine color, it should be shaken up with 5 or 10 drops of such solution.

Ethylene Bromide Merck.—(*Not Ethyl Bromide*).

Brownish, volatile, emulsifiable liq.; chloroform odor.—MISCIBLE with alcohol; insol. in water.—Anti-epileptic and Sedative.—USES: Epilepsy, delirium tremens, nervous headache, etc.—**Dose:** 1—2 ℳ, 2—3 t. daily, in emulsion or capsules.

Eucaine, Alpha-, Hydrochlorate.

Wh. powd.—SOL. in 10 parts water.—Local Anesthetic, like cocaine.—APPLIED to mucous surfaces in 1—5% solut.—SUBCUT. 15—60 min. of 6% solut.

Eucaine, Beta-, Hydrochlorate.

Wh. powd.—SOL. in 28 parts water.—Local Anesthetic, specially intended for ophthalmologic use.—APPLIED in 2% solut.

Eucalyptol Merck.—U. S. P.—C. P.

Dose: 5—16 ℳ, 4 or 5 t. daily, in capsules, sweetened emulsion, or sugar.—INJECTION: 8—16 ℳ of mixture of 2—5 eucalyptol and 10 liq. paraffin.

Eucalyptus—U. S. P.

Preparation: F. E. (D., 5-20 min.).—See also, Oil Eucalyptus and Eucalyptol.

Eudoxine.

Bismuth salt of nosophen.—Odorl., tastel., insol. powd.; 52.9% iodine.—Intest. Antiseptic and Astringent.—**Dose:** 4—10 grn., 3—5 t. daily.

Eugallol.

PYROGALLOL MONOACETATE, *Knoll.*—Syrupy, transparent, dark-yellow mass.—SOL. in water readily.—Succedaneum for Pyrogallol in obstinate chronic psoriasis; very vigorous in action.—APPLIED like pyrogallol.

Eugenol Merck.—Pure.

EUGENIC ACID; CARYOPHYLLIC ACID.—Colorl., oily liq.; spicy odor; burning taste.—SOL. in alcohol, ether, chloroform, solut. caustic soda.—Antiseptic, Antitubercular, Local Anesthetic.—USES: *Extern.*, oint. with adeps lanæ in eczema and other skin diseases, local anesthetic in dentistry, etc.; *intern.*, tuberculosis, chronic catarrhs, etc.—**Dose:** 8—30 ℳ.—MAX. D.: 45 ℳ.

Euonymus—U. S. P.

WAHOO.—*Preparation:* Ext. (D., 2-5 grn.).

Eupatorium—U. S. P.

BONESET.—*Preparation:* F. E. (D., 20-60 min.).

Euonymin, American, Brown, Merck.

Brownish powd.—USES: Cholagogue and drastic purgative, similar to podophyllin.—**Dose:** 1½—6 grn.

Euphorin Merck.—(*Not Europhen.*)

PHENYL-ETHYL URETHANE.—Colorl. needles; slight aromatic odor; clove taste.—SOL. in alcohol, ether, slightly in water.—Antirheumatic, Anodyne, Antiseptic, Antipyretic.—USES: *Intern.*, rheumatism, tuberculosis, headache, and sciatica; *extern.*, dusting-powd. in venereal and other skin diseases, ulcers.—**Dose:** 8—16 grn., 2—3 t. daily.

Euquinine.

QUININE ETHYL-CHLOROCARBONATE.—Slightly bitter powd.—SOL. in alcohol, ether, chloroform; slightly sol. in water.—Succedaneum for other quinine salts, internally.—**Dose:** About 1½ times that of quin. sulph., in powd. or cachets, or with soup, milk, or cacao.

Euresol.

RESORCIN MONOACETATE, *Knoll.*—Viscid, transparent, yellow mass, readily pulverizable.—Succedaneum for Resorcin.—USES: Chiefly acne, sycosis simplex, seborrhea, etc.—EXTERN. in 5 to 20% oint.

Eurobin.

CHRYSAROBIN TRIACETATE, *Knoll.*—SOL. in chloroform, acetone, ether; insol. in water.—Succedaneum for Chrysarobin; very active reducer or "reactive".—EXTERN. in 1 to 20% solut. in acetone, with 5 to 10% of saligallol.

Europhen.

Yellow powd.; 27.6% iodine.—SOL. in alcohol, ether, chloroform, fixed oils; insol. in water or glycerin.—Antisyphilitic, Surgical Antiseptic.— **Dose:** (by inject.): ½–1½ grn. once daily, in oil.—EXTERN. like iodoform.

Extract, Bone-Marrow, (Armour's).

Not completely defined.—(Stated: "Medullary glyceride, containing all the essential ingredients of fresh red bone-marrow.—Hematinic, Nutrient. —USES: Anemia, chlorosis, etc.—**Dose:** 1–2 fl. drs., in water, milk, or wine, 3 t. daily.")

Extract Cod-Liver Oil,—see GADUOL.

Extract, Ergot, Aqueous, Soft,—see ERGOTIN.

Extract, Licorice, Purified, Merck.—U. S. P.—Clearly soluble.

PURE EXTRACT GLYCYRRHIZA.—Used to cover taste of bitter mixtures, infusions, or decoctions; also as pill-excipient. Enters into Comp. Mixt. Glycyrrhiza.

Extract, Male Fern,—see OLEORESIN, MALE FERN.

Extract, Malt, Merck.—Dry, Powd.

Contains maximum amount diastase, dextrin, dextrose, protein bodies, and salts from barley.—Tonic, Dietetic.—USES: Children, scrofulous patients, dyspeptics, etc.—**Dose:** 1–4 drams.

Extract, Monesia, Merck.—Aqueous, Dry

Alterative, Instestinal Astringent.—USES: Chronic diarrhea, catarrh, scrofula, scurvy, etc.—**Dose:** 2–5 grn.

Extract, Muira-puama, Fluid, Merck.

Aphrodisiac, Nerve-stimulant.—USES: Sexual debility, senile weakness, etc.—**Dose:** 15–30 ℳ.

Extract, Opium, Aqueous, Merck.—U. S. P.—Dry.

Dose: ¼–1 grn.—MAX. D.: 2 grn. single; 5 grn. daily.

Ferropyrine.

FERRIC-CHLORIDE-ANTIPYRINE,*Knoll;* FERRIPYRINE.—64% antipyrine, 12% iron, 24% chlorine.—Orange-red non-hygroscopic powd.—SOL. in 5 parts water, 9 parts boiling water; in alcohol, benzene, slightly in ether.—Hematinic, Styptic, Astringent, Antineuralgic.—USES: *Intern.*, anemia, chlorosis, migraine, headache, neuralgia; *extern.*, gonorrhea, nosebleed, etc. **Dose:** 5–15 grn., with peppermint-oil sugar, or in solut.—EXTERN. in 1–1½% solut. for gonorrhea; 20% solut. or pure for hemorrhages.

Firwein (Tilden's).

Not completely defined.—(Stated: "Each fl. dr. contains $1/100$ grn. phosphorus, $1/6$ grn. iodine, $1/6$ grn. bromine.—Alterative, Anticatarrhal.—USES: Chronic bronchitis, phthisis, catarrh, etc.—**Dose:** 1–2 fl. drs., before meals.")

Fluorescein Merck.

Orange-red powd.—Sol. in ether, alkaline solut.—Uses: Diagnosis of corneal lesions and impervious strictures of nasal duct. Solut. 10 grn., with 15 grn. sodium bicarbonate, in ounce water.

Foods and Dietetic Products.

Bovinine.—"Unaltered bovine blood."

Carnrick's Soluble Food.

·Eskay's Albumenized Food.

Globon.—A chemically pure albumin.—See under "G."

Hemaboloids.—"Iron-bearing nucleo-albumins, reinforced by bone-marrow extract, and antiseptically treated with nuclein."

Horlick's Food.—"Containing in 100 parts 3.39 water, 0.08 fat, 34.99 glucose, 12.45 cane sugar, 6.71 albuminoids, 1.28 mineral constituents, but no starch."

Imperial Granum.—"Unsweetened food, prepared from the finest growths of wheat; contains no glucose, cane sugar, or malt."

Infant Food, Keasbey & Mattison's.

Liebig's Soluble Food.

Malted Milk, Horlick's.

Maltine.—"Extraction of all the nutritive and digestive properties of wheat, oats, and malted barley."—Maltine M'f'g Co., Brooklyn, N. Y.

Maltzyme.—See under "M".

Mellin's Food.—"Consists of dextrin, maltose, albuminates, and salts."

Nestle's Food.—"40% sugar, 5% fat, 15% proteids, 30% dextrin and starch."

Nutrose.—"Casein-sodium."

Panopepton.—"Bread and beef peptone; containing the entire edible substance of prime, lean beef, and of best wheat flour."

Peptogenic Milk Powder.—"For modifying cow's milk to yield a food for infants, which, in physiological, chemical and physical properties, is almost identical with mother's milk."

Peptonized Milk.—See Peptonizing Tubes.

Peptonoids, Beef.—"From beef and milk, with gluten."

Peptonoids, Liquid.—"Beef Peptonoids in cordial form."

Saccharin.—Antidiabetic and Hygienic Substitute for Sugar.—See under "S."

Sanose.—"80% purest casein, 20% purest albumose."

Somatose.—"Deutero- and hetero-albumoses."

Trophonine.—"Containing the nutritive elements of beef, egg albumen, and wheat gluten."

Formalbumin.

Formaldehyde-Proteid, *Merck;* from Casein.—Yellowish powd., almost odorl. and tastel.—Protective Vulnerary, forming a film from which formaldehyde is gradually liberated, thus persistently disinfecting the wound-surface.

Formaldehyde Merck.

Aqueous solut. formaldehyde gas; about 35%.—Colorl., volatile liq.; pungent odor.—Non-corrosive Surgical and General Antiseptic (in wounds, abscesses, etc., for clothing, bed-linen, walls, etc.); preservative of collyria and anatomical or botanical specimens.—Applied in vapor or solut.: In surgery, $\frac{1}{4}$-$\frac{1}{2}$% solut.; general antisepsis, $\frac{1}{4}$ - 2% solut. or in vapor: for collyria, $1/_{10}$% solut.; for hardening anatomical specimens. 4-10% solut. [Other brands of this preparation are sold under special names, such as "Formalin," "Formol," etc. The Merck article is sold under its true chemical name: "Formaldehyde".]

Formaldehyde, Para-,—see PARAFORMALDEHYDE.

Formalin or Formol,—see FORMALDEHYDE.

Formin.

HEXAMETHYLENE-TETRAMINE, *Merck*.—Alkaline cryst. powd.—SOL. in water, slightly in alcohol.—Uric-acid Solvent and Genito-urinary Antiseptic.—USES: Gout, cystitis, etc.—**Dose:** 15–30 grn. daily, taken in the morning, or morning and evening, in lithia water or carbonated water.

Frangula—U. S. P.

BUCKTHORN.—Laxative.—*Preparation:* F. E. (D., 15–30 min.).

Fuchsine, Medicinal, Merck.

Fuchsine free from arsenic.—SOL. in water.—Antiseptic, Antinephritic. —USES: *Intern.*, nephritis, cystitis; said to reduce anasarca and arrest albuminuria.—**Dose:** ⅓–3 grn., several t. daily, in pills.—CAUTION : Do not confound with Fuchsine *Dye!*

Gaduol.

ALCOHOLIC EXTRACT COD-LIVER OIL, *Merck*.—Brown, oily liq.; bitter, acrid taste; contains the therapeutically active principles of cod-liver oil (iodine, bromine, phosphorus, and alkaloids), without any of the inert ballast of the oil.—Alterative, Nutrient.—USES: Instead of cod-liver oil.—**Dose:** 5–30 ℳ, as elixir or wine.—[Further information in "Merck's Digest" on "GADUOL," containing detailed information, formulas, etc.]

Gall, Ox, Inspissated, Merck.—Purified, Clearly Sol.

Laxative, Digestive.—USES: Typhoid fever, deficiency of biliary secretion, etc.—**Dose:** 2–5 grn., several t. daily, in capsules or pills.

Gallanol Merck.

GALLIC ACID ANILIDE.—Wh. or grayish powd.—SOL. in alcohol, ether; sl. in water, chloroform.—Antiseptic Dermic.—USES: *Extern.*, instead of chrysarobin or pyrogallol: acute or chronic eczema, 1–7 parts in 30 parts ointment; psoriasis, 20% solut. in chloroform or traumaticin; moist eczema, 25% with talcum; favus, prurigo and tricophyton, 20% solut. in alcohol with little ammonia.

Gallobromol Merck.

DIBROMO-GALLIC ACID.—Small, grayish cryst.—SOL. in alcohol, ether, 10 parts water.—Sedative, Antiseptic, Astringent.—USES: *Intern.*, instead of potassium bromide; *extern.*, cystitis, gonorrhea, gleet, and other skin diseases.—**Dose:** 10–30 grn.—EXTERN. in 1–4% solut., powd., or paste.

Gamboge—U. S. P.

Dose: 1–5 grn.—Enters in Comp. Cathartic Pills.

Gelanthum.

Lauded by Unna as an ideal water-soluble vehicle for the application of dermics. Forms a smooth, homogeneous covering without any tendency to stickiness. Does not stain the skin or the linen. Readily takes up 50% ichthyol, 40% salicylic acid, resorcin, or pyrogallol, 5% carbolic acid, and 1% mercuric chloride. Keeps insoluble drugs well suspended.

Gelseminine (Alkaloid) Merck.—C. P.

White microscopic cryst.—SOL. in alcohol, ether, chloroform.—(The *hydrochlorate* and *sulphate* are sol. in water.)—Antineuralgic, Antispasmodic.—USES: Neuralgia, rheumatism, dysmenorrhea, etc.; also antidote to strychnine.—**Dose:** $^1/_{120}$–$^1/_{30}$ grn.—MAX. D.: $^1/_{30}$ grn. single, $^1/_6$ grn. daily.—ANTIDOTES: Emetics early, atropine, strophanthin, artificial respiration, external stimulation.

(The salts of Gelseminine are not described because used substantially as the above.)

Gelsemium—U. S. P.

YELLOW JASMINE.—*Preparations:* F. E. (D., 2-5 min.), Tr. (D., 10-30 min.).—See also, Gelseminine.

Gentian—U. S. P.

Dose: 10-30 grn.—*Preparations:* Ext. (D., 2-6 grn.); F. E. (D., 10-30 min.); Comp. Tr. (D., 1-2 drams).

Geranium—U. S. P.

CRANESBILL.—**Dose:** 30-60 grn.—*Preparation:* F. E. (1:1).

Ginger—U. S. P.

Dose: 5-20 grn.—*Preparations:* F. E. (1:1); Oleores. (D., ½-2 min.); Tr. (1:5); Troches (3 min. Tr.); Syr. (3 per cent. F. E.).

Globon.

Chemically pure albumin.—Yellowish, dry, odorl., tastel. powd.— INSOL. in water.—Albuminous Nutritive and Reconstructive; more nutritious than meat, milk, or any other aliment; very easily assimilated.— USED in acute diseases and during convalescence therefrom; in anemia, gastric affections, diabetes, and gout; also in children.—**Dose:** ½—1 dram several t. daily, best taken with amylaceous food; children ¼-½ as much.

Glycerin—U. S. P.

Dose: 1-4 drams.—*Preparation:* Suppos. (95 per cent.).

Glycerin Tonic Compound (Gray's).

Not completely defined.—(Stated: " Combination of glycerin, sherry, gentian, taraxacum, phosphoric acid, and carminatives.—Alterant Tonic [especially in diseases of chest and throat].—**Dose:** ½ fl. oz., before meals, in water.")

Glyco-thymoline.

Not completely defined.—(Stated: " Alkaline, antiseptic, cleansing solut. for treatment of diseased mucous membrane, especially nasal catarrh.— USED chiefly *extern.;* generally in 20% solut."—**Dose:** 1 fl. dr., diluted.)

Glycozone.

Not completely defined.—(Stated: " Result of the chemical reaction when glycerin is subjected to the action of 15 times its own volume of ozone. under normal atmospheric pressure at 0°C.—Colorl, viscid liq.: sp. gr. 1.26.—Disinfectant, Antizymotic.—**Dose:** 1-2 fl. drs., after meals, in water.—ENEMA: ½-1 fl. oz. in 1-2 pints water.")

Glycyrrhiza—U. S. P.

LICORICE ROOT.—*Preparations:* Ext. and F. E. (vehicles); Comp. Powd. (D., 1-2 drams).; Comp. Mixt. (D., 2-4 fl. drs.); Troches Glyc. and Opium (one-twelfth grn. Op.).—See also, Glycyrrhizin, Ammoniated.

Glycyrrhizin, Ammoniated, Merck.—Clearly Soluble.

Dark-brown or brownish-red, sweet scales.—SOL. in water, alcohol.—Expectorant, Demulcent.—USES: Chiefly with bitter or neutral medicines, to cover taste; also as cough remedy.—**Dose:** 5-15 grn.—INCOMPATIBLE with acids.

Gold Bromide, Auric, Merck.

GOLD TRIBROMIDE.—Dark-brown powd.—SOL. in water, ether.—USES, **Dose,** ETC.: same as of Gold Bromide, Aurous.

Gold Bromide, Aurous, Merck.

GOLD MONOBROMIDE.—Yellowish-gray, friable masses.—INSOL. in water. Anti-epileptic, Anodyne, Nervine.—USES: Epilepsy, migraine, etc.; said to act, in small doses, quickly and continuously, without bromism.— **Dose:** *Anti-epileptic,* $1/10$–$1/5$ grn. 2 or 3 t. daily, in pills; *anodyne,* $1/20$ grn. 2 t. daily. Children, half as much.

Gold Chloride Merck.

AURIC CHLORIDE.—Brown, very deliquescent, cryst. masses.—SOL. in water, alcohol.—Antitubercular, Alterative.—USES: Phthisis and other tubercular affections; lupus.—**Dose:** $1/50$–$1/15$ grn.—CAUTION: Keep dry, from light!

Gold Cyanide, Auric, Merck.

GOLD TRICYANIDE.—Colorl. hygroscopic plates.—SOL. in water, alcohol. USES: Antitubercular.—**Dose:** $1/20$–$1/10$ grn.—ANTIDOTES: As Gold Cyanide, Aurous.

Gold Cyanide, Aurous, Merck.

GOLD MONOCYANIDE.—Yellow cryst. powd.—INSOL. in water, alcohol, or ether.—**Dose:** $1/16$–$1/4$ grn., several t. daily, in pills.—ANTIDOTES: Emetics, stomach siphon, artificial respiration, ferric or ferrous sulphate, ammonia, chlorine, hot and cold douche, etc.

Gold Iodide Merck.

AUROUS IODIDE.—Greenish or yellow powd.—Alterative.—USES: Scrofula and tuberculosis.—**Dose:** $1/64$–$1/8$ grn.

Gold and Sodium Chloride Merck.—U. S. P.

Dose: $1/24$–$1/4$ grn.—INCOMPATIBLES: Silver nitrate, ferrous sulphate, oxalic acid.

Grindelia—U. S. P.

Preparation: F. E., (D., 20–60 min.).

Guaiac—U. S. P.

RESIN GUAIAC.—**Dose:** 5–15 grn.—*Preparations:* Tr. (D., 20–60 min.); Ammon. Tr. (D., 1–2 drams).

Guaiacol Merck.

Colorl., limpid, oily liq.; characteristic aromatic odor.—SOL. in alcohol; ether, 200 parts water.—Antitubercular, Antiseptic, Antipyretic, Local Analgesic.—USES: *Intern.,* phthisis, lupus, and intestinal tuberculosis, febrile affections.—**Dose:** 2 ♏ 3 t. daily, gradually increased to 16 ♏, in pills, or in 1–2% solut. brandy, wine, etc., after meals.—EXTERN. (Analgesic and Antipyretic): 16—32 ♏, pure or with equal parts glycerin or oil.

Guaiacol Benzoate,—see BENZOSOL.

Guaiacol Carbonate.

DUOTAL.—Small, wh., odorl., tastel. cryst.—INSOL. in water.—Antitubercular.—**Dose:** 4–8 grn. 2 or 3 t. daily, gradually increased to 90 grn. a day if necessary, in powd.

Guaiacol Phosphite.

GAIACOPHOSPHAL.—Oily liq.: 92% guaiacol.—SOL. in alcohol, glycerin, oils.—Antitubercular, etc., like guaiacol.—**Dose:** Same as of guaiacol; in pills, elixir, or wine.

Guaiacol Salol Merck.

GUAIACOL SALICYLATE.—White, insipid cryst.; salol odor.—SOL. in alcohol; insol. in water.—Intestinal Antiseptic, Antitubercular, Antirheumatic.—USES: Phthisical diarrhea, dysentery, rheumatism, marasmus, chorea, etc.—**Dose:** 15 grn., several t. daily.—MAX. D.: 150 grn. daily.

Guaiaquin.

QUININE GUAIACOL-BISULPHONATE.—Yellowish, acrid, bitter powd.; 61.36% quinine, 23.48% guaiacol.—SOL. in water, alcohol, dil. acids.—Antiperiodic, Intest. Antiseptic.—**Dose:** 5—10 grn., 3 t. daily, before meals.

Guarana—U. S. P.

Dose: 15–60 grn.—*Preparation:* F. E. (1:1).

Guethol Merck.

GUAIACOL-ETHYL.—Oily liq., congealing in the cold.—SOL. in alcohol, ether, chloroform; insol. in water or glycerin.—Local Anesthetic, Topical and Internal Antitubercular.—USES: Chiefly as succedaneum for guaiacol: *extern.*, in neuralgia, tubercular cystitis, etc.; *intern.* in phthisis.—**Dose:** 5—10 ℳ 3 t. daily, in sweetened hydro-alcoholic solut. —EXTERN. as paint with equal part chloroform, or in 10—20% oint.

Haema-, Haemo-,—see under HEMA-, HEMO-, etc.

Hamamelis—U. S. P.

WITCHHAZEL.—*Preparation:* F. E. (D., 15–60 min.).

Hedeoma—U. S. P.

PENNYROYAL.—*Preparations:* Oil (D., 3–10 min.); Spt. (10 per cent oil).

Hematoxylon—U. S. P.

LOGWOOD.—*Preparations:* Ext. (D., 10–20 grn.).

Hemogallol.

HEMOGLOBIN REDUCED BY PYROGALLOL, *Merck.*—Reddish-brown powd. containing iron in condition for easy assimilation.—Hematinic, Constructive, Tonic.—USES: Anemia, chlorosis, chronic nephritis, diabetes, and in convalescence; readily transformed into blood coloring-matter in debilitated people, and uniformly well borne; much superior to inorganic preparations of iron.—**Dose:** 4—8 grn., 3 t. daily, ½ hour before meals, in powd. with sugar, or in pills or chocolate tablets.

Hemoglobin Merck.

Brownish-red powd. or scales.—SOL. in water.—Hematinic.—USES: Anemia, chlorosis, etc.—**Dose:** 75—150 grn., daily, in wine or syrup.

Hemol.

HEMOGLOBIN REDUCED BY ZINC, *Merck.*—Dark-brown powd. containing easily assimilable iron, with slight traces of zinc oxide.—Hematinic, Antichlorotic.—USES: Anemia and chlorosis, neurasthenia, etc.—**Dose:** 2—8 grn., before meals, in powd. with sugar, or in wafers.

Hexamethylene-tetramine,—see FORMIN.

Hexamethylene-tetramine Salicylate,—see SALIFORMIN.

Holocaine.

Wh. needles.—SOL. in 40 parts water; undecomposed on boiling.—Local Anesthetic, like cocaine.—USES: Chiefly in eye diseases in 1% solut.

Homatropine Hydrobromate Merck.

Small white cryst.—SOL. in 10 parts water, 133 parts alcohol.—USES: Mydriatic in ophthalmic surgery; in night-sweats of phthisis, and as Sedative. Mydriatic effect commences in ¼ to ½ hour, reaches maximum in 1 hour, and disappears in 6 hours. Accommodation paresis ceases earlier. **Dose:** 1⁄120—1⁄60 grn.—EXTERN., to the eye, in 1% solut.

Honey—U. S. P.

Preparations. Clarified Honey; Honey of Rose; Confect. Rose—all vehicles.

Hops—U. S. P.

Preparation: fr. (D., 1–3 drams).

Hydrastine (Alkaloid) Merck.—C. P.

White prisms.—Sol. in alcohol, ether, chloroform; slightly in water.—Alterative, Tonic, Antiperiodic.—**Dose:** ¼—1 grn.

Hydrastine Hydrochlorate Merck.—C. P.—(*Not Hydrastinine, etc.*)

Amorph., white powd.—Sol. in water.—Astringent, Dermic, Tonic, Hemostatic.—Uses: *Intern.*, uterine hemorrhage, dyspepsia, hemorrhoids, etc.; *extern.*, gonorrhea, conjunctivitis, endometritis, leucorrhea, cervical erosions, acne, hyperidrosis, seborrhea, etc.—**Dose:** ½—1 grn., every 2 hours if necessary.—Extern. as *astringent*, $\frac{1}{10}$—½% solut.; in *skin diseases*, 1% oint'ts or lotions.

Hydrastinine Hydrochlorate Merck.—U. S. P.—C. P.—(*Not Hydrastine, etc.*)

Yellow, cryst. powd.—Sol. in water.—Uterine Hemostatic, Emmenagogue, Vaso-constrictor.—Uses: Hemorrhages, congestive dysmenorrhea, metrorrhagia, epilepsy, hemoptysis, etc.—**Dose:** ¼—½ grn., 3—4 t. daily, in capsules.

Hydrastis—U. S. P.

GOLDEN SEAL.—*Preparations:* F. E. (D., 10–30 min.): Glycerite (1:1 [extern.]): Tr. (D., 30–120 min.).

Hydrastis (Lloyd's).

Not completely defined.—(Stated: "Solution in glycerin and water of the valuable properties of hydrastis.—Colorl. liq.—Astringent, Tonic.—Used chiefly *extern.* (gonorrhea, leucorrhea, sore throat, etc.), in 1—2:16 dilut. —**Dose:** 10—40 min., 3 t. daily.")

Hydrogen Peroxide Solution,—see SOLUTION, HYDROGEN PEROXIDE.

Hydroleine.

Not completely defined.—(Stated: "2 fl. drs. contain 80 min. cod-liver oil, 35 min. dist. water, 5 grn. pancreatin, ⅓ grn. soda, ¼ grn. salicylic acid.—**Dose:** ¼—½ fl. oz., after each meal.")

Hydrozone.

Not completely defined.—(Stated: "30 vols. preserved aqueous solut. of H_2O_2.—Clear liq., acid taste.—Disinfectant, Cicatrizant.—**Dose:** 1 fl. dr., well dil., before meals.—Extern. in 2 or 3% solut.")

Hyoscine Merck.—True, Amorph.

From Hyoscyamus niger.—Thick, colorl. syrup.—Sol. in alcohol, ether; slightly in water.—Hypnotic, Sedative.—Uses: To quiet and give sleep to insane and others.—**Dose:** For *insane*, $\frac{1}{32}$ grn., cautiously increased or repeated until effect is produced; for *sane*, $\frac{1}{210}$—$\frac{1}{100}$ grn.—Injection: For *insane*, $\frac{1}{120}$—$\frac{1}{60}$ grn.; for *sane*, $\frac{1}{460}$—$\frac{1}{260}$ grn.—Antidotes: Emetics, stomach pump, muscarine, tannin, animal charcoal, emetics again; heat or cold externally; cathartics, etc.

Hyoscine Hydrobromate Merck.—U. S. P.—True, Cryst.

Colorl. cryst.—Sol. in water, alcohol.—Uses and Doses, same as Hyoscine.

(Other salts of Hyoscine are not described because used substantially as the above.)

Hyoscyamine, True, Merck.—C. P., Cryst.—(*Much stronger than Amorph.!*)

From Hyoscyamus niger.—White, silky, permanent cryst.—Sol. in alcohol, ether, chloroform, acidulated water; slightly in water.—Hypnotic,

Sedative.—Uses: To quiet insane and nervous; ease cough in consumption; asthma, etc.—**Dose:** $\frac{1}{128}$-$\frac{1}{32}$ grn., several t. daily, in pill or solut.; as *hypnotic* for insane, $\frac{1}{8}$-$\frac{1}{4}$ grn.—Antidotes: As for Atropine.

Hyoscyamine, True, Merck.—Pure, Amorph.—(*Much weaker than Cryst.!*)

Brown, syrupy liq.—**Dose:** $\frac{1}{8}$-$\frac{1}{4}$ ℿ.

Hyoscyamine Sulphate, True, Merck.—U. S. P.—C. P., Cryst.

White, deliquescent, microscopic needles; acrid taste.—Sol. in water, alcohol.—Uses, Dose, etc.: As of Hyoscyamine, True, *Cryst.*

Hyoscyamine Sulphate, True, Merck.—Pure, Amorph.

Yellowish, hygroscopic powd.—Sol. in water, alcohol.—**Dose:** $\frac{1}{8}$-$\frac{1}{4}$ grn.

Other salts of Hyoscyamine are not described because (used substantially as the above.)

Hyoscyamus—U. S. P.
Henbane.—*Preparations:* Ext. (D., 1-3 grn.); F. E. (D., 5-15 min.); Tr. (D., 20-60 min.)

▌chthalbin.

Ichthyol Albuminate, *Knoll.*—Gray-brown, odorl.,almost tastel. powd.; 4 parts equal 3 parts ichthyol.—Sol. in alkaline fluids (such as intestinal secretion); insol. in ordinary solvents and in diluted acids (as gastric juice).—Succedaneum for Ichthyol *internally* as an Alterant, Antiphlogistic, and Assimilative.—Uses: Phthisis, scrofula, rheumatism, skin diseases, etc.—**Dose:** 15—30 grn., 2 or 3 t. daily, before meals.—[Further information in "Merck's Digest" on "Ichthalbin," containing clinical reports and detailed information.]

Ichthyol.

Ammonium Sulpho-ichthyolate, *Ichthyol Co.*,—$(NH_4)_2C_{28}H_{36}S_3O_6$.— Thick, brown liq.; bituminous odor; 15% easily assimilable sulphur.—Sol. in water, mixture alcohol and ether; miscible with glycerin, oils.—Antiphlogistic, Anodyne, Alterative, Antigonorrhoic, Dermic.—Uses: *Intern.*, skin diseases, rheumatism, scrofula, nephritis: *extern.*, 5 to 50% oint., solut., etc.: uterine and vaginal inflammation, urticaria, erosions, pruritus, gout, boils, carbuncles, acne, eczema, herpes, burns, catarrh, etc.; 2% solut. in gonorrhœa.—**Dose:** 3—10 ℿ, in pills, capsules, or water.—(See "Ichthalbin,"—a preferable form for *internal* use.)

Ingluvin.

Digestive ferment obtained from gizzard of chicken.—Yellowish powd.— **Dose:** 5—20 grn.

Iodia.

Not completely defined.—(Stated: "Combination of active principles from green roots of stillingia, helonias, saxifraga, menispermum; with 5 grn. potass. iodide per fl. dr.—Alterative, Uterine Tonic.—**Dose:** 1—3 fl. dr., 3 t. daily.")

Iodine Merck.—U. S. P.—Resublimed.

Dose: $\frac{1}{4}$—1 grn.—*Preparations:* Oint. (4%); Comp. Solut. (5%, with 10% KI); Tr. (7%).—Antidotes: Emetics, stomach pump; starchy food in abundance.—Incompatibles: Oil turpentine, starch, tannin.

Iodipin.

Iodine addition-product of sesame oil.—Yellow fluid, of purely oleaginous taste; 10% iodine.—Alterative Tonic; carried even to remotest parts of body.—Uses: Syphilis, scrofula, etc.—**Dose:** 1—2 fl. drs., 3 or 4 t. daily, in emulsion with peppermint water and syrup; children in proportion.

Iodo–bromide of Calcium Comp. (Tilden's).

Not completely defined.—(Stated: "Each fl. oz. contains 72 grn. combined salts of bromine, iodine, and chlorine with calcium, magnesium, iron, sodium, potassium; together with combined constituents of 1 oz. mixed stillingia, sarsaparilla, rumex, dulcamara, lappa, taraxacum, menispermum.—Alterative, Tonic.—Uses: Scrofula, cancer, chronic coughs, eczema, etc.—**Dose:** 1—2 fl. drs., in water, before meals.")

Iodoform Merck.—U. S. P.—C. P., Cryst. or Powd.

Dose: 1–3 grn.—*Preparation:* Oint. (10%).—INCOMPATIBLE: Calomel.

Iodoformogen.

IODOFORM ALBUMINATE, *Knoll.*—Brown-yellow, fine, dry, non-conglutinating powd.; about 3 times as voluminous as iodoform, more pervasive, and free from its odor.—Especially convenient, economical, and efficient form of Iodoform; liberates the latter, on contact with wound surfaces, gradually and equably, and hence is more persistent in action.

Iodole.

TETRAIODO-PYRROLE, *Kalle.*—Light, fine, grayish-brown powd.; 89% iodine.—SOL. in alcohol, chloroform, oils; 3 parts ether; slightly in water.—Antiseptic, Alterative.—USES: *Intern.*, syphilis, scrofula, etc.; *extern.*, 5 to 10% oint. in chronic ulcers, lupus, chancre, etc.; powd. or solut. on mucous membranes, as in ozena, tonsillitis, etc.—**Dose:** 8—15 grn., daily, in wafers.

Iodothyrine.

THYROIODINE.—Dry preparation of thyroid gland.—Alterative, Discutient. —USES: Goiter, corpulency, myxedema, etc.—**Dose:** 15—40 grn. per day.

Ipecac—U. S. P.

Dose: *Stomachic,* ⅙-1 grn.; *emetic,* 10–20 grn.—*Preparations:* F. E. (1·1); Powd. of Ipecac and Opium (1.10 each); Troches (¼ grn.); Syr. (7 per cent. F. E.); Tr. Ipecac and Opium (D., 5–15 min.); Troches w. Morphine (one-twelfth grn. ipecac, one-fortieth grn. morph.); Wine (10 per cent. F. E.).

Iris—U. S. P.

BLUE FLAG.—*Preparations:* Ext. (D., 2–6 grn.); F. E. (D., 10–30 min.).

Iron, by Hydrogen, Merck.—(*Reduced Iron, U.S.P.*).

QUEVENNE'S IRON.—**Dose:** 2–5 grn.

Iron Acetate Merck.—Scales.

Dose: 3—10 grn.

Iron Albuminate Merck.—Scales or Powd.

Brown; very stable.—SOL. in water.—Hematinic.—**Dose:** 3–10 grn.

Iron Arsenate Merck.

Yellowish-green, insol. powd.—**Dose:** $\frac{1}{16}$—⅛ grn., in pill

Iron Carbonate, Mass—U. S. P.
    VALLET'S MASS.—50 per cent. Fe CO_3.—Dose: 3-5 grn., in pill

Iron Carbonate, Mixture—*Compound Iron Mixture, U. S. P*
    GRIFFITH'S MIXTURE.—Dose: 1-2 fl. oz.

Iron Carbonate, Saccharated, Merck, (*Saccharated Ferrous Carbonate, U. S. P.*).

Dose: 5—30 grn.

Iron Citrate Merck (*Ferric Citrate, U. S. P.*).—Scales.

Dose: 3—10 grn.

Iron Glycerino-phosphate Merck.

Yellowish scales.—SOL. in water, dil. alcohol.—USES: Deficient nerve-nutrition, neurasthenia, etc.—**Dose:** 2–5 grn., 3 t. daily, in cinnamon water.

Iron Hydrate with Magnesia—U. S. P.

ARSENIC ANTIDOTE.—(I) Solut. ferric sulphate 50 Cc., water 100 Cc.; (II) magnesia 10 Gm., water to make 750 Cc. (in a 1000 Cc. bottle). For immediate use, add I to II.

Iron Hypophosphite Merck (*Ferric Hypophos., U.S.P.*).—C. P.

Whitish powd.—Insol. in water.—**Dose:** 5—10 grn.

Iron Iodide, Saccharated, Merck, (*Saccharated Ferrous Iodide, U. S. P.*).

Dose: 2–5 grn.—CAUTION: Keep dark, cool, and well-stoppered!

Iron Lactate Merck (*Ferrous Lactate, U. S. P.*).—Pure.

Dose: 1–5 grn.

Iron Oxalate, Ferrous, Merck.

Pale-yellow, odorl., cryst. powd. –INSOL. in water.—**Dose:** 2–6 grn.

Iron Oxide, Red, Saccharated, Merck.—Soluble.

IRON SACCHARATE.—2.8% iron.—Brown powd.—SOL. in water.—USES: Antidote for arsenic; also in chlorosis, anemia, etc.—**Dose:** 10–30 grn.

Iron, Peptonized, Merck.—Powd. or Scales.

5% iron oxide, with peptone.—SOL. in water.—USES: Mild, easily assimilable chalybeate.—**Dose:** 5–20 grn.

Iron Phosphate, Soluble, Merck, (*Soluble Ferric Phosphate, U. S. P.*).

Dose: 5–10 grn.

Iron Pyro-phosphate, with Sodium Citrate, Merck, (*Soluble Ferric Pyro-phosphate, U. S. P.*).

Dose: 5–10 grn.

Iron Succinate Merck.

Amorph., reddish-brown powd.—SOL. slightly in cold water; easily in acids. -Tonic, Alterative. –USES: Solvent biliary calculi.—**Dose:** 10 grn., gradually increased to 60 grn. if necessary, after meals; associated with 10 drops of chloroform, 4 to 6 t. daily.

Iron Sulphate, Basic, Merck.—Pure.

MONSEL'S SALT; IRON SUBSULPHATE. –**Dose:** 2–5 grn

Iron Sulphate, Ferrous, Merck, (*Ferrous Sulphate, U. S. P.*)

Dose: 1–3 grn.

Iron Sulphate, Ferrous, Dried, Merck.

Best form for pills.—**Dose:** ½–2 grn.

Iron Tartrate, Ferric, Merck.

Brown scales.—SOL. in water.—**Dose:** 5–10 grn

Iron Valerianate Merck.—(*Ferric Valerianate, U. S. P.*).

Brick-red powd.; valerian odor; styptic taste.—Tonic, Nervine, Emmenagogue.—USES: Anemia or chlorosis. with hysteria or nervous exhaustion; epilepsy, chorea, etc.—**Dose:** 3–15 grn.

Iron and Ammonium Citrate Merck.—U. S. P.—Brown Scales.

SOL. in water.—**Dose:** 3–10 grn.—*Preparation:* Wine (4%).

Iron and Ammonium Sulphate, Ferric, Merck.—U. S. P.

AMMONIO-FERRIC ALUM.—**Dose:** 5–15 grn.

Iron and Manganese, Peptonized, Merck.

Brown powd.—SOL. in water.—**Dose:** 5–20 grn.

Iron and Potassium Tartrate Merck.—U. S. P.

Dose: 5–10 grn.

Iron and Quinine Citrate, Soluble, Merck.—U. S. P.

Dose: 3–10 grn.—*Preparation:* Bitter Wine Iron (5%).

Iron and Quinine Citrate, with Strychnine, Merck.

1% strychnine.—Green scales.—SOL. in water.—**Dose:** 2–5 grn.

Iron and Strychnine Citrate Merck.—U. S. P.

1% strychnine.—**Dose:** 2–5 grn.

Itrol,—see SILVER CITRATE.

Jalap—U. S. P.

Dose: 10–30 grn.—*Preparations:* Ext. (D., 2–5 grn.); Comp. Powd. (D., 20–60 grn.); Resin (D., 2–5 grn.).

Juice, Cineraria, Merck.

USES: *Extern.*, cataract of the eye; 2 drops 3 t. daily.

Kamala—U. S. P.

Dose: 1–2 drams, with hyoscyamus, in honey.

Kefir Fungi Merck.

USES: In making Kefir (" Kumyss ").—[Further information in descriptive circular.]

Keratin, Pepsinized, Merck.

Horn-substance purified by pepsin.—Yellowish-brown powd.—USES: Coating enteric pills.—[Further information in descriptive circular.]

Kermes Mineral,—see ANTIMONY, SULPHURATED.

Kino—U. S. P.

Dose: 10–20 grn.—*Preparation:* Tr. (1:10).

Koussein Merck.—Amorph.

BRAYERIN, KUSSEÏN.—Yellowish-brown powd.—SOL. in alcohol, ether, chloroform; slightly in water.—Anthelmintic.—**Dose:** 15–30 grn., divided into 4 parts, intervals of half hour; followed by castor oil. Children, half this quantity.

Kousso—U. S. P.

BRAYERA.—*Preparation:* F. E. (D., 1–4 drams).—See also, Koussein.

Krameria—U. S. P.

RHATANY.—**Dose:** 5–30 grn.—*Preparations:* Ext. (D., 2–10 grn.); F. E. (1:1); Syr. (45 per cent.); Tr. (1:5); Troches (1 grn. ext.).

Kryofine.

METHOXY-ACET-PHENETIDIN.—Colorl., odorl., powd.; faint bitter-pungent taste.—SOL. in 600 parts water; freely in alcohol, chloroform, ether.—Analgesic, Antipyretic.—**Dose:** 8–15 grn. in tabl. or powd.

Lactopeptine.

Not completely defined.—(Stated : " Contains pepsin, pancreatin, ptyalin, lactic and hydrochloric acids.—Grayish powd.—Digestant.—**Dose:** 10–20 grn., in powd. or tabl.")

Lactophenin.

LACTYL-PHENETIDIN.—Wh., odorl., slightly bitter powd.—SOL. in 500 parts water, 9 alcohol.—Antipyretic and Analgesic.—**Dose:** 8–15 grn.

Lactucarium Merck.—U. S. P.

Dose: *Hypnotic* and *anodyne*, 5—20 grn.; *sedative*, 3—8 grn.—*Preparations:* Tr. (1:2); Syr. (1:20).

Lanolin.

Wool-fat, analogous to Adeps Lanæ, which see.

Lappa—U. S. P.

BURDOCK.—Alterative.—*Preparation:* F. E. (D., 30–60 min.)

Largin.

Silver-albumin compound; 11% silver.—Gray powd.—SOL. in 9 parts water, also in glycerin.—Powerful Bactericide and Astringent, like silver nitrate but non-irritating and not precipitable by sodium chloride or albumin.—USES: Chiefly gonorrhea, in ¼—1½% solut. (according to stage), 3 t. daily.

Lead Acetate Merck.—U. S. P.—C. P., Cryst. or Powd.

Dose: 1—4 grn.—ANTIDOTES: Emetics. stomach siphon; sulphate of sodium or potassium or magnesium; milk, albumen, opium (in pain).—INCOMPATIBLES: Acids; soluble sulphates, citrates, tartrates, chlorides, or carbonates; alkalies, tannin, phosphates.

Lead Carbonate Merck.—C. P.

Not used internally.—*Preparation:* Oint. (10%).

Lead Iodide Merck.—U. S. P.—Powd.

Dose: 1—4 grn.—*Preparation:* Oint. (10%).

Lead Nitrate Merck.—U. S. P.—Pure, Cryst.

Dose: 1—4 grn.

Lemon Juice—U. S. P.

Preparation: Acid. Citric (q. v.).

Lemon Peel—U. S. P.

Preparations: Oil; Spt.; Syr.—all flavorings.

Lenigallol.

PYROGALLOL TRIACETATE, *Knoll.*—White powd.—INSOL. in water; sol. with decomposition in warm aqueous solut's of alkalies.—Mild succedaneum for Pyrogallol: non-poisonous, non-irritating, and non-staining. —APPLIED in ½ –5% oint.

Lenirobin.

CHRYSAROBIN TETRACETATE, *Knoll.*—INSOL. in water.—Mild "reactive" or "reducing" Dermic; succedaneum for Chrysarobin especially in herpes; non-poisonous, non-irritating, non-staining.—EXTERN. like chrysarobin.

Leptandra—U. S. P.

CULVER'S ROOT.—Dose: 20–60 grn.—*Preparations:* Ext. (D., 3–10 grn.); F. E. (1:1).

Leptandrin Merck.—Pure.

Dose: *Cholagogue* and *alterative*, 1—3 grn.; *purgative*, 8 grn.

Levico Water,—see AQUA LEVICO.

Lime Merck.—U. S. P.

CALCIUM OXIDE; BURNT LIME.—Escharotic, in cancers, etc.

Lime, Sulphurated, Merck.

(So-called "CALCIUM SULPHIDE".)—**Dose:** ¼—2 grn.

Lime Water,—see SOLUTION, CALCIUM HYDRATE.

Liquor,—see SOLUTION.

Listerine.

Not completely defined.—(Stated: "Essential antiseptic constituents of thyme, eucalyptus, baptisia, gaultheria, and mentha arvensis, with 2 grn. benzo-boric acid, in each fl. dr.—Clear, yellow liq. of arom. odor.—Antiseptic, Deodorant, Disinfectant.—**Dose:** 1 fl. dr., diluted.—EXTERN. generally in solut. up to 20%.")

Lithium Benzoate Merck.—U. S. P.

Dose: 5—20 grn.

Lithium Bromide Merck.—U. S. P.

Dose: 10—30 grn.

Lithium Carbonate Merck.

Dose: 5—15 grn.

Lithium Citrate Merck.

Dose: 5—15 grn.

Lithium Hippurate Merck.—C. P.

White powd.—SOL., slightly in hot water.—**Dose:** 5—15 grn.

Lithium Iodide Merck.

SOL. in water.—**Dose:** 1—5 grn.

Lithium Salicylate Merck.—U. S. P.—C. P.

Dose: 10—30 grn.

Lobelia—U. S. P.

Preparations: F. E. (D., 2-10 min.); Tr. (D., 10-40 min.).—See also, Lobeline.

Lobeline Sulphate (fr. Seed) Merck.

Very deliquescent, yellow, friable pieces.—SOL. in water, alcohol.—USES: Chiefly asthma; also dyspnea, whooping-cough, and spasmodic neuroses.—**Dose:** (*Spasmodic Asthma*): 1 grn. daily, gradually increasing to 3–6 grn. daily.—Children $\frac{1}{6}$–$\frac{3}{4}$ grn. daily.—ANTIDOTES: Stomach siphon, emetics, tannin; later brandy, spirit ammonia; morphine.

Loretin.

Yellow, odorl., insol. powd. Forms emulsions with ethereal and oily fluids (especially w. collodion).—Succedaneum for iodoform externally.—APPLIED like the latter.

Losophan.

TRI-IODO-CRESOL.—Colorl. needles, peculiar odor; 80% iodine.—SOL. in ether, chloroform; insol. in water.—Antiseptic, Vulnerary, Dermic.—EXTERN. in 1% solut. in 75% alcohol, or in 1—3% oint.

Lupulin—U. S. P.

Dose: 3-8 grn.—*Preparations:* F. E. (1:1); Oleores. (D., 2-5 grn.).

Lycetol.

DIMETHYL-PIPERAZINE TARTRATE.—Wh. powd.—SOL. in water.—Uric-acid Solvent, Diuretic.—USES: Gout, lithiasis, etc.—**Dose:** 4—10 grn.

Lycopodium—U. S. P.

Used only extern., as dusting-powd.

Lysidine.

50% solut. Ethylene-ethenyl-diamine.—Pinkish liq.; mousy odor.—MISCIBLE with water.—Uric-acid Solvent, Diuretic.—USES: Gout, lithiasis, etc.—**Dose:** 15—30 grn., in carbonated water.

Magnesium Carbonate.—U. S. P.

Antacid, Antilithic.—**Dose:** 30—120 grn.

Magnesium Citrate Merck.—Soluble.

Dose: 30—120 grn.

Magnesium Oxide, Light, Merck, (*Magnesia, U. S. P.*).

LIGHT or CALCINED MAGNESIA.—Light, white powd.: slightly alkaline taste.—SOL. in diluted acids, carbonic-acid water.—Antacid, Laxative, Antilithic.—USES: *Intern.,* sick headache, heartburn, gout, dyspepsia, sour stomach, constipation, gravel, and as antidote to arsenous acid. *Extern.,* ulcers and abraded surfaces: dusting-powd. for babies; and in tooth powders.—**Dose:** 10—30—60 grn. Small doses are antacid or antilithic; large are laxative.

Magnesium Oxide, Heavy, Merck, (*Heavy Magnesia, U. S. P.*).
Dose: 10—60 grn.

Magnesium Salicylate Merck.—C. P.

SOL. in water.—**Dose:** 15—60 grn.

Magnesium Sulphate Merck.—U. S. P.—C. P.

EPSOM SALT.—**Dose:** ½—1 oz.

Magnesium Sulphite Merck

USES: Instead of sodium sulphite: has less disagreeable taste.—**Dose:** 10—60 grn.

Maltzyme.

Not completely defined.—(Stated: "A concentrated, diastasic essence of malt.— Nutritive, Digestant —USES : Malnutrition, starchy indigestion, etc.—**Dose:** ½—1 fl. oz., during meals; children in proportion.")

Manganese Dioxide Merck.

MANGANESE PEROXIDE; BLACK OXIDE OF MANGANESE.—Containing over 90% MnO_2.—**Dose:** 2—10 grn.

Manganese Hypophosphite Merck.

Permanent rose-red cryst.—**Dose:** 10—30 grn.

Manganese Iodide Merck.

Brown, deliquescent masses. –SOL. in water, with decomposition.—USES: Anemia, chlorosis, scrofula, syphilis, and enlargement of spleen.—**Dose:** 1—3 grn.

Manganese, Peptonized, Merck.

Brown powd.; 4% manganic oxide.—SOL. in water.—USES: Anemia and chlorosis.—**Dose:** 10—30 grn.

Manganese Peroxide,—see MANGANESE DIOXIDE.

Manganese Sulphate Merck.—U. S. P.—Pure, Cryst.

SOL. in 1 part water.—**Dose:** 5—15 grn.

Manna—U. S. P.

Dose: ½—1 oz.

Marrubium—U. S. P.

HOREHOUND.—Used chiefly as infus. (1:16) taken hot, or as confectionery; in coughs, colds, etc.

Mastic—U. S. P.

MASTICHE.—*Preparations:* Pills Aloes and Mastic (2 grn. A., ⅔ grn. M.).

Matico—U. S. P.

Dose: 30–60 grn.—*Preparations:* F. E. (1:1); Tr. (1:10).

Matricaria—U. S. P.

GERMAN CHAMOMILE.—Used chiefly as tea, in colds.

Melachol.

Not completely defined.—(Stated: "Liquefied combination of sodium phosphate with sodium nitrate; 1 fl. dr. = 85 grn. sod. phosphate — Laxative, Nervine.—**Dose:** *Lax.*, 1–6 fl. drs., in water, before meals; *nerv*., ½ fl. dr., 3 t. daily.")

Melissa—U. S. P.

BALM.—Carminative.—See also, Spt. Melissa.

Menispermum—U. S. P.

YELLOW PARILLA.—**Dose:** 10–20 grn., in F. E. (1:1) or infus.

Menthol Merck.—U. S. P.—C. P., Recryst.

Dose: 3–5 grn.—For toothache: put a crystal into cavity.—Tampons, 1 in 5 of oil.

Mercauro.

Not completely defined.—(Stated: "10 min. contain $\frac{1}{32}$ grn. each gold, arsenic, and mercury bromides.—Alterative, Antisyphilitic.—**Dose:** 5–15 min., in water, after meals.")

Mercuro-iodo-hemol.

Brown powd.; 12.35% mercury, 28.68% iodine, with hemol.—Antisyphilitic (chiefly); without untoward action.—**Dose:** 2–5 grn., 3 t. daily, in pills.

Mercury—U. S. P.

Preparations: Mass (33 per cent.): Mercury with Chalk (D., 3–10 grn.); Oint. (50 per cent.); Plaster (18 per cent.); Ammoniac and Mercury Plaster (30 per cent. Hg.).

Mercury, Ammoniated,—see MERCURY-AMMONIUM CHLORIDE.

Mercury Benzoate, Mercuric, Merck.

White cryst.—SOL. in alcohol, solut. sodium chloride; slightly in water.—USES: Syphilis and skin diseases.—**Dose:** $\frac{1}{32}$–⅛ grn., in pills or hypodermically.

Mercury Bichloride Merck (*Corrosive Mercuric Chloride, U. S. P.*).—Recryst.

Dose: $\frac{1}{32}$–½ grn.—MAX. D.: ⅙ grn. single; ⅙ grn. daily.—ANTIDOTES: Zinc sulphate, emetics, stomach siphon, white of egg, milk in abundance, chalk mixture, castor oil, table salt, reduced iron, iron filings. White of egg and milk 2 or 3 t. daily for a week.—INCOMPATIBLES: Reduced iron, sulphurous acid, albumin, alkalies, carbonates.

Mercury Chloride, Mild, Merck.—U. S. P.

CALOMEL.—INCOMPATIBLES: Sulphurous acid, hydrocyanic acid; alkali chlorides, bromides, iodides, sulphites, carbonates, hydrates; organic acids, lime water, etc.

Mercury Cyanide Merck.

Dose: $\frac{1}{16}$—$\frac{1}{8}$ grn., in solut.—EXTERN. (gargle) 1:10000.

Mercury Imido-succinate,—see MERCURY SUCCINIMIDE.

Mercury Iodide, Red, Merck.

MERCURY BINIODIDE.—**Dose:** $\frac{1}{16}$—$\frac{1}{4}$ grn., in pills.

Mercury Iodide, Yellow, Merck.—U. S. P.

MERCURY PROTO-IODIDE.—**Dose:** $\frac{1}{2}$—2 grn. CAUTION: Never prescribe this with a soluble iodide, since mercury biniodide (highly poisonous) is formed!

Mercury Oxide, Black (Hahnemann), Merck.

HAHNEMANN'S SOLUBLE MERCURY.—Grayish-black powd.; decomposes on exposure to light.—**Dose:** $\frac{1}{4}$—3 grn.

Mercury Oxide, Red, Merck.—U. S. P.—Levigated.

Not used internally.—*Preparation:* Oint. (10%).—INCOMPATIBLES: Chlorides.

Mercury Oxide, Yellow, Merck.

Not used internally.—*Preparation:* Oint. (10%).

Mercury Oxycyanide Merck.

White, cryst. powd.—SOL. in water.—Antiseptic.—USES: *Extern.*, diphtheria, erysipelas, and skin diseases; said superior as antiseptic dressing to mercuric chloride because more active as germicide and less easily absorbed.—APPLIED in 0.6% solut. to wounds and in surgical operations.

Mercury Salicylate Merck.

White powd.; about 59% mercury.—SOL. in solut. of sodium chloride, dilute alkalies.—USES: *Extern.*, chancre, gonorrhea, and venereal affections; 1% powd. or oint.; *injection* in urethra, 1–5% water.—Reported easily borne by the stomach, and to produce no salivation.—**Dose:** $\frac{1}{3}$—1 grn.

Mercury Succinimide Merck.

MERCURY IMIDO-SUCCINATE.—White powd.—SOL. in 25 parts water; slightly in alcohol.—Antisyphilitic, Alterative.—Said to be free from disagreeable local and secondary effects.—**Dose:** $\frac{1}{8}$ grn., hypodermically.

Mercury Sulphate, Basic, Merck.

MERCURY SUBSULPHATE; TURPETH MINERAL.—**Dose:** *Emetic*, 2—5 grn.; *alterative*, $\frac{1}{4}$—$\frac{1}{2}$ grn.; in pills or powd.

Mercury Tannate Merck.

Greenish-gray powd.; about 50% mercury.—Antisyphilitic.—**Dose:** 1—2 grn., in pills.

Mercury-Ammonium Chloride Merck.—U. S. P.

WHITE PRECIPITATE; AMMONIATED MERCURY.—Not used internally.—*Preparation:* Oint. (10%).

Methyl Salicylate Merck.—U. S. P.

SYNTHETIC OIL GAULTHERIA (WINTERGREEN).—**Dose:** 5—30 ℳ.

Methylene Blue Merck.—C. P., Medicinal.

Bluish cryst., or blue powd.—SOL. in 50 parts water.—USES: Rheumatism, malaria, cystitis, nephritis, etc.—**Dose:** 2—4 grn., in capsules.—INJECTION: 1 grn.—MAX. D.: 15 grn., single or daily.—[Further information in "Merck's Digest" on "Methylene Blue," containing clinical reports.]

Mezereum—U. S. P.

MEZEREON.—Alterative.—**Dose:** 5–10 grn.—*Preparations:* F. E. (irritant). Enters into Comp. Decoct. Sarsaparilla, and Comp. F. E. Sarsaparilla.

Milk Sugar—U. S. P.

LACTOSE.—Nutritive, Diuretic.—**Dose:** 1–6 oz. a day, in milk.

Monsel's Salt,—see IRON SULPHATE, BASIC.

Morphine Merck.—U. S. P.—Pure, Cryst.

Almost insol. in water.—**Dose:** ⅛–½ grn.—ANTIDOTES: Emetics, stomach tube, permanganate potassium, paraldehyde, picrotoxin, atropine, strychnine, caffeine, cocaine, exercise, electric shock, etc.—INCOMPATIBLES: Alkalies, tannic acid, ⁴potassium permanganate, etc.

Morphine Hydrochlorate Merck.—U. S. P.

SOL. in 24 parts water.—**Dose:** ⅛–½ grn.

Morphine Meconate Merck.

MORPHINE BIMECONATE.—Yellowish-white powd.—SOL. in alcohol; 25 parts water.—Said to have less disagreeable effect on brain, stomach, and intestines than other morphine salts.—**Dose:** Same as Morphine.

Morphine Sulphate Merck.—U. S. P.

SOL. in 21 parts water.—**Dose:** ¼–½ grn.—*Preparations:* Comp. Powd. (1:60); Troches Morph. and Ipecac ($\frac{1}{40}$ grn. M., $\frac{1}{12}$ grn. I.).

(Other salts of Morphine are not described because used substantially as the above.)

Muscarine Nitrate Merck.

Brown, deliquescent mass.—SOL. in water, alcohol.—Antihidrotic, Antispasmodic.—USES: Night-sweats, diabetes insipidus; antidote to atropine, etc.—**Dose:** $\frac{1}{32}$–$\frac{1}{16}$ grn.

Muscarine Sulphate Merck.

USES and DOSES: Same as the Nitrate.

Musk—U. S. P.

Stimulant, Antispasmodic.—Dose: 3–10 grn.—*Preparation:* Tr. (i :20).

Mydrine Merck.

Combination of ephedrine and homatropine hydrochlorates (100:1).—Wh. powd.—SOL. in water.—Mydriatic.—USES: Where evanescent mydriasis is desired; especially valuable in diagnosis.—APPLIED in 10% solut.

Myrrh—U. S. P.

Astringent, Carminative. Cathartic, Emmenagogue.—**Dose:** 5–20 grn.—*Preparations:* Tr. (1:20); Tr. Aloes and Myrrh (each 10 per cent.); Pills Aloes and Myrrh (2 grn. A., 1 grn. M.).

Myrtol Merck.

Constituent of essential oil of Myrtus communis, L.—Clear, colorl. liq.; agreeable, ethereal odor.—SOL. in alcohol.—Antiseptic, Sedative, Stimulant.—USES: Chronic bronchitis, tonsillitis, cystitis.—**Dose:** 1–2 ℳ.

Naftalan.—(*Not Naphtalin!*)

NAPHTALAN.—Obtained by fractional distillation of a natural naphta from Armenia.—Blackish-green, unctuous, neutral mass; empyreumatic odor.—SOL. in fats, oils, ether, chloroform; insol. in water, glycerin.—Analgesic, Antiphlogistic, Parasiticide.—USES: Succedaneum for oil cade or oil tar in skin diseases; also in burns, contusions, epididymitis, etc.—CONTRA-INDICATED in very irritated conditions; ineffectual in psoriasis.—APPLIED pure, and well covered. The stains it may make readily disappear on immersion in kerosene or benzin.—Keep from air !

Naphtalin Merck.—U. S. P.—C. P., Medicinal.

Uses: *Intern.*, intestinal catarrhs, worms, cholera, typhoid fever, etc.; *extern.*, skin diseases.—**Dose**: 2–8 grn., in powd. or capsule: for tapeworm, 15 grn., followed some hours later by castor oil.—Max. D.: 30 grn.

Naphtol, Alpha-, Merck.—Recryst., Medicinal.

Colorl. or pinkish prisms; disagreeable taste.—Sol. in alcohol, ether; slightly in water.—Antiseptic, Antifermentative.—Uses: Diarrhea, dysentery, typhoid fever, and summer complaint.—**Dose**: 2–5 grn.

Naphtol, Beta-, Merck.—U. S. P.—Recryst., Medicinal.

Dose: 3–8 grn.—Max. D.: 10 grn. single; 30 grn. daily.

Naphtol, Beta-, Benzoate, Merck.—Pure.

Benzo-naphtol.—Whitish powd.; darkens with age.—Sol. in alcohol, chloroform. — Intestinal Disinfectant. — Uses: Diarrhea, dysentery, typhoid fever, cholera, etc.—**Dose**: 5–15 grn.

Narceine-sodium and Sodium Salicylate,—see Antispasmin.

Neurodin.

Acetyl-para-oxyphenyl-urethane, *Merck.*—Colorl., inodorous cryst.—Sol. slightly in water.—Antineuralgic, Antipyretic.—Uses: Sciatica, rheumatic pains, migraine, various forms of fever.—**Dose**: 15–25 grn. as *antineuralgic*; 5–10 grn. as *antipyretic*.

Neurosine.

Not completely defined.—(Stated: " Each fl. dr. represents 5 grn. each potass., sod., and ammon. bromides; zinc bromide ⅛ grn., ext. bellad. and ext. cannab. ind. each $\frac{1}{64}$ grn.; ext. lupuli 4 grn.; fl. ext. cascara 5 min.; with aromatic elixirs.—Neurotic, Anodyne, Sedative.—**Dose**: 1—2 fl. drs.")

Nickel Bromide Merck.

Greenish-yellow powd.—Sol. in water, alcohol, ether.—Nerve Sedative.—Uses: Epilepsy, etc.—**Dose**: 5–10 grn.

Nosophen.

Tetraiodo-phenolphtalein.—Yellow, odorl., tastel., insol. powd.; 60% iodine.—Surgical Antiseptic, like iodoform.

Nutgall—U. S. P.
  Galls.—*Preparations:* Tr. (D., 30–60 min.); Oint. (1:5).

Nutmeg—U. S. P.
  Aromatic, Carminative.—**Dose**: 5–20 grn.—*Preparations:* Oil (D., 1–5 min.); Spt. (5 per cent. oil).—Enters into Aromatic Powder, and Comp. Tr. Lavender.

Nux Vomica—U. S. P.
  Stomachic, Tonic, Respir. Stimulant.—**Dose**: 1–5 grn.—*Preparations:* Ext. (D., ⅛-½ grn.); F. E. (1:1); Tr. (2 per cent. ext.).—See also, Strychnine.

Oil, Almond, Bitter-, Merck.—U. S. P.

Dose: ⅙–½ ℳ.—Antidotes: Emetics, stomach siphon, ammonia, brandy, iron persulphate.—Caution: Poison !

Oil, Cade, Merck.—U. S. P.

Juniper Tar.—Uses: Only *extern.*, in psoriasis, favus, etc.

Oil, Cajuput—U. S. P.
  Stimulant, Diaphoretic.—**Dose**: 5–20 min.

Oil, Castor—U. S. P.
  Dose:—½–1 fl. oz., with saccharin or in emuls.

Oil, Cod-Liver—U. S. P.
  Dose: 1–4 drams.—See also, Gaduol.

Oil, Croton, Merck.—U. S. P.—Colorless.

USES : *Intern.*, obstinate constipation ; amenorrhea, dropsy ; *extern.*, rheumatism, neuralgia, and indolent swellings; hypodermically to nævi. — **Dose :** 1—2 ℳ, in pills.—ANTIDOTES : Stomach siphon, oils, mucilage, opium, cocaine, etc.—CAUTION : Poison !

Oil, Eucalyptus, Australian, Merck.

USES : *Intern.*, intermittent and remittent fever, bronchitis, cystitis, and dysentery, and by inhalation in asthma or catarrh; *extern.*, skin diseases. —**Dose :** 5—15 ℳ.

Oil, Gaultheria—U. S. P.
OIL WINTERGREEN.—Dose: 5-20 min.—*Preparation:* Spt. (5 per cent.).

Oil, Juniper Berries, Merck, (*Oil of Juniper, U. S. P.*).

Diuretic.—**Dose :** 5—15 ℳ.—*Preparations* : Spt. (5%); Comp. Spt. (0.4%).

Oil, Mustard, Natural, Merck, (*Volatile Oil of Mustard, U. S. P.*)—Rectified.

Dose : ⅛—¼ ℳ, with much water.—*Preparation :* Comp. Lin. (3%).

Oil, Olive—U. S. P.
Emollient, Nutrient, Laxative.—Dose: ¼-1 oz. ; in hepatic colic, 3-6 oz.

Oil, Pinus Pumilio, Merck.

OIL MOUNTAIN PINE.—Fragrant oil ; terebinthinous taste.—SOL. in alcohol, ether, chloroform.—Antiseptic, Expectorant.—USES : *Inhalation* in pectoral affections ; *intern.*, as stimulating expectorant ; *extern.*, lately employed in glandular enlargements, boils, and skin diseases.—**Dose :** 5—10 ℳ, in capsules.

Oil, Pinus Sylvestris, Merck.

OIL SCOTCH FIR ; OIL PINE NEEDLES.—Antiseptic, Antirheumatic.— USES : By *inhalation*, chronic pulmonary diseases ; *extern.*, in chronic rheumatism.

Oil, Rosemary—U. S. P.
Stimulant, Diuretic, Carminative, Emmenagogue.—Dose: 2-5 min.

Oil, Santal—U. S. P.
OIL SANDAL WOOD.—Internal Antiseptic, Anticatarrhal.—Dose: 5-20 min. in emuls. or capsules.

Oil, Tar—U. S. P.
Dose: 2-5 min.—Used chiefly extern.

Oil, Thyme—U. S. P.
Dose: 3-10 min.—Used chiefly extern.—See also, Thymol.

Oil, Turpentine, Rectified, Merck.—U. S. P.

For *internal* use only the *rectified* oil answers.—**Dose :** 5—30 ℳ; for tapeworm, 1—2 drams.—*Preparation :* Lin. (35%, with 65% resin cerate).

Ointment, Mercuric Nitrate—U. S. P.
CITRINE OINTMENT.—Stimulative and Alterative Dermic.—APPLIED in 10-50 per cent. dilution with fatty vehicle.

Ointment, Rose Water—U. S. P.
COLD CREAM.—18 per cent. borax.—Astringent Emollient.

Oleate, Cocaine, Merck.—5% and 10%.

Local Anesthetic.

Oleate, Mercury, Merck.—20% and 40%.

USES : *Extern.*, skin diseases, pediculi. Also for endermic administration of mercury.

Oleoresin, Capsicum, Merck.—U. S. P.

Sol. in alcohol, ether.—Rubefacient, Stimulant.—Uses : *Intern.*, flatulence, and to arouse appetite ; *extern.*, diluted with soap liniment or olive oil, in lumbago, neuralgia, and rheumatic affections.—**Dose :** ¼–1 ℳ, highly diluted, in beef tea or other hot liq.

Oleoresin, Male Fern, Merck, (*Oleoresin of Aspidium, U. S. P.*).

"Extract" Male Fern.—Thick, brown liq.; bitter, unpleasant taste. Efficacious and safe Anthelmintic. — **Dose:** In *Tænia solium* (the *usual* kind of tapeworm), 2½–3 drams, in *Tænia mediocanellata* 3–4 drams; in capsules, followed if necessary in 1–2 hours by calomel and jalap.

Merck's Oleoresin of Male Fern *exceeds* the requirements of the U. S. P., and conforms to the stricter demands of the Ph. G. III. Merck's preparation is made from rhizomes of a *pistachio-green* color inside, and only the crop of each current year is used.

Opium, Merck.—U. S. P.

Not less than 9 per cent. morphine.

Opium, Powdered, Merck.—U. S. P.

13-15 per cent morphine.—**Dose :** ½–2 grn.—Antidotes : Emetics, stomach-pump, warm coffee ; atropine or strychnine hypodermically, potass. permanganate, exercise.—*Preparations :* Deodorized (Denarcotized) Opium ; Ext. (D., ¼–1 grn.) ; Pills (1 grn.) ; Dover's Powder (Ipecac and Opium, ea. 10 per cent.) ; Tr. (1:10) ; Camph. Tr. (4:1000) ; Troches Liquorice and Opium (one-twelfth grn. O.) ; Vinegar (1:10) ; Wine (1:10).

Orange Peel, Bitter—U. S. P.

Preparations: F. E. (1:1); Tr. (1:5)—both flavorings.

Orange Peel, Sweet—U. S. P.

Preparations: Syr. (1.20); Tr. (1:5)—both flavorings.

Orexine Tannate.

Phenyl-dihydro-quinazoline Tannate, *Kalle.*—Yellowish-white, odorl. powd., practically tasteless.—Appetizer, Anti-emetic, Stomachic.—Uses : Anorexia in phthisis, chlorosis, cardiac diseases, surgical operations ; also for vomiting of pregnancy. Contra-indicated in excessive acidity of stomach and in gastric ulcers.—**Dose :** 4–8 grn., 2 t. daily; with chocolate.

Orphol,—see Bismuth Beta-Naphtolate.

Orthoform.

Methyl Ester of Meta-amido-para-oxybenzoic Acid.—Wh. odorl. powd.—Sol. slightly in water.—Local and intern. Anodyne, Antiseptic.—Uses : Chiefly extern., on painful wounds, burns, etc.—Applied pure or in trituration or oint.—**Dose:** 8—15 grn.

Ovariin Merck.

Dried ovaries of the cow.—Coarse, brownish powd.—Uses: Molimina climacterica and other ills referable to the ovaries.—**Dose:** 8—24 grn., 3 t. daily, in pills flavored with vanillin, or in tablets.

Pancreatin Merck.—Pure, Powd. or Scales.

Dose: 5—15 grn.

Papain Merck.

Papayotin.—Concentrated active principle of juice Carica Papaya, L. (Papaw).—An enzyme similar to pepsin, but acting in alkaline, acid, or neutral solut.—Whitish, hygroscopic powd.—Sol. in water, glycerin.—

USES: For dissolving false membrane, and for aiding digestion.—**Dose:** 2—5 grn.—EXTERN. in 5% solut. equal parts glycerin and water, for diphtheria and croup.—CAUTION: Not to be confounded with the vastly weaker preparations from papaw, known by various names.

Papine.

Not completely defined.—(Stated: "Anodyne principle of opium, without the narcotic and convulsive elements.—1 fl. dr. represents ⅛ grn. morphine.—**Dose:** 1—2 fl. drs.")

Paraformaldehyde Merck.

PARAFORM; TRIOXY-METHYLENE.—White, cryst. powd.—SOL. in water.—Antiseptic, Astringent.—USES: *Intern.*, cholera nostras, diarrhea, etc.; *extern.*, to generate (by heating) formaldehyde, for impregnating antiseptic bandages and surgical dressings, and for disinfecting atmosphere of rooms.—**Dose:** 8—15 grn., several t. daily.

Paraldehyde Merck.—U. S. P.—C. P.

Colorl. fluid; cryst. below 10.5° centigrade; peculiar, aromatic, suffocating odor and warm taste.—SOL. in alcohol, ether, oils, chloroform; about 10 parts water.—Hypnotic, Antispasmodic, Stimulant.—USES: Insomnia, and as antidote for morphine.—**Dose:** 30—90 ℳ, well diluted, with elixir, sweet water, brandy, or rum.

Pareira—U. S. P.

Diuretic, Laxative, Tonic.—**Dose:** 30–60 grn.—*Preparation:* F. E. (1:1).

Pelletierine Sulphate Merck.

PUNICINE SULPHATE.—Brown, syrupy liq.—SOL. in water, alcohol.—Anthelmintic.—**Dose:** 6 grn., with 8 grn. tannin, in 1 ounce water.—Give brisk cathartic in half an hour.

Pelletierine Tannate Merck.

Grayish-brown, hygroscopic, tastel. powd.—SOL. in 800 parts alcohol, 700 parts water.—Anthelmintic. Principal and most efficacious salt of Pelletierine.—**Dose:** 8—24 grn., in 1 ounce water, followed in 2 hours by cathartic.

Pepper—U. S. P.

Dose: 3–15 grn.—*Preparation:* Oleores. (D., ¼–1 min.).—See also, Piperin.

Peppermint—U. S. P.

Preparations: Oil (D., ·5 min.); Spt. (10 per cent oil); Troches (one-sixth min. oil); Water (one-fifth per cent. oil).—See also, Menthol.

Pepsin Merck.—U. S. P.—1:3,000; Powd., Granular, or Scales.

Dose: 5—15 grn.—INCOMPATIBLES: Alcohol, tannin, or alkali carbonates.

Pepsin, Saccharated, Merck.—U. S. P.--1:3oo.

Dose: 60—150 grn.

Peptenzyme.

Not completely defined.—(Stated: "Contains the digestive principles of the stomach, pancreas, liver, spleen, salivary and Brunner's glands, and Lieberkuhn's follicles.—Digestant.—**Dose:** 3—10 grn., 3 t. daily, in tabl., powd., or elix.")

Pepto-Mangan (Gude).

Not completely defined.—(Stated: "Aromatized solut. peptonized iron and manganese.—Hematinic.—**Dose:** 1—4 fl. drs., before meals.")

Peptonizing Tubes.

Each containing 25 grn. of peptonizing powder (pancreatin 1, sod. bicarb. 4) sufficient to peptonize 1 pint milk.

Peronin.

BENZYL-MORPHINE HYDROCHLORATE, *Merck.*—White powd.—SOL. readily in water; insol. in alcohol, chloroform, and ether.—Substitute for Morphine as a Sedative and Anodyne.—USES: Coughs, catarrhs, rheumatic and neuralgic pains, etc.; almost wholly free from the by-effects of morphine.—**Dose:** ⅓—1 grn., in pill or sweetened solut.

Phenacetin.

PARA-ACETPHENETIDIN.—Wh., tastel., cryst. powd.—SOL. in 1500 parts water, 16 alcohol.— Antipyretic, Antineuralgic, Analgesic.—**Dose:** *Antipyr.*, 8--10 grn.; *analg.*, 15—24 grn.; *children*, up to 5 grn.

Phenalgin.

Not completely defined.—(Stated: "AMMONIO-PHENYLACETAMIDE.—Wh. powd., of ammoniacal odor and taste.—Antipyretic, Analgesic.—**Dose:** *Antipyr.*, 5—10 grn.; *analg.*, 10—20 grn.; in tabl., caps., or cachets.")

Phenocoll Hydrochlorate.

Colorl. needles.—SOL. in 16 parts water.—Antipyretic, Analgesic, Antiperiodic.—**Dose:** 5—15 grn.

Phosphorus—U. S. P.

SOL. in oils.—**Dose:** one one-hundredth to one-thirty-second grn.—*Preparations:* Elix. (21 per cent. Spt. Phosph.); Oil (1 per cent.); Pills (one one-hundredth grn.); Spt. (¾ per cent.).—ANTIDOTES: Emetics, stomach-pump; 1 per cent. solut. potass. permang.; avoid oils.—INCOMPATIBLES: Sulphur, iodine, oil turpentine, potass. chlorate, etc.—CAUTION: Inflammable ! Keep under water.

Physostigma—U. S. P.

CALABAR BEAN.—*Preparations:* Ext. (D., one-twelfth to ¼ grn.); Tr. (D., 5-15 min.).—See also, Eserine (Physostigmine).

Physostigmine,—see ESERINE.

Phytolacca Root—U. S. P.

POKE ROOT.—Alterative, Antifat.—Dose: 1-5 grn.—*Preparation:* F. E. (1:1).

Picrotoxin Merck.—U. S. P.

COCCULIN.—Antihidrotic, Nervine, Antispasmodic.—USES: Night-sweats of phthisis; also paralysis, epilepsy, chorea, flatulent dyspepsia, dysmenorrhea; also antidote to chloral.—**Dose:** $\frac{1}{150}$—$\frac{1}{30}$ grn.—MAX. D.: $\frac{1}{10}$ grn.—ANTIDOTES: Emetics, stomach siphon, chloral hydrate, and stimulants.

Pilocarpine Hydrochlorate Merck.—U. S. P.

Sialagogue, Myotic, Diaphoretic, Diuretic.—USES: *Intern.*, dropsy, coryza, laryngitis, bronchitis, asthmatic dyspnea, uremic convulsions, croup, pneumonia, etc.; as antidote to atropine; contra-indicated in heart failure and during fasting; *extern.*, 1—2% aqueous solut. for collyrium.—**Dose:** ⅙—¼ grn. in water, hypodermically, or by mouth.—MAX. D.: ⅓ grn.—ANTIDOTES: Emetics, stomach siphon, atropine, ammonia, brandy.—INCOMPATIBLES: Silver nitrate, corrosive sublimate, iodine, alkalies.

(Other salts of Pilocarpine are not described because used substantially as the above.)

Pilocarpus—U. S. P.

JABORANDI.—Dose: 10-30 grn.—*Preparation:* F. E. (1:1).—See also, Pilocarpine.

Pimenta—U. S. P.

ALLSPICE.—Aromatic, Stomachic.—Dose: 10-40 grn.—*Preparation:* Oil (D., 2-5 min.).

Piperazine.

DIETHYLENE-DIAMINE.—Colorl., alkaline cryst.—SOL. freely in water.—Antipodagric, Antirheumatic.—**Dose:** 5—10 grn. 3 t. a day, well diluted.

Piperin Merck.—U. S. P.

Stomachic and Antiperiodic.—USES: Feeble digestion, and as substitute for quinine in remittent and intermittent fevers.—**Dose:** *Stomachic*, ½—1 grn.; *antiperiodic*, 6—8 grn., both in pills.

Pitch, Burgundy—U. S. P.

Used only extern., as counterirritant.—*Preparations.* Plaster (80 per cent.); Canthari-dal Pitch Plaster (8 per cent. cerate cantharides, 92 per cent. pitch).

Podophyllin,—see RESIN, PODOPHYLLUM.

Podophyllum—U. S. P.

MAY APPLE.—*Preparations:* Ext. (D., 2-5 grn.); F. E. (D., 10-30 min.); Resin (D., ⅙—½ grn.).—See also, Resin Podophyllum.

Pomegranate—U. S. P.

Dose: 1-2 drams, as decoct. (1:4) or fl. ext. (1:1).—See also, Pelletierine.

Potassa,—see POTASSIUM HYDRATE.

Potassa, Sulphurated, Merck.—U. S. P.—Pure.

USES: *Intern.*, small doses increase frequency of pulse: large doses: rheumatism, gout, scrofula, painter's colic, skin diseases, catarrh, croup; antidote in lead and mercury poisoning; *extern.*, lotion in parasitic skin diseases.—**Dose:** 2—10 grn.—ANTIDOTES: Emetics, stomach siphon, lead or zinc acetate, brandy.—INCOMPATIBLES: Acids, alcohol, carbonated waters, etc.

Potassa, Sulphurated, Merck.—Crude.

USES: For baths in skin affections, 2—4 ounces to one bath.—CAUTION: Avoid metal bath-tubs, metal spoons, and water with much carbon dioxide.

Potassium Acetate Merck.—C. P.

Very deliquescent.—SOL. in 0.36 part water, 1.9 parts alcohol.—**Dose:** 10—60 grn.

Potassium Antimonate Merck.—Purified, Washed.

DIAPHORETIC ANTIMONY; "WHITE OXIDE ANTIMONY".—White powd.—Diaphoretic, Sedative.—USES: Pneumonia, puerperal fever, etc.—**Dose:** 8—24 grn.

Potassium Arsenite Merck.—Pure.

White powd.—SOL. in water.—**Dose:** $\frac{1}{32}$—$\frac{1}{16}$ grn.

Potassium Bicarbonate Merck.—U. S. P.—C. P., Cryst. or Powder.

SOL. in water.—Diuretic, Antilithic, Antacid.—USES: Dyspepsia, dropsy, lithiasis, sour stomach, jaundice, etc. Usually taken effervescent with tartaric or citric acid.—**Dose:** 20—60 grn.

Potassium Bichromate Merck.—U. S. P.—C. P., Cryst.

SOL. in 10 parts water.—Corrosive, Astringent, Alterative.—USES: *Intern.*, syphilis: *extern.*, sweating feet, tubercular nodules, syphilitic vegetations, and warts.—**Dose:** $\frac{1}{16}$—$\frac{1}{4}$ grn.—EXTERN. in 5% solut. for sweating feet; 10% solut. as caustic.—ANTIDOTES: Emetics and stomach pump, followed by soap, magnesia, or alkali carbonates.

Potassium Bisulphate Merck.—C. P., Cryst.

Colorl., more or less moist, plates.—SOL. in water.—Aperient, Tonic.—USES: Constipation with weak appetite.—**Dose:** 60–120 grn., with equal weight sodium carbonate.

Potassium Bitartrate Merck.—C. P., Cryst. or Powd.

CREAM OF TARTAR.—**Dose:** 1–8 drams.

Potassium Bromide.—U. S. P.

Dose: 15–60 grn.

Potassium Cantharidate Merck.

White, amorph. powd., or cryst. mass.—SOL. in water.—USES: Hypodermically in tuberculosis (Liebreich).—INJECTION: 3–6 ♏ of 3:5000 solut.

Potassium Carbonate Merck.—U. S. P.—C. P.

Dose: 10–30 grn.

Potassium Chlorate Merck.—U. S. P.—C. P.

Dose: 10–20 grn.—*Preparation:* Troches (4½ grn.).—INCOMPATIBLES: Iron iodide, tartaric acid.—CAUTION: Do not triturate with sulphur, phosphorus, or organic or combustible compounds. Inflames or explodes with sulphuric acid and any organic powd. Do not administer on empty stomach!

Potassium Citrate Merck.—U. S. P.—Pure.

SOL. in 0.6 part water; slightly in alcohol.—USES: Rheumatism, lithiasis, fevers.—**Dose:** 20–25 grn.

Potassium Cyanide Merck.—C. P.

SOL. in 2 parts water; slightly in alcohol.—Sedative, Antispasmodic, Anodyne. USES: *Intern.,* dyspnea, asthma, phthisis, catarrh, whooping-cough, etc.; *extern.,* 0.2–0.8% aqueous solut. in neuralgia and local pains; 0.6–1.2% aqueous solut. removes silver-nitrate stains from conjunctiva.—**Dose:** ⅛ grn.—ANTIDOTES: Chlorine water, chlorinated-soda solut., ammonia, cold affusion, 10 grn. iron sulphate with 1 dram tincture of iron in ounce of water.—INCOMPATIBLES: Morphine salts, acid syrups, and silver nitrate.

Potassium Glycerino-phosphate Merck.—50% Solut.

Thick liq.—SOL. in water.—Nerve-tonic.—USES: Neurasthenia, phosphaturia, convalescence from influenza, etc.—INJECTION: 3–4 grn. daily, in water containing sodium chloride.

Potassium Hydrate Merck.—C. P.

CAUSTIC POTASSA.—SOL. in water, alcohol.—Escharotic, Antacid, Diuretic.—**Dose:** ¼–1 grn., highly diluted with water.—*Preparation:* Solut. (5%).—ANTIDOTES: Vinegar, lemon juice, orange juice, oil, milk; opium if pain; stimulants in depression.

Potassium Hydrate with Lime (*Potassa with Lime, U. S. P.*). —Powder.

VIENNA CAUSTIC; POTASSA-LIME.—USES: *Extern.,* cautery, in paste with alcohol.

Potassium Hypophosphite Merck.

SOL. in 0.6 part water, 7.3 parts alcohol.—**Dose:** 10–30 grn.—CAUTION: Explodes violently on trituration or heating with any nitrate, chlorate, or other oxidizer.

Potassium Iodide Merck.—C. P.

Sol. in 0.75 part water, 2.5 parts glycerin, 18 parts alcohol.—Incompatibles: Chloral hydrate, tartaric acid, calomel, silver nitrate. potassium chlorate, metallic salts, acids.—*Preparation:* Oint. (12%).

Potassium Nitrate Merck.—U. S. P.—C. P.

Saltpeter; Niter.—Sol. in 3.8 parts water.—**Dose:** 10—60 grn.—*Preparation:* Paper (fumes inhaled in asthma).

Potassium Nitrite Merck.—C. P.

White, deliquescent sticks.—Sol. in water.—Uses: Asthma, epilepsy, hemicrana.—**Dose:** ¼—2 grn. several t. daily.

Potassium Permanganate Merck.—U. S. P.

Sol. in 16 parts water.—Disinfectant, Deodorant, Emmenagogue.—**Dose:** 1—2 grn., in solut. or pills made with kaolin and petrolatum, or with cacao butter, after meals.—Incompatibles: All oxidizable substances, particularly organic ones, such as glycerin, alcohol, etc.—Remove stains with oxalic, or hydrochloric, acid.

Potassium Phosphate, Dibasic, Merck.—C. P.

Deliquescent, amorph., white powd.—Sol. in water.—Alterative.—Uses: Scrofula, rheumatism, phthisis, etc.—**Dose:** 10—30 grn.

Potassium Salicylate Merck.

White, slightly deliquescent powd.—Sol. in water, alcohol.—Antirheumatic, Antipyretic, Analgesic.—Uses: Rheumatism, pleurisy, pericarditis, lumbago, muscular pains, etc.—**Dose:** 6—15 grn.

Potassium Sulphate Merck.—U. S. P.—C. P.

Sol. in 9.5 parts water.—Uses: Constipation, and as antigalactic.—**Dose:** 20—120 grn., several t. daily, in solut.

Potassium Sulphite Merck.—Pure.

White, opaque cryst., or slightly deliquescent, white powd.—Sol. in 4 parts water, slightly in alcohol.—Antizymotic.—Uses: Acid fermentation of stomach, and gastric ulceration.—**Dose:** 15—60 grn.

Potassium Tartrate Merck.—Pure.

Soluble Tartar.—Colorl. cryst.—Sol. in 1.4 parts water.—Diuretic, Laxative.—**Dose:** *Diuretic*, 15—30 grn., *laxative*, 1—3 drams.

Potassium Tellurate Merck.—C. P.

White cryst.—Sol. in water.—Antihidrotic.—Uses: Night-sweats of phthisis.—**Dose:** ½—¾ grn., at night, in pills or alcoholic julep.

Potassium and Sodium Tartrate Merck.—U. S. P.—C. P.

Rochelle, or Seignette, Salt.—**Dose:** 2—8 drams.—*Preparation:* Seidlitz Powder.

Powder, Antimonial—U. S. P.

James's Powder.—33 per cent. antimony oxide.—Alterative, Diaphoretic, Antipyretic.—Dose: 2-10 grn.

Propylamine, so-called,—see Solution, Trimethylamine.

Prostaden.

Standardized Dried Extract Prostate Gland, *Knoll.*—Uses: Hypertrophy of prostate.—**Dose:** Up to 40 grn., daily, in tablets or powder.

Protargol.

Proteid compound of silver; 8% silver.—Yellow powd.—Sol. in water.—Antigonorrhoic.—Applied in ¼—1% solut.

Protonuclein.

Not completely defined.—(Stated: "Obtained from the lymphoid structures of the body by direct mechanical and physiological processes."—Brownish powd.—Antitoxic, Invigorator, Cicatrizant.—**Dose:** 3–10 grn., 3 t. daily.—EXTERN. [to cancers] pure.")

Ptyalin Merck.

Amylolytic ferment of saliva.—Yellowish powd.—SoL. in glycerin; partly in water.—USES: Amylaceous dyspepsia.—**Dose:** 10–30 grn.

Pulsatilla—U. S. P.

Antiphlogistic, Sedative, Antispasmodic.—Used chiefly in 1·10 tinct., the dose of which is 2–10 min.

Pumpkin Seed—U. S. P.

Anthelmintic.—**Dose:** 1–2 drams.

Pyoktanin, Blue.—Powder.—Also, Pencils.

PENTA- and HEXA-METHYL-PARAROSANILINE HYDROCHLORATE, *Merck.*—Non-poisonous, violet, cryst. powd.; nearly odorl.; solut. very diffusible in animal fluids.—SoL. in 12 parts 90% alcohol, 50 glycerin, 75 water; insol. in ether.—Antiseptic, Disinfectant, Analgesic.—USES: Surgery, ophthalmiatric and otiatric practice, diseases of throat and nose, gonorrhea, leucorrhea, varicose ulcers, burns, wounds, malignant and syphilitic neoplasms, conjunctivitis, etc. Stains removed by soap, rubbing well and washing with alcohol.—**Dose:** In pyloric carcinoma, 1–5 grn., in caps.: at first once daily, then 2, finally 3 t. a day.—MAX. D.: 10 grn.—EXTERN. pure, or 1:100—1:100 solut.

Pyoktanin, Yellow.—Powder.—Also, Pencils.

IMIDO-TETRAMETHYL-DIAMIDO-DIPHENYL-METHANE HYDROCHLORATE, *Merck*; APYONINE; C. P. AURAMINE.—Yellow powd.—SoL. in water, alcohol.—Antiseptic, Disinfectant.—USES: Considerably weaker than the blue, and principally employed in diseases of skin and in ophthalmiatric practice.

Pyrethrum—U. S. P.

PELLITORY.—Topical Sialagogue; not used internally.—*Preparation:* Tr. (1:5).

Pyridine Merck.—C. P.

Colori., limpid, hygroscopic liq.; empyreumatic odor; sharp taste.—MISCIBLE with water, alcohol, ether, fatty oils, etc.—Respiratory Sedative, Antigonorrhoic, Antiseptic.—USES: Asthma, angina pectoris, dyspnea, gonorrhea, etc. Contra-indicated in heart weakness.—**Dose:** 2–10 drops, several t. daily in water. Usually by *inhalation;* 45—75 ℳ, evaporated spontaneously in room. As urethral *injection,* ⅜% solut.; as *paint,* 10% solut.

Pyrogallol,—see ACID, PYROGALLIC.

Quassia—U. S. P.

Dose: 10–30 grn.—*Preparations:* Ext. (D., 2–5 grn.); F. E. (1:1), Tr. (1·10), Infus. (1:60).—Used by enema as teniacide.

Quassin, Merck.—C. P.

Intensely bitter cryst. or powd.—SoL. in alcohol, chloroform: slightly in water.—Tonic, Stimulant.—USES: Invigorate digestive organs.—**Dose:** ¹⁄₃₀–⅓ grn.

Quillaja—U. S. P.

SOAP BARK.—Expectorant, Antiparasitic, Antihidrotic —**Dose:** 10–30 grn.—*Preparation:* Tr. (1:5).

Quinalgen.

ANALGEN.—Derivative of quinoline.—Wh., tastel., insol. powd.—Anodyne.—USES: Sciatica, migraine, gout, rheumatism, etc.—**Dose:** 5—15 grn.

Quinidine Merck.

CHINIDINE; CONCHININE.—From some species of Cinchona bark.—Colorl. prisms; effloresce on exposure.—SOL. in 20 parts alcohol, 30 parts ether, 2000 water.—Antiperiodic, Antipyretic, Antiseptic, Tonic.—USES: Substitute for quinine. Salts less agreeable to take, but more prompt in action.—**Dose:** *Tonic,* ½—3 grn.; *antiperiodic,* 20—30 grn.; for a *cold,* 5—10 grn. in syrup, capsule, or pill.—MAX. D.: 40 grn.

Quinidine Sulphate Merck.—U. S. P.

SOL. in 8 parts alcohol, 100 water.—**Dose:** As of quinidine.

Quinine (Alkaloid) Merck.—U. S. P.

The salts are usually prescribed. For hypodermic use, the bisulphate, dihydrochlorate, or carbamidated hydrochlorate is to be preferred.—**Dose:** *Tonic,* ½—2 grn. 3 t. daily; *antiperiodic,* 8—15 grn. 6—12 hrs. before paroxysm; *antipyretic,* 15—30 grn. in the course of an hour.

Quinine Bisulphate.—U. S. P.

SOL. in 10 parts water, 32 parts alcohol; eligible for subcutaneous use.—NASAL INJECTION (in hay fever): 0.2% aqueous solut.—**Dose :** Same as of quinine alkaloid.

Quinine Dihydrochlorate Merck.

Well adapted to subcutaneous injection, on account of solubility.—**Dose:** Same as of quinine alkaloid.

Quinine Glycerino-phosphate Merck.

Colorl. needles; 68% quinine.—SOL. in water, alcohol.—Nervine, chiefly in malarial neurasthenia, malnutrition, or neuralgia.—**Dose:** 2—5 grn., 3 t. daily, in pills.

Quinine Hydrobromate Merck.

Dose: Same as of quinine alkaloid.

Quinine Hydrochlorate Merck.

SOL. in 3 parts alcohol, 9 parts chloroform, 34 parts water.—**Dose:** Same as of quinine alkaloid.

Quinine Salicylate Merck.

White, bitter cryst.—SOL. in 20 parts alcohol, chloroform, 120 parts ether, 225 parts water.—USES: Typhoid, rheumatism, lumbago, and muscular pain from cold.—**Dose:** 2—30 grn., in pill or caps.

Quinine Sulphate.—U. S. P.

SOL. in dil. acids; 740 parts water, 65 alcohol, 40 glycerin.—**Dose:** Same as of quinine alkaloid.—INCOMPATIBLES : Ammonia, alkalies, tannic acid, iodine, iodides, Donovan's solution, etc.

Quinine Tannate Merck.—Neutral and Tasteless.

Light-brown, insol. powd.—USED chiefly for children.—**Dose** (Children): 5—15 grn., with chocolate, in powd. or tablets.

Quinine Valerianate Merck.—U. S. P.

Slight odor of valerian.—Sol. in 5 parts alcohol, 100 parts water.—Nerve tonic, Antipyretic, etc.—Uses: Hemicrania and debilitated or malaria condition with a nervous state or hysteria.—**Dose:** 2—6 grn.

Quinine & Urea Hydrochlorate Merck.

Carbamidated Quinine Dihydrochlorate.—Colorl. cryst.—Sol. freely in water, alcohol.—Used by Injection: 2—8 grn.

    (Other salts of Quinine are not described because used substantially as the above.)

Quinoidine Merck.

Chinoidine.—Very bitter, brownish-black mass.—Sol. in diluted acids, alcohol, chloroform.—Antiperiodic, Tonic, etc.—Uses: Intermittent and remittent fevers. Best taken between paroxysms.—**Dose:** 2—15 grn.

Resin—U. S. P.

    Rosin; Colophony.—Vulnerary; Irritant.—*Preparations:* Cerate (35 per cent.) Plaster (14 per cent.).

Resin, Jalap, Merck.—U. S. P.—True, Brown.

Heavy Jalap Resin.—Sol. in alcohol; partly solut. in ether.—**Dose:** 2—5 grn.

Resin, Podophyllum, Merck.—Perfectly and Clearly Sol. in Alcohol and in Ammonia.

Podophyllin.—In habitual constipation, small continued doses act best. —**Dose:** ⅛—½ grn.; in *acute* constipation, ¾—1½ grn.

Resin, Scammony, Merck.—White, and Brown.

Dose: 3—8 grn.

Resinol.—(*Not Retinol!*)

Unguentum Resinol.—Not completely defined.—(Stated: " Combination of active principle of Juniperus oxycedrus and a synthetical derivative of the coal-tar series, with lanolin-petrolatum base.—Antipruritic, Antiphlogistic, Dermic.—Extern.: pure, night and morning.")

Resorcin Merck.—C. P., Resublimed or Recryst.

Resorcinol.—White cryst.: reddish on exposure; unpleasant sweet taste.—Sol. in 0.5 part alcohol, 0.6 part water; ether, glycerin.—Antiseptic, Antispasmodic, Antipyretic, Anti-emetic, Antizymotic.—Uses: *Intern.*, for vomiting, seasickness, asthma, dyspepsia, gastric ulcer, cholera infantum, hay-fever, diarrhea, whooping-cough, cystitis, and diphtheria; *extern.*, inflammatory diseases of skin, eyes, throat, nose, mouth, urethra, vagina, etc.—**Dose:** *Seasickness*, chronic gastric catarrh, cholera nostras, or cholera morbus, 2—3 grn. every 1—2 hours, in solut. or powder; *ordinary*, 5—10 grn. several t. daily; *antipyretic*, 15—30 grn.—Max. D.: 45 grn.—Extern. in 5—30% solut.

Retinol Merck.

Rosin Oil.—Viscid, yellow, oily liq.—Sol. in ether, oils, alcohol, oil turpentine, glycerin.—Antiseptic.—Uses: *Intern.*, venereal affections; *extern.*, oint. or liniment in skin diseases, and injection for gonorrhea; also solvent of phosphorus, salol, camphor, naphtol, carbolic acid, etc. Recommended as excipient for phosphorus.—**Dose:** 5—10 ♏, 4—6 t. daily, in capsules.—Extern.: 10—50% oint.

Rhubarb—U. S. P.

Dose: *Tonic*, 3-10 grn., *lax.*, 10–20 grn.: *purg.*, 20–40 grn.—*Preparations:* Ext. (D., 1-3 –10 grn.); F. E. (1:1); Pills (3 grn.); Comp. Pills (rhub., aloes, myrrh); Tr. (1:10); Arom. Tr. (1:5); Sweet Tr. (1:10, with liquorice and glycerin); Syr. (10 per cent. F. E.); Arom. Syr. (15 per cent. arom. tr.); Comp. Powd. (rhub., 25; magnes., 65; ginger, 10).

Rhus Glabra—U. S. P.
  SUMACH BERRIES.—Astringent.—*Preparation*: F. E. (D., 30–60 min.).

Rhus Toxicodendron—U S P
  POISON IVY; POISON OAK.—Alterative, Cerebral and Spinal Stimulant.—Used mostly as 20 per cent. tr., 5–30 min. per dose.

Rochelle Salt,—see POTASSIUM & SODIUM TARTRATE.

Rose, Red—U. S. P.
  Astringent.—*Preparations*: F. E. (30–60 min.); Confect. (8:100), Honey (12 per cent. F. E.); Syr. (12½ per cent. F. E.).

Rubidium Iodide Merck.

White cryst.—SOL. in water.—Alterative.—USES: As potassium iodide. Does not derange stomach.—**Dose**: 1–5 grn.

Rubidium & Ammonium Bromide Merck.

White, or yellowish-white, powd.; cooling taste; saline after-taste.——SOL. in water.—Anti-epileptic, Sedative, Hypnotic.—USES: Epilepsy, and as soporific, instead of potassium bromide.—**Dose**: *Anti-epileptic*, 60—100 grn. daily, in solut.; *hypnotic*, 60—75 grn.

Rubus—U. S. P.
  BLACKBERRY.—Astringent.—*Preparations*: F. E. (D., 30–60 min.); Syr. (25 per cent. F. E.).

Rumex—U. S. P.
  YELLOW DOCK.—Alterative, Antiscorbutic.—*Preparation*: F. E. (D., 15–60 min.).

Saccharin Tablets Merck.

Each tablet equal in sweetness to a large lump of sugar.—USES: For sweetening tea, coffee, and other beverages.

Saccharin.—Refined.

BENZOYL-SULPHONIC IMIDE, *Fahlberg;* GLUSIDE.—White powd.; over 500 times as sweet as cane sugar.—SOL. in 50 parts ether, 30 parts alcohol, 230 parts water. Alkaline carbonates increase solubility in water.—Non-fermentable Sweetener.—USES: Sweeten food of diabetics and dyspeptics; cover taste of bitter and acrid remedies.

Saffron—U. S. P.
  Dose: 10–20 grn.—*Preparation*: Tr. (1:10).

Salicin Merck.

SOL. in 28 parts water, 30 parts alcohol.—Tonic, Antiperiodic, Antirheumatic.—USES: Rheumatism, malaria, general malaise, and chorea.—**Dose**: 20—30 grn.—MAX. D.: 150 grn. daily.

Saliformin.

HEXAMETHYLENE-TETRAMINE SALICYLATE, *Merck;* FORMIN SALICYLATE.—White, cryst. powd., of agreeable acidulous taste.—SOL. easily in water or alcohol.—Uric-acid Solvent and Genito-urinary Antiseptic.—USES: Gout, gravel, cystitis, etc.—**Dose**: 15—30 grn. daily.

Salligallol.

PYROGALLOL DISALICYLATE, *Knoll.*—Resinous solid.—SOL. in 6 parts acetone, 15 parts chloroform.—Skin varnish, of weak pyrogallol effect.—USES: Chiefly as vehicle for eugallol, eurobin, and other dermics applicable as varnish.—EXTERN.: 2—15% solut. in acetone.

Salipyrine.

ANTIPYRINE SALICYLATE.—Wh. powd.; odorl.; sweetish taste.—SOL. in 250 parts water in alcohol, chloroform, ether.—Antirheumatic, Analgesic. —**Dose**: 10—30 grn., in cachets.

Salol Merck.—U. S. P.

PHENOL SALICYLATE.—SOL. in 0.3 part ether; chloroform, 10 parts alcohol; fatty oils; almost insol. in water.—Antiseptic, Antirheumatic, Antipyretic, etc.—USES: *Intern.*, typhoid fever, diarrhea, dysentery, fermentative dyspepsia, rheumatism, grip, and cystitis; *extern.*, wounds, burns, sores, etc. Coating for enteric pills; such pills should be taken one hour or more after meals, and no oil with them.—**Dose**: 3—15 grn.; as *antipyretic*, 30—45 grn.

Salophen.

ACETYL-PARA-AMIDOPHENOL SALICYLATE.—Wh., odorl., tastel. leaflets or powd.; 51% salicylic acid.—SOL. in alcohol, ether; insol. in water.—Antirheumatic.—**Dose**: 15—20 grn.

Salt, Epsom,—see MAGNESIUM SULPHATE.

Salt, Glauber,—see SODIUM SULPHATE.

Salt, Rochelle,—see POTASSIUM AND SODIUM TARTRATE.

Saltpeter,—see POTASSIUM NITRATE.

Salvia—U. S. P.

SAGE.—Tonic, Astringent, Stimulant.—**Dose**: 10–30 grn., as infus. (1:30) or fl. ext. (1:1).

Sambucus—U. S. P.

ELDER.—Stimulant, Diuretic, Diaphoretic.—**Dose**: 4–8 drams, in infus. drank hot.

Sanguinaria—U. S. P.

BLOOD ROOT.—Expectorant, Emetic.—**Dose**: 3–20 grn.—*Preparations:* F. E. (1:1); Tr. (15:100).—See also, Sanguinarine.

Sanguinarine Merck.—C. P.

Small, white needles; acrid, burning taste.—SOL. in chloroform, alcohol, ether.—Expectorant, Alterative, Emetic.—USES: Chiefly as expectorant; also in dyspepsia, debility, etc.—**Dose:** *Expectorant,* $\frac{1}{12}$—$\frac{1}{8}$ grn., in solut.; *alterative,* $\frac{1}{8}$—$\frac{1}{4}$ grn.; *emetic,* $\frac{1}{2}$—1 grn.

Sanguinarine Nitrate Merck.

Red powd.—SOL. in water, alcohol.—USES, DOSES, ETC., same as alkaloid.

Sanguinarine Sulphate Merck.

Red powd.—SOL. in water, alcohol.—USES, DOSES, ETC., same as alkaloid

Santonin Merck.—U. S. P.

ANHYDROUS SANTONINIC ACID.—SOL. in 4 parts chloroform, 40 parts alcohol, 140 parts ether, 5000 parts water.—**Dose:** 2–4 grn.; children of 2 years, $\frac{1}{4}$—$\frac{1}{2}$ grn.—*Preparation:* Troches ($\frac{1}{2}$ grn.).

Sarsaparilla—U. S. P.

Preparations: Comp. Decoct. (D., 1–4 oz.); F. E. (30–120 min.); Comp. F. E. (D., 30–120 min.); Comp. Syr. (flavoring).

Sassafras—U. S. P.

Carminative, Aromatic Stimulant.—*Preparation:* Oil (D., 1–3 min.)

Sassafras Pith—U. S. P.

Demulcent, Emollient.—*Preparation:* Mucilage (1:50).

Savine—U. S. P.
  Rubefacient, Emmenagogue.—*Preparations·* F. E. (D., 5–20 min.); Oil (D., 1–5 min.).

Scammony—U. S. P.
  Dose: 5–15 grn.—*Preparation:* Resin (D., 3–8 grn.).

Scoparius—U. S. P.
  BROOM.—Diuretic, Purgative.—**Dose:** 30–60 grn., as fl. ext. (1:1) or infus. (1:20).—See also, Sparteine.

Scopolamine Hydrobromate Merck.

  Colorl., hygroscopic cryst.—SOL. in water, alcohol.—Mydriatic, Sedative. —USES: *Extern.,* in ophthalmology, $\frac{1}{10}$–$\frac{1}{2}$% solut.; *subcutaneously* for the insane.—INJECTION: $\frac{1}{250}$–$\frac{1}{64}$ grn.—ANTIDOTES: Emetics, stomach pump, muscarine, tannin, animal charcoal, cathartics, etc.

Scutellaria—U. S. P.
  SCULLCAP.—Sedative, Antispasmodic.—*Preparation:* F. E. (D., 30–60 min.).

Senega—U. S. P.
  Dose: 5–20 grn.—*Preparation:* F. E. (1:1); Syr. (20 per cent. F. E.).

Seng.

  Not completely defined.—(Stated: "Active constituents of **Panax Schin-seng** in an aromatic essence.—Stomachic.—**Dose:** 1 fl. dr.")

Senna—U. S. P.
  Dose: 1–4 drams.—*Preparations:* Confect. (D., 1–2 drams), F. E. (1:1); Comp. Infus. (D., 1–3 oz.); Syr. (1.4).—Enters into Comp. Liquorice Powd.

Serpentaria—U. S. P.
  VIRGINIA SNAKEROOT.—Tonic, Antiperiodic, Diaphoretic.—**Dose:** 10–30 grn.—*Preparations:* F. E. (1:1); Tr. (1.10).—Enters into Comp. Tr. Cinchona.

Serum, Antituberculous, Maragliano.—(Only in 1 Cc. [16 min.] tubes.)

  Antitoxin against Pulmonary Tuberculosis.—**Dose** (subcutaneous): In *apyretic* cases, 16 ℳ (1 cubic centimetre) every other day for 10 days, then daily for 10 days, and 30 ℳ twice a day thereafter until sweats have entirely subsided, when 16 ℳ are injected for a month every other day, and finally once a week for a year. In *febrile* cases, if the fever be slight and intermittent, dosage the same as above; if continuous and intense, inject 160 ℳ; and if there be a marked fall of temperature repeat in a week, and so continue until fever is gone, then inject 16—32 ℳ daily.

Silver Chloride Merck.

  White powd.; blackens on exposure to light.—SOL. in ammonia, potassium thiosulphate, potassium cyanide.—Antiseptic, Nerve-sedative.— USES: Chorea, gastralgia, epilepsy, pertussis, diarrhea, and various neuroses.—**Dose:** ½—1½ grn., in pills.—MAX. D.: 3 grn.

Silver Citrate Merck.

  White, dry powd.—SOL. in about 4000 parts water.—Antiseptic Astringent. —USES: Wounds, gonorrhea, etc.—APPLIED in 1—2% oint., or 1—2:8000 solut.—Always prepare solut. fresh!

Silver Cyanide Merck.—U. S. P.

  SOL. in solut's of potassium cyanide, ammonia, sodium thiosulphate.— Antiseptic, Sedative.—USES: Epilepsy, chorea.—**Dose:** $\frac{1}{60}$—$\frac{1}{20}$ grn., in pills.—ANTIDOTES: Ammonia, chlorine, mixture of ferric and ferrous sulphates, artificial respiration, stomach siphon.

Silver Iodide Merck.—U. S. P.

  SOL. in solut. potassium iodide or cyanide, ammonium thiosulphate.— Alterative.—USES: Gastralgia and syphilis.—**Dose:** ¼—1 grn., in pills.

Silver Lactate Merck.

Small needles or powd.—SOL. in 20 parts water.—Antiseptic Astringent.—USES: Sore throat, gonorrhea, etc.—APPLIED in 1—2:4000 solut.

Silver Nitrate Merck.—U. S. P.—Cryst.

SOL. in 0.6 part water, 26 parts alcohol.—**Dose:** ⅛–½ grn.—ANTIDOTES: Solut. common salt, sal ammoniac, mucilaginous drinks, emetics, stomach siphon, white of egg, milk, etc.—INCOMPATIBLES: Organic matter, hydrochloric acid, chlorides, phosphates, arsenites, opium, extracts, resins, essential oils, tannin, etc.

Silver Nitrate, Moulded (Fused), Merck.—U. S. P.

LUNAR CAUSTIC.

Silver Nitrate, Diluted, Merck.—U. S. P.

MITIGATED CAUSTIC.—33⅓% silver nitrate.

Silver Oxide Merck.—U. S. P.

Dose: $1/12$–$1/6$–¼ grn., best mixed with some chalk and put up in capsules.—INCOMPATIBLES: Ammonia, creosote, tannin, acids.—CAUTION: Do not triturate with oxidizable matter; may cause explosion!

Soap—U. S. P.

WHITE CASTILE SOAP.—Detergent, Laxative.—**Dose:** 3–10 grn.—*Preparations:* Lin. Plaster.

Soap, Soft—U. S. P.

GREEN SOAP.—Not used internally.—*Preparation:* Lin.

Sodium Acetate Merck.—U. S. P.—C. P.

SOL. in 1.4 parts water, 30 parts alcohol.—Diuretic.—**Dose:** 15–120 grn.

Sodium Arsenate Merck.—U. S. P.

SOL. in 4 parts water, 2 parts glycerin.—**Dose:** $1/24$–⅛ grn.—*Preparation:* Solut. (1%).—ANTIDOTES: Emetics, stomach siphon, fresh ferric hydrate, dialyzed iron, ferric hydrate and magnesia, demulcents, stimulants, warmth, etc.

Sodium Benzoate Merck.—U. S. P.

SOL. in about 2 parts water, 45 parts alcohol.—Antirheumatic, Antipyretic. Antiseptic.—USES: Rheumatism, gout, uremia, cystitis, lithemia, tonsillitis, colds, etc.—**Dose:** 10–40 grn.

Sodium Bicarbonate Merck.—U. S. P.—C. P.

Dose: 10–40 grn.—*Preparation:* Troches (3 grn.).—CAUTION: Should not be given as acid-antidote, as it evolves large quantities of carbon dioxide gas.

Sodium Bisulphite Merck.—U. S. P.

SOL. in 4 parts water, 72 parts alcohol.—Antiseptic.—USES: *Intern.,* sore mouth, diphtheria, yeasty vomiting; *extern.,* skin diseases.—**Dose:** 10–30 grn.

Sodium Borate Merck.—U. S. P.

BORAX; SODIUM PYROBORATE; so-called "SODIUM BIBORATE" or "TETRABORATE".—SOL. in 16 parts water; 1 part glycerin.—USES: *Intern.,* amenorrhea, dysmenorrhea, epilepsy, uric-acid diathesis; *extern.,* sore mouth, conjunctivitis, urethritis, etc.—**Dose:** 30–40 grn.

Sodium Borate, Neutral, Merck.

Erroneously designated as "SODIUM TETRABORATE".—Transparent, fragile, splintery, glass-like masses.—SOL. in water.—Antiseptic, Astringent.—USES: *Extern.*, chiefly in diseases of nose and ear; a cold saturated solut. used for bandages.

Sodium Borobenzoate Merck.—N. F.

White, cryst. powd.—SOL. in water.—Antiseptic, Antilithic, Diuretic.—USES: Rheumatism, gravel, and puerperal fever.—**Dose:** 30—120 grn.

Sodium Bromide.—U. S. P.

SOL. in 1.2 parts water, 13 parts alcohol.—**Dose:** 10—60 grn.

Sodium Carbonate Merck.—U. S. P.

SOL. in 1.6 parts water, 1 part glycerin.—**Dose:** 5—20 grn.—ANTIDOTES: Acetic acid, lemon juice, olive oil, etc.

Sodium Chlorate Merck.—U. S. P.

Colorl. cryst.; odorl.; cooling, saline taste.—SOL. in 1.1 parts water, 5 parts glycerin, 100 parts alcohol.—Deodorant, Antiseptic, Alterative.—USES: *Intern.*, diphtheria, tonsillitis, pharyngeal and laryngeal inflammation, stomatitis, gastric cancer, mercurial ptyalism, etc.; *extern.*, as wash, gargle or injection.—**Dose:** 5—15 grn.—INCOMPATIBLES: Organic matters, easily oxidizable substances.—CAUTION: Do not triturate with sulphur or phosphorus. or any combustible substance; severe explosion may occur!

Sodium Choleate Merck.

DRIED PURIFIED OX-GALL.—Yellowish-white, hygroscopic powd.—SOL. in water, alcohol.—Tonic, Laxative.—USES: Deficient biliary secretion, chronic constipation, etc.—**Dose:** 5—10 grn.

Sodium Cinnamate Merck.—C. P.

White powd.—SOL. in water.—Antitubercular, like cinnamic acid.—INJECTION (intravenous or parenchymatous): ⅛—1 grn. in 5% solut., twice a week.

Sodium Dithio-salicylate, Beta-, Merck.

Grayish-white, hygroscopic powd.—SOL. in water.—Antineuralgic, Antirheumatic.—USES: *Intern.*, sciatica, gonorrheal rheumatism, etc.—**Dose:** 2—10 grn.

Sodium Ethylate, Liquid, Merck.

Colorl. syrupy liq.; turns brown on keeping.—Escharotic.—USES: Warts, nævi, etc.—APPLIED with glass rod, pure. Chloroform arrests caustic action.

Sodium Ethylate, Dry, Merck.

White or brownish, hygroscopic powd.—ACTION AND USES: As above.—APPLIED in solut. 1:3 absolute alcohol.

Sodium Fluoride Merck.—Pure.

Clear cryst.—SOL. in water.—Antispasmodic, Antiperiodic, Antiseptic.—USES: *Intern.*, epilepsy, malaria, tuberculosis; *extern.*, antiseptic dressing for wounds and bruises, as mouth-wash, in vaginitis, etc. Does not attack nickel-plated instruments.—**Dose:** $^1/_{12}$—$^1/_6$ grn., in solut. with sodium bicarbonate.—APPLIED: *Wounds*, in $^1/_{20}$—$^1/_{10}$% solut.; *mouth-wash*, etc., in ½—1% solut.

Sodium Formate Merck.

White, deliquescent cryst.—SOL. in water, glycerin.—USES: Hypodermically in surgical tuberculosis.—INJECTION (parenchymatous): *Children*, ½—1 grn. in solut., every 7—10 days; *adults*, 3 grn., every 7—10 days.

Sodium Glycerino-phosphate Merck.—50% Solut.

Yellowish liq.—Sol. in water.—Uses: Deficient nerve-nutrition, neuras thenia, phosphaturia, convalescence from influenza, etc.—Injection 3—4 grn. daily, in physiological solut. sodium chloride.

Sodium Hippurate Merck.

White powd.—Uses: In cachexias, and diseases due to uric-acid diathesis —**Dose:** 10—20 grn.

Sodium Hydrate Merck.—U. S. P.—C. P.

Sodium Hydroxide; Caustic Soda.—**Dose:** ½—1 grn., freely diluted —*Preparation:* Solut. (5%).—Antidotes: Water, and then vinegar, o lemon juice.

Sodium Hypophosphite Merck.—Purified.

Sol. in 1 part water, 30 parts alcohol.—**Dose:** 10—30 grn.

Sodium Hyposulphite,—see Sodium Thiosulphate.

Sodium Iodide Merck.—U. S. P.

Sol. in about 1 part water, 3 parts alcohol.—Uses: Rheumatism, pneu monia, tertiary syphilis, asthma, chronic bronchitis, scrofula, etc.— **Dose:** 5—60 grn.

Sodium Naphtolate, Beta-, Merck.

Microcidin.—Yellowish to white powd.—Sol. in 3 parts water.—Uses Surgical antiseptic on bandages, etc.—Applied in 3—5% aqueous solut.

Sodium Nitrate Merck.—U. S. P.—C. P.

Chili Saltpeter.—Sol. in 1.3 parts water, 100 parts alcohol.—Uses Intern., inflammatory condition of intestines, dysentery, etc.; extern. rheumatism, 1:3 aqueous solut.—**Dose:** 10—60 grn.

Sodium Nitrite Merck.—C. P.

White cryst. or sticks; mildly saline taste.—Sol. in 1.5 parts water slightly in alcohol.—Antispasmodic, Diaphoretic, Diuretic.—Uses: An gina pectoris, dropsy, and diseases of genito-urinary organs.—**Dose:** 1—3 grn.

Sodium Paracresotate Merck.

Microcryst. powd.; bitter taste.—Sol. in 24 parts warm water.—Anti pyretic, Intestinal Antiseptic, Analgesic.—Uses: Acute gastric catarrh acute rheumatism, pneumonia, typhoid fever, etc.—**Dose:** 2—20 grn. according to age, 3 t. daily, in aqueous solut. with extract licorice.

Sodium Phosphate Merck.—C. P.

Colorl. cryst.—Sol. in about 20 parts water.—Uses: Chronic rheumatism, stimulant of biliary secretion, mild laxative, and vesical calculi.—**Dose:** 5—40 grn.; as laxative, ½—1 ounce.

Sodium Pyrophosphate Merck.—U. S. P.

Sol. in 10 parts water.—Uses: Lithiasis.—**Dose:** 5—40 grn.

Sodium Salicylate Merck.—U. S. P.

Sol. in 1 part water, 6 parts alcohol; glycerin.—**Dose:** 5—40 grn.—Max. D.: 60 grn.—Incompatibles: Ferric salts.

Merck's Sodium Salicylate is *the only brand* which yields a clear and *colorless* solut.

Sodium Salicylate Merck.—From Oil Wintergreen.

Uses, etc., as above.

Sodium Santoninate Merck.

Stellate groups of needles: mildly saline and somewhat bitter taste: turn yellow on exposure to light.—SOL. in 3 parts water, 12 parts alcohol.—Anthelmintic.—USES: Instead of santonin; less powerful.—**Dose** (adult): 2—6 grn., in keratinized pills: children 4—10 years old, 1—3 grn.

Sodium Silico-fluoride Merck.

White cryst. or granular powd.—SOL. in 200 parts water.—Antiseptic, Germicide, Deodorant, Styptic.—USES: *Extern.*, wounds, carious teeth, cystitis, gonorrhea, for irrigating cavities, and in gynecological practice.—APPLIED in $^1/_5\%$ solut.

Sodium Sulphate Merck.—C. P., Cryst. or Dried.

GLAUBER'S SALT.—SOL. in 3 parts water; glycerin.—**Dose:** *Cryst.*, 2—8 drams; *dried*, 1—4 drams.

Sodium Sulphite Merck.—U. S. P.

SOL. in 4 parts water, sparingly in alcohol.—USES: Skin diseases, sore mouth, diphtheria, sarcina ventriculi, and chronic mercurial affections.—**Dose:** 10—60 grn.

Sodium Sulpho-carbolate Merck.—U. S. P.

SOL. in 5 parts water, 132 parts alcohol.—Antiseptic, Disinfectant.—USES: *Intern.*, dyspepsia, phthisis, typhoid fever, dysentery, etc.: *extern.*, gonorrhea, putrid wounds, etc.—**Dose:** 8—30 grn.—EXTERN.: ½—1% solut.

Sodium Tartrate Merck.—C. P.

White cryst.—SOL. in water.—USES: Tastel. substitute for Epsom salt.—**Dose:** 4—8 drams.

Sodium Tellurate Merck.

White powd.—SOL. in water.—Antihidrotic, Antiseptic, Antipyretic.—USES: Night-sweats of phthisis; gastric ulcerations, rheumatism, and typhoid fever.—**Dose:** ¼—¾ grn., in alcoholic mixture or elixir.

Sodium Thiosulphate Merck (*Sodium Hyposulphite, U. S. P.*)

SOL. in 1 part water.—USES: Parasitic skin diseases, sore mouth, sarcina ventriculi, diarrhea, flatulent dyspepsia, etc.—**Dose:** 5—20 grn.—INCOMPATIBLES: Iodine, acids.

Solanin Merck.—Pure.

Colorl., lustrous, fine needles; bitter taste.—Analgesic, Nerve-sedative.—USES: Neuralgia, vomiting of pregnancy, bronchitis, asthma, painful gastric affections, epileptoid tremors, locomotor ataxia, etc.—**Dose:** ¼—1 grn.—MAX. D.: 1½ grn. single, 8 grn. daily.

Solution, Aluminium Acetate, Merck.

8% basic aluminium acetate.—Clear, colorl. liq.—Antiseptic, Astringent.—USES: *Intern.*, diarrhea and dysentery; *extern.*, lotion for putrid wounds and skin affections, mouth-wash.—**Dose:** 3—15 ℳ.—EXTERN., solut. 1 : 15; as mouth-wash or enema, 1 : 150.

Solution, Ammonium Acetate—U. S. P.

SPIRIT MINDERERUS.—Diaphoretic, Antipyretic, Diuretic.—**Dose:** 2-8 drams.

Solution, Arsenic and Mercuric Iodides, Merck.—U. S. P.

DONOVAN'S SOLUTION.—**Dose:** 5—10 ℳ.—ANTIDOTES: Same as for arsenous acid.—INCOMPATIBLES: Alkalies and alkaloids or their salts.

Solution, Calcium Bisulphite, Merck.

Liq.; strong sulphurous odor.—Disinfectant, Antiseptic.—USES: *Extern.*, diluted with 4—8 t. weight water, in sore throat, diphtheria, vaginitis, endometritis, wounds, etc.

Solution, Calcium Hydrate—U. S. P.
LIME WATER.—Antacid, Astringent.—Dose: 1-4 oz.—*Preparation:* Liniment.

Solution, Fowler's, Merck, (*Solut. Potassium Arsenite, U. S. P.*).

Never give on an empty stomach !—**Dose:** 1—5 M.—ANTIDOTES : Emetics, stomach siphon; freshly precipitated ferric hydrate; or ferric hydrate with magnesia; or saccharated ferric oxide; etc.

Solution, Hydrogen Peroxide.—U. S. P.

3% H_2O_2 (= 10 vols. available O).—SOL. in all proportions water or alcohol.—Disinfectant, Deodorant, Styptic, Antizymotic.—USED chiefly *extern.;* in diphtheria, sore throat, wounds, gonorrhea, abscesses, etc.; *rarely intern.;* in flatulence, gastric affections, epilepsy, phthisical sweats, etc.—**Dose:** 1—4 fl. drs., well dil. EXTERN.: in 20% solut. to pure.—CAUTION: Keep cool and quiet. It rapidly deteriorates !

Solution, Iodine, Compound—U. S. P
LUGOL'S SOLUTION.—5 per cent. iodine, 10 per cent. potass. iodide.—Alterative.—**Dose:** 1-10 min.

Solution, Iron Acetate—U. S. P.
31 per cent. (=7.5 per cent. iron).—Chalybeate, Astringent.—**Dose:** 2-10 min.

Solution, Iron Albuminate, Merck.

Brown liq.—0.4% iron.—Hematinic; easily assimilable.—USES : Anemia, chlorosis, etc.—**Dose:** 1—4 drams, with milk, before meals.
MERCK'S Solution of Iron Albuminate is superior to other makes in point of palatability and stability, besides being perfectly free from acidity and astringency and hence not injuring the teeth or stomach.

Solution, Iron Chloride, Ferric—U. S. P.
37.8 per cent.—Styptic (chiefly in post-partum hemorrhage: 1 dram to pint water).

Solution, Iron Citrate, Ferric—U. S. P.
7.5 per cent. iron.—Hematinic.—**Dose:** 5-15 min.

Solution, Iron Nitrate—U. S. P.
6.2 per cent. ferric nitrate.—Tonic, Intern. Astringent.—**Dose:** 5-15 min.

Solution, Iron Subsulphate, Ferric—U. S. P.
MONSEL'S SOLUTION.—Styptic. Astringent.—USED chiefly extern.· pure or in strong solut.—**Dose:** 2-10 min.

Solution, Iron and Ammonium Acetate—U. S. P.
BASHAM'S MIXTURE.—Hematinic, Astringent.—**Dose :** ½-1 fl oz

Solution, Lead Subacetate—U. S. P.
GOULARD'S EXTRACT.—25 per cent.—Astringent, Antiseptic.—USED chiefly to make the *Diluted Solution* (lead water), and the *Cerate* (20 per cent.).

Solution, Magnesium Citrate—U. S. P.
Laxative, Refrigerant.—**Dose:** 6-12 fl. oz.

Solution, Mercury Nitrate, Mercuric—U. S. P.
60 per cent.—Caustic.— USED only extern.· pure.

Solution, Potassium Hydrate—U. S. P
POTASSA SOLUTION.—5 per cent. KOH.—Antacid, Antilithic, Diuretic.—**Dose:** 5-20 min., well diluted.—INCOMPATIBLES: Organic matter, alkaloids, ammonium salts.—ANTIDOTES Mild acids. oils. milk.

Solution, Soda, Chlorinated—U. S. P
LABARRAQUE'S SOLUTION.—2.6 per cent. available chlorine.—Disinfectant, Antizymotic.—**Dose:** 20-60 min., diluted.—EXTERN. in 3-10 per cent. solut.

Solution, Sodium Arsenate—U. S. P.

1 per cent.—Alterative, Antiperiodic.—**Dose:** 3–10 min.

Solution, Sodium Hydrate—U. S. P.

SODA SOLUTION.—5 per cent. Na OH.—ACTION, USES, DOSE, etc.: As of Solut. Potass. Hydr.

Solution, Sodium Silicate—U. S. P.

20 per cent. silica, 10 per cent. soda.—USED only for surgical dressings.

Solution, Trimethylamine, Merck.—10%.—Medicinal.

So-called "PROPYLAMINE".—Colorl. liq.; strong fishy and ammoni-acal odor.—Antirheumatic, Sedative.—USES: Rheumatism, chorea, etc.—**Dose:** 15–45 ♏; in chorea as much as 1½ ounces daily may be given, in sweetened, flavored water.

Solution, Zinc Chloride—U. S. P.

50 per cent.—Disinfectant, Astringent.

Sozoiodole-Mercury.

MERCURY DIIODO-PARAPHENOL-SULPHONATE, *Trommsdorff.* — O r a n g e powd.—SOL. in solut. of sodium chloride or potassium iodide.—Anti-syphilitic, Antiseptic, Alterative. USES: Syphilitic eruptions and ulcers, enlarged glands, parasitic skin diseases, and diseased joints.—APPLIED in 2–20% oint. or powd.; *Injection* (hypodermically), 1–3 grn., in solut. of potassium iodide.

Sozoiodole-Potassium.

POTASSIUM DIIODO-PARAPHENOL-SULPHONATE, *Trommsdorff.*—White, odorl., cryst. powd.; 52.8% of iodine; 20% of phenol; and 7% sulphur.—SOL. slightly in cold water; insol. in alcohol.—Antiseptic Vulnerary ; Non-poi-sonous Succedaneum for Iodoform.—USES: *Extern.,* scabies, eczema, herpes tonsurans, impetigo, syphilitic ulcers, diphtheria, burns, and scalds; ozena, otitis, and rhinitis; injection for gonorrhea.—APPLIED in 10–25% oint's or dusting-powders, which are as effective as iodoform pure.—INCOMPATIBLES: Mineral acids, ferric chloride, silver salts.

Sozoiodole-Sodium.

SODIUM DIIODO-PARAPHENOL-SULPHONATE, *Trommsdorff.* — Colorl. needles.—SOL. in 44 parts water; alcohol, 20 parts glycerin.—Antiseptic, Astringent, Antipyretic.—USES: *Intern.,* as intestinal antiseptic, and in diabetes: *extern.,* gonorrhea, cystitis, nasal catarrh, ulcers, whooping-cough, etc.—**Dose:** 5–30 grn. daily.—EXTERN.: 10% oint., with adeps lanæ, 1% solut. in water, or 2% solut. in paraffin. In whooping-cough, 3 grn. daily, blown into nose.

Sozoiodole-Zinc.

ZINC DIIODO-PARAPHENOL-SULPHONATE, *Trommsdorff.*—Colorl. needles.—SOL. in 25 parts water, in alcohol, glycerin.—Antiseptic Astringent.—USES: Gonorrhea, nasal and pharyngeal catarrhs, etc.—APPLIED: *Rhinitis,* 5–10% trituration with milk sugar by insufflation, or 3–5% paint; *gonorrhea,* ½–1% solut.; *skin diseases,* 5–10% oint.; *gargle,* 1–2% solut.

Sparteine Sulphate Merck.—U. S. P.

SOL. in water, alcohol.—Heart-stimulant, Diuretic.—USES: Best where digitalis fails or is contra-indicated.—**Dose:** ¼—1 grn.

Spearmint—U. S. P.

Preparations: Oil (D., 2–5 min.); Spt. (10 per cent. oil); Water (one-fifth per cent. oil).

71

Spermine, Poehl.—Sterilized.

2% solut. of spermine hydrochlorate with sodium chloride.—Nervine.—
Uses: Nervous diseases with anemia, neurasthenia, hystero-epilepsy,
angina pectoris. locomotor ataxia, asthma, etc.; usually hypodermically.
—Injection: 15 ℔, usually given on the lower extremities or near the
shoulder-blade, once daily, for 8 or 10 days.—Incompatible with potassium
iodide treatment.

Spermine Poehl.—Essence.

4% aromatized alcoholic solut. of the double-salt spermine hydrochlorate-
sodium chloride.—Uses: *Intern.*, for same diseases as the preceding.—
Dose: 10—30 ℔, in alkaline mineral water, every morning.

Spigelia—U. S. P.

Pinkroot.—Anthelmintic.—Dose: 1–2 drams.—*Preparation:* F. E. (1:1).

Spirit, Ants, True, Merck.

From ants.—Rubefacient.—Uses: Counter-irritant in painful local affections.—Applied undiluted.

Spirit Glonoin—U. S. P.

Spirit (Solution) of Nitroglycerin (Trinitrin).—1 per cent.—Anti-spasmodic,
Vaso-dilator.—Dose: 1–3 min.

Spirit, Melissa, Concentrated, Merck.

Rubefacient, Stimulant, Carminative.—Uses: *Extern.*, as counter-irritant; *intern.*, in cardialgia, colic, and diarrhea.—**Dose:** ½—1 dram on
sugar.

Spirit, Nitrous Ether—U. S. P.

Dose: 30–90 min.—Incompatibles: Antipyrine, tannin, acetanilid, phenacetin, iodides,
fl. ext. buchu, tr. guaiac, and morphine salts.

Squill—U. S. P.

Dose: 1–3 grn.—*Preparations:* F. E. (1:1); Syr. (45 per cent. vinegar squill); Comp. Syr.
(F. E. squill, 8 per cent. ; F. E. senega, 8 per cent. ; tartar emetic, one-fifth per cent.);
Tr. (15:100); Vinegar (1:10).

Staphisagria—U. S. P.

Stavesacre.—Parasiticide.—Used extern., in substance or 1:16 solut. of fl. ext. in dil.
acetic acid.

Starch—U. S. P.

Preparation: Glycerite (1:10).

Starch, Iodized, Merck.

2% iodine.—Bluish-black powd.—Disinfectant, Antiseptic.—Uses: *Intern.*,
diarrhea, typhoid fever, etc.; *extern.*, with adeps lanæ, as substitute for
tincture of iodine.—**Dose :** 3—10 grn.

Stillingia—U. S. P.

Queen's Root.—Alterative, Resolvent.—*Preparation:* F. E. (D., 15–60 min.).

Storax—U. S. P.

Stimulant, Antiseptic, Expectorant.—Dose: 5–20 grn.—Enters into Comp. Tr. Benzoin.

Stramonium Leaves—U. S. P.

Dose: 2–5 grn.

Stramonium Seed—U. S. P.

Dose: 1–3 grn.—*Preparations:* Ext. (D., ¼–½ grn.); F. E. (1:1); Oint. (10 per cent.
Ext.); Tr. (15:100).

Strontium Arsenite Merck.

White powd.—Almost insol. in water.—Alterative, Tonic.—Uses: Skin
diseases and malarial affections.—**Dose:** 1/30—1/12 grn., in pills.

Strontium Bromide Merck.—Cryst.

Deliquescent, colorl., odorl. needles; bitter-saline taste.—Sol. in alcohol; 1—2 parts water.—Gastric Tonic, Nerve-sedative, Anti-epileptic, Anti-nephritic.—Uses: Hyperacidity of stomach; rheumatism, gout, epilepsy, nervousness, hysteria, headache, etc.—**Dose:** 10—40 grn. In epilepsy as much as 150 grn. may be given daily.

Strontium Iodide Merck.

White or yellowish, deliquescent powd. or plates; bitterish-saline taste.— Sol. in alcohol, ether; 0.6 parts water.—Alterative, Sialagogue.—Uses: Substitute for potassium iodide in heart disease, asthma, rheumatism, scrofula, etc.—**Dose:** 10—20 grn.

Strontium Lactate Merck.—U. S. P.—C. P.

White, granular powd.; slightly bitter taste.—Sol. in alcohol, 4 parts water.—Anthelmintic, Antinephritic, Tonic.—Uses: Nephritis, worms, rheumatism, gout, and chorea. Decreases albumin in urine, without diuresis.—**Dose:** 10—20 grn.; for worms, 30 grn. twice daily for 5 days.

Strontium Salicylate Merck.—Cryst.

Sol. in about 20 parts water, in alcohol.—Antirheumatic, Tonic.—Uses: Rheumatism, gout, chorea, muscular pains, and pleurisy. **Dose:** 10—40 grn.

Strophanthin Merck.—C. P.

White powd.; very bitter taste.—Sol. in water, alcohol.—Heart Tonic, *not* Diuretic.—Uses: Similar to digitalin.—**Dose:** $\frac{1}{205}$—$\frac{1}{80}$ grn.—Anti-dotes: Emetics, stomach siphon, muscarine, atropine, camphor, picrotoxin.

Strophanthus—U. S. P.

Cardiac Tonic, like digitalis.—*Preparation:* Tr. (D., 3–10 min.).

Strychnine (Alkaloid) Merck.—U. S. P.

Sol. in 7 parts chloroform, 110 parts alcohol, 6700 parts water.—**Dose:** $\frac{1}{60}$—$\frac{1}{20}$ grn.—Antidotes: Stomach pump, tannin, emetics, charcoal, paraldehyde, urethane, potassium bromide, chloroform, chloral hydrate, artificial respiration, etc.

Strychnine Arsenate Merck.

White powd.; very bitter taste.—Sol. in about 15 parts water.—Alterative, Antitubercular.—Uses: Tuberculosis, skin diseases, malarial affections, etc.; usually hypodermically, 0.5% in liq. paraffin; of this 4—10 ♍ may be injected daily.—**Dose:** $\frac{1}{64}$—$\frac{1}{16}$ grn.

Strychnine Arsenite Merck.

White powd.—Sol. slightly in water.—Uses, Doses, Etc., as of the Arsenate.

Strychnine Hypophosphite Merck.

White cryst. powd. — Sol. in water. — Uses: Tubercular affections, scrofula, and wasting diseases generally.—**Dose:** $\frac{1}{32}$—$\frac{1}{12}$ grn.

Strychnine Nitrate Merck.

Groups of silky needles.—Sol. in 50 parts water, 60 parts alcohol.—Uses, Doses, Etc.: About as the Alkaloid. Most frequently used in *dipsomania*.

Strychnine Sulphate Merck.—U. S. P.

Sol. in 50 parts water, 109 parts alcohol. Uses, Doses, Etc., same as of the Alkaloid.

Stypticin.

COTARNINE HYDROCHLORATE, *Merck.*—Yellow cryst.—SOL. in water.—Hemostatic, Uterine Sedative.—USES: Uterine hemorrhage, dysmenorrhea, fibroids, subinvolution, climacteric disorders, etc.—**Dose:** 2—5 grn. 4 t. daily, in pearls.—INJECTION (urgent cases): 2—3 grn., in 10% solut.

Sulfonal.

DIETHYLSULPHONE-DIMETHYL-METHANE.—Colorl., tastel. cryst.—SOL. 500 parts in water; 135 ether; 110 dil. alcohol.—Hypnotic, Sedative.—**Dose:** 15—45 grm., in powd.

Sulphur Merck.—Precipitated.

LAC SULPHURIS; MILK OF SULPHUR.—**Dose:** ⅙—2 drams.

Sulphur, Sublimed—U. S. P.

FLOWERS OF SULPHUR.—Intended for external use only.

Sulphur, Washed—U. S. P.

Dose: 1–3 drs.—*Preparation:* Oint. (30 per cent.).—Enters into Comp. Liquorice Powd.

Sulphur Iodide Merck.

80% iodine.—Grayish-black masses.—SOL. in 60 parts glycerin.—Antiseptic, Alterative.—USES: *Intern.*, scrofula, and chronic skin diseases; *extern.*, in 5—10% oint., for eczema, psoriasis, prurigo, etc.—**Dose:** 1—4 grn.

Sumbul—U. S. P.

MUSK ROOT.—Antispasmodic, Sedative.—*Preparation:* Tr. (D., 15–60 min.).

Svapnia.

Not completely defined.—(Stated: "Purified opium; 10% morphine; contains the anodyne and soporific alkaloids codeine and morphine, but excludes the convulsive alkaloids thebaine, narcotine, and papaverine.—**Dose:** Same as of opium.")

Syrup, Hydriodic Acid—U. S. P.

1 per cent. absol. HI.—Alterative.—Dose: 30–60 min.

Syrup, Hypophosphites—U. S. P.

Ea. fl. dr. contains 2¼ grn. calc. hypophos., 1 grn. ea. of pot. and sod. hypophos.—Alterative, Tonic.—Dose: 1–2 fl. drams.

Syrup, Hypophosphites, Fellows'.

Not complt ly defined.—(Stated: "Contains hypophosphites of potash, lime, iron, manganese; phosphorus, quinine, strychnine.—Alterative, Reconstructive.—**Dose:** 1—2 fl. drs., 3 t. daily, in wineglassful water.")

Syrup, Hypophosphites, McArthur's.

Not completely defined.—(Stated: "Contains chemically pure hypophosphites of lime and soda; prepared acc. to formula of Dr. Churchill, Paris.—Alterative, Reconstructive.—**Dose:** 2—4 fl. drs., in water, after meals.")

Syrup, Hypophosphites, with Iron—U. S. P.

Ea. fl. dr. contains 2¼ grn. calc. hypophos., 1 grn. ea. of pot. and sod. hypophos., ⅜ grn. iron lactate.—Alterative, Hematinic.—Dose: 1–2 fl. drams.

Syrup, Iron Iodide—U. S. P.

10 per cent. ferrous iodide.—Alterative, Hematinic.—Dose: 15–30 min.

Syrup, Iron, Quinine, and Strychnine Phosphates—U. S. P.

EASTON'S SYRUP.—Ea. fl. dr. contains 1 grn. ferric phosph., 1¾ grn. quinine, one-ninetieth grn. strychnine.—Nervine, Hematinic.—Dose: 1–2 fl. drs.

Syrup, Lime—U. S. P.

Antacid, Antidote to Carbolic Acid.—Dose: 30–60 min.

Taka-Diastase.

(Diastase Takamine).—Brownish powd.; alm. tastel.—SOL. in water; insol. in alcohol.—Starch-digestant (1 part stated to convert over 100 parts dry starch).—USED in amylaceous dyspepsia.—**Dose:** 1—5 grn.

Tannalbin.

TANNIN ALBUMINATE, EXSICCATED, *Knoll.*—Light-brown, odorl., tastel. powd.; contains 50% tannin.—SOL. in alkaline, insol. in acid fluids.— Intestinal Astringent and Antidiarrheal. Not acted upon in stomach, but slowly and equably decomposed in the intestines; thus causing no gastric disturbance, while gently yet firmly astringent on entire intestinal mucosa. Innocuous, and without by- or after-effects.—**Dose:** 45—150 grn. daily, in 15—30 grn. portions. In urgent acute cases repetition in 2- or even 1-hourly intervals has proved useful for promptly creating the first impression, the frequency being decreased with the improvement. The dose for *nurslings* is 5—8 grn.; for *children,* up to 15 grn.—["Merck's Digest" on "TANNALBIN" contains clinical reports and detailed information.]

Tannigen.

ACETYL-TANNIN.—Gray, slightly hygrosc. powd.; alm. odorl. and tastel. —SOL. in alkaline fluids, alcohol; insol. in water.—Intestinal Astringent. **Dose:** 5—15 grn.

Tannin,—see ACID, TANNIC.

Tannoform.

TANNIN-FORMALDEHYDE, *Merck.*—Loose, reddish powd.—SOL. in alkaline liqs.; insol. in water.—SICCATIVE ANTISEPTIC and DEODORANT.— USES: Hyperidrosis, bromidrosis, ozena, etc.—APPLIED pure or in 25—50% triturations.—[Further information in "Merck's Digest" on "TANNOFORM," containing clinical reports.]

Tannopine.

HEXAMETHYLENE-TETRAMINE-TANNIN.—Brown, sl. hygrosc. powd.; 87% tannin.—SOL. in dil. alkalies; insol. in water, alcohol, or dil. acids.—Intestinal Astringent.—**Dose:** 15 grn., several t. a day; children 3-8 grn.

Tar—U. S. P.

Preparations: Oint. (50 per cent.); Syr. (D., 1-4 drams).

Taraxacum—U. S. P.

DANDELION.—Bitter Tonic, Hepatic Stimulant.—*Preparations:* Ext. (D., 10-30 grn.); F. E. (D., 1-2 drams).

Tartar Emetic,—see ANTIMONY AND POTASSIUM TARTRATE.

Tartar, Soluble,—see POTASSIUM TARTRATE.

Terebene Merck.

Colorl. or slightly yellowish liq.; resinifies when exposed to the light; thyme-like odor.—SOL. in alcohol, ether; slightly in water.—Expectorant, Antiseptic, Antifermentative.—USES: *Intern.,* in chronic bronchitis, flatulent dyspepsia, genito-urinary diseases, emphysema, phthisis, bronchitis, dyspnea, etc.; *extern.,* uterine cancer, gangrenous wounds, skin diseases, etc. In phthisical affections it is given by inhalation (about 2 oz. per week).—**Dose:** 4—20 ℧, with syrup or on a lump of sugar.

Terpin Hydrate Merck.—U. S. P.

Colorl., lustrous prisms; slightly bitter taste.—SOL. in 10 parts alcohol, 100 parts ether; 200 parts chloroform, 250 parts water.—Expectorant, Antiseptic, Diuretic, Diaphoretic.—USES: Bronchial affections, whooping-cough, throat affections, tuberculosis, genito-urinary diseases, etc.— **Dose:** *Expectorant,* 3—6 grn.; *diuretic,* 10—15 grn.; several t. daily.

Terpinol Merck.

Oily liq., hyacinthine odor.—SOL. in alcohol, ether.—Bronchial Stimulant, Antiseptic, Diuretic.—USES: To diminish expectoration and lessen odor in phthisis; also for tracheal and bronchial catarrhs.—**Dose:** 8—15 ℧.

Testaden.

STANDARDIZED DRIED EXTRACT TESTICULAR SUBSTANCE, *Knoll.*—1 part represents 2 parts fresh gland.—Powd.—USES: Spinal and nervous diseases, impotence, etc.—**Dose:** 30 grn., 3 or 4 t. daily.

Tetraethyl-ammonium Hydroxide Merck.—10% Solut.

Alkaline, bitter, caustic liq.—Solvent of Uric acid.—USES: Rheumatism, gout, etc.—**Dose:** 10—20 ♏ 3 t. daily, well diluted.—CAUTION: Keep well-stoppered !

Thalline Sulphate Merck.—(*Not Thallium!*)

Yellowish needles, or cryst. powd.; cumarin-like odor; acid-saline-bitterish, aromatic taste; turns brown on exposure.—SOL. in 7 parts water, 100 parts alcohol.—Antiseptic, Antipyretic.—USES: *Intern.*, typhoid fever, malarial fever, etc.; *extern.*, 1—2% injection for gonorrhea; in chronic gonorrhea a 5% solut. in oil is best.—**Dose:** 3—8 grn.—MAX. D.: 10 grn. single, 30 grn. daily.

Thalline Tartrate Merck.—(*Not Thallium!*)

Cryst., or cryst. powd.—SOL. in 10 parts water, 300 parts alcohol.—USES, DOSES, ETC., as the Sulphate.

Thallium Acetate Merck.—(*Not Thalline!*)

White, deliquescent cryst.—SOL. in water, alcohol.—USES: Recently recommended in phthisical night-sweats.—**Dose:** 1½—3 grn., at bedtime.

Theine,—see CAFFEINE.

Theobromine Merck.—C. P.

White powd.; bitter taste.—SOL. in ether; insol. in water or chloroform. —Diuretic, Nerve-stimulant.—**Dose:** 5—15 grn.

Theobromine Salicylate Merck.—True Salt.

Small, white, acid, permanent needles; not decomposable by water.—SOL. slightly in water.—USES: Powerful Diuretic and Genito-urinary Antiseptic; similar in action to diuretin, but perfectly stable.—**Dose:** 15 grn., several t. daily, in wafers, or in powd. with saccharin.

Theobromine and Lithium Benzoate,—see UROPHERIN B.

Theobromine and Lithium Salicylate,—see UROPHERIN S.

Theobromine and Sodium Salicylate Merck.

DIURETIN.—White, fine powd., odorl.; containing 49.7% theobromine, 38.1% salicylic acid; decomposes on exposure.—Diuretic.—USES: Heart disease; nephritis, especially of scarlet fever.—**Dose:** 15 grn., 5—6 t. daily, in powd., or capsules, followed by water.

Thermodin.

ACETYL-PARAETHOXY-PHENYLURETHANE, *Merck.*—Colorl., odorl. cryst.—SOL. slightly in water.—Antipyretic, Analgesic.—USES: Typhoid, pneumonia, influenza, tuberculosis, etc. Temperature reduction begins in 1 hour after taking and reaches its lowest in four hours.—**Dose:** *Antipyretic,* 5—10 grn.; *anodyne,* 15—20 grn.

Thiocol.

POTASSIUM GUAIACOLSULPHONATE, *Roche.*—White, odorl. powd., of faint bitter, then sweet, taste; 30% guaiacol.—SOL. freely in water.—ANTITUBERCULAR and ANTICATARRHAL; reported non-irritating to mucosæ of digestive tract, readily assimilated, uniformly well borne even by the most sensitive, and perfectly innocuous.—USES: Phthisis, chronic coughs and catarrhs, scrofulous disorders, etc.—**Dose:** 8 grn., gradually increased to 30 or 40 grn., 3 t. daily; preferably in solut. with orange syrup.

Thiosinamine Merck.

ALLYL SULPHO-CARBAMIDE.—Colorl. cryst.; faint garlic odor; bitter taste.—SOL. in water, alcohol, or ether.—Discutient, Antiseptic.—USES: *Extern.*, lupus. chronic glandular tumors; and for removing scar tissue. Possesses the power of softening cicatricial tissue, also tumors of the uterine appendages.—**Dose:** ½ grn., grad. increased to 1½ grn., twice daily, in diluted alcohol; *hypodermically*, 2–8 grn. in glycerino-aqueous solut., once every 3 or 4 days.

Thymol Merck.—U. S. P.—Cryst.

THYMIC ACID.—SOL. in alcohol, ether, chloroform; 1200 parts water.— USES: *Intern.*, rheumatism, gout, chyluria, worms, gastric fermentation, etc.; *extern.*, inhaled in bronchitis, coughs, coryza, etc.; for toothache and mouth-wash, and for wounds, ulcers, and skin diseases.—**Dose:** 1–10 grn.

Thyraden.

STANDARDIZED DRIED EXTRACT THYROID GLAND, *Knoll.*—1 part represents 2 parts fresh gland. Light-brownish, sweet, permanent powd., free from ptomaines.—Alterative.—USES: Diseases referable to disturbed function of the thyroid gland (myxedema, cretinism, struma, certain skin diseases, etc.).—**Dose:** 15–25 grn. daily, gradually increased if necessary; children, ¼–½ as much.

Tincture, Aconite, Merck.—U. S. P.

Dose: 1–3 ℳ.—ANTIDOTES: Emetics, stomach siphon, stimulants, strychnine, or digitalis.—CAUTION: Tincture Aconite, U. S. P., is 3½ times as powerful as that of the German Pharmacopœia.

Tincture, Adonis Æstivalis, Merck.

Anti-fat.—**Dose:** 10–30 ℳ, after meals, in lithia water.—CAUTION: Do not confound with Tincture Adonis Vernalis!

Tincture, Adonis Vernalis, Merck.

Cardiac Stimulant, Diuretic; said to act more promptly than digitalis.— **Dose:** 3–20 ℳ.—ANTIDOTES: Emetics, stomach siphon, tannin, brandy, ammonia, opium.—CAUTION: Do not confound with Tincture Adonis Æstivalis!

Tincture, Arnica Flowers, Merck.—U. S. P.

Antiseptic, Antipyretic.—USES:—*Intern.*, to check fever; *extern.*, chiefly in bruises and other injuries.—**Dose:** 10–30 ℳ.

Tincture, Bursa Pastoris, Merck.

TINCTURE SHEPHERD'S PURSE.—USES: Chiefly in vesical calculus.— **Dose:** 30 ℳ three t. daily.

Tincture, Cactus Grandiflorus, Merck.

Heart-tonic; claimed free from cumulative action.—**Dose:** 15 ℳ, every 4 hours.—MAX. D.: 30 ℳ.

Tincture, Hydrastis, Merck.—U. S. P.

Hemostatic, Astringent, Alterative.—USES: Uterine hemorrhages, chronic catarrh, hemorrhoids, leucorrhea, gonorrhea, etc.—**Dose:** 30–60 ℳ.

Tincture, Hyoscyamus, Merck.—U. S. P.

Dose: 10–60 ℳ.—ANTIDOTES: Animal charcoal followed by emetic; opium; pilocarpine hypodermically, artificial respiration, brandy, ammonia, etc.

Tincture, Iron Chloride.—U. S. P.

Dose: 5–20 ℳ., diluted.—INCOMPATIBLES : Alkalies, alkali benzoates and carbonates, antipyrine, most vegetable infusions and tinctures, mucilage acacia, etc.

Tincture, Nerium Oleander, from leaves, Merck.

Succedaneum for Digitalis.—**Dose:** 20 ℳ, three t. daily.

Tincture, Nux Vomica, Merck.—U. S. P.

Assayed.—Containing 0.3 gramme of combined alkaloids of nux vomica
in 100 cubic centimetres.—Tonic, Stimulant.—Uses : Atonic indigestion;
stimulant to nervous system; in chronic bronchitis, adynamic pneumonia; in poisoning by opium, chloral, or other narcotics; in all affections with impaired muscular nutrition; anemia, etc.—**Dose:** 5—15
ℳ.—Antidotes: Emetics, stomach pump, tannin, potassium iodide,
chloroform, amyl nitrite, opium, absolute repose, etc.

Tincture, Pulsatilla, Merck.

Antispasmodic, Sedative, Anodyne.—Uses: *Intern.*, asthma, whooping-
cough, spasmodic dysmenorrhea, orchitis, etc.; *extern.*, leucorrhea (1:10
water).—**Dose :** 3—20 ℳ.

Tincture, Rhus Toxicodendron, Merck.

Uses : Chronic rheumatism, incontinence of urine, skin diseases.—
Dose: 15 ℳ.

Tincture, Simulo, Merck.

Nervine, Anti-epileptic.—Uses: Hysteria, nervousness, and epilepsy.—
Dose: 30—60 ℳ, two or three t. daily, in sweet wine.

Tincture, Stramonium Seed, Merck.—U. S. P.

Dose: 5—10 ℳ.—Antidotes: Emetics, stomach siphon, animal charcoal, tannin, opium; pilocarpine hypodermically.

Tincture, Strophanthus, Merck.—U. S. P.—1:20.

Dose: 3—10 ℳ.—Antidotes: Emetics, stomach siphon, cathartics,
tannin, opium, coffee, brandy, etc.

Tincture, Veratrum Viride, Merck.—U. S. P.

Dose: 1—5 ℳ.—Antidotes: Emetics, stomach siphon, tannic acid,
stimulants, external heat, stimulation by mustard or friction.

Toluene Merck.

Toluol.—Colorl., refractive liq.; benzene-like odor.—Sol.: Alcohol, ether,
chloroform; slightly in water.—Uses: *Topically*, in diphtheria, as "Loef-
fler's Solution" = Toluene 18 cubic centimetres, Solut. Iron Chloride
2 cubic centimetres, Menthol 5 grammes, Alcohol 30 cubic centimetres.

Tongaline.

Not completely defined.—(Stated: "Each fluid dram represents 30 grn.
tonga, 2 grn. ext. cimicifuga, 10 grn. sod. salicylate, ₁⁄₁₀ grn. pilocarpine
salicylate, 1/500 grn. colchicine.—Antirheumatic, Diaphoretic.—**Dose:**
1—2 fl. drs.")

Traumaticin Merck.

10% solut. gutta-percha in chloroform.—Thick, viscid, dark-brown liq.—
Uses : *Extern.*, in dentistry and surgery, as a protective covering for
bleeding surfaces, cuts, etc.; also as a vehicle for application of chrysarobin or other antiseptics, in skin diseases.

Tribromphenol Merck.

Bromol.—White cryst.; disagreeable, bromine odor; sweet, astring. taste.
—Sol. in alcohol, ether, chloroform, glycerin, oils; insol. in water.—External and Internal Antiseptic.—Uses: *Intern.*, cholera infantum, typhoid
fever, etc.; *extern.*, purulent wounds, diphtheria, etc.—**Dose:** 3—8 grn.
daily.—Extern. in 1:30 oily solut., or 1:8 oint.; in diphtheria, 4% solut.
in glycerin.

Trimethylamine Solution, Medicinal, — see SOLUTION, TRIMETHYLAMINE.

Trional.

Colorl., odorl. plates; peculiar taste.—SOL. in 320 parts water; also in alcohol or ether.—Hypnotic, Sedative.—**Dose :** 15—30 grn.—MAX. DOSE: 45 grn.

Triphenin.

PROPIONYL-PHENETIDIN, *Merck*.—Colorl. cryst.—SOL. in 2000 parts water.—Antipyretic and Antineuralgic, like Phenacetin; prompt, and without by- or after-effect.—**Dose:** *Antipyretic,* 4—10 grn.; *antineuralgic,* 15—20 grn.—[Further information in "Merck's Digest" on "TRIPHENIN", containing clinical reports.]

Triticum—U. S. P.

COUCH-GRASS.—Demulcent, Diuretic.—**Dose:** 1-4 drams, in F. E. (1:1) or infus. (1:20).

Tritipalm.

Not completely defined.—(Stated: " Comp. Fld. Ext. Saw Palmetto and Triticum. Ea. fl. dr. represents 30 grn. fresh saw palmetto berries and 60 grn. triticum.—Genito-urinary Tonic.—**Dose :** 1 fl. dr., 4 t. daily.")

Tropacocaine Hydrochlorate Merck.

BENZOYL-PSEUDOTROPEINE HYDROCHLORATE.—Colorl. cryst.—SOL. in water.—Succedaneum for Cocaine. According to Drs. Vamossy, Chadbourne, and others, tropacocaine is not half as toxic as cocaine. Anesthesia from it sets in more rapidly and lasts longer than with cocaine. It causes much less hyperemia than does cocaine. Mydriasis does not always occur, and when it does, is much less than with cocaine. The activity of its solution is retained for two to three months. Tropacocaine may replace cocaine in every case as an anesthetic.—APPLIED in 3% solut., usually in 0.6% sodium-chloride solut.—[Further information in "Merck's Digest" on "TROPACOCAINE ", containing clinical reports.]

Turpentine, Chian, Merck.

Thick, tenacious, greenish-yellow liq.; peculiar, penetrating odor.—Antiseptic.—USES: *Extern.,* cancerous growths.

Turpentine, Canada—U. S. P.

BALSAM OF FIR.—Used chiefly extern.—**Dose:** 5-30 grn., in pill

Unguentine.

Not completely defined.—(Stated: " Alum ointment, with 2% carbolic acid, 5% ichthyol.—Antiseptic. Astringent, Antiphlogistic.—USES: Burns and other inflam. diseases of skin.")

Uranium Nitrate Merck.—C. P.

Yellow cryst.—SOL. in water, alcohol, ether.—USES: Diabetes.—**Dose:** 1—2 grn., gradually increasing to 15 grn., two or three t. daily.

Urea Merck.—Pure.

CARBAMIDE.—White cryst.—SOL. in water, alcohol.—Diuretic.—USES: Cirrhosis of liver, pleurisy, renal calculus, etc.—**Dose:** 150—300 grn. a day, in hourly instalments, in water.

Urethane Merck.—C. P.

ETHYL URETHANE.—Colorl. cryst.; faint, peculiar odor; saltpeter-like taste.—SOL. in 0.6 part alcohol, 1 part water, 1 part ether, 1.5 part chloroform, 3 parts glycerin, 20 parts olive oil.—Hypnotic, Antispasmodic, Sedative.—USES: Insomnia. eclampsia. nervous excitement, tetanus; and as antidote in strychnine, resorcin, or picrotoxin poisoning. Does not interfere with circulation; no unpleasant after-effects. In eclampsia it should be given per enema.—**Dose:** *Sedative,* 10—20 grn., 1—4 t. daily; *hypnotic,* 30—45 grn., in 3 portions at ½—1 hour intervals, in 10% solut.—MAX. D.: 80 grn.—INCOMPATIBLES: Alkalies, acids.

Uricedin.

Not completely defined.—(Stated: "Uniform combination of sodium sulphate, sodium chloride, sodium citrate, and lithium citrate.—Wh. granules.—SOL. freely in water.—Antilithic.—**Dose:** 15—30 grn., in hot water, 3 t. daily.")

Uropherin B.

THEOBROMINE AND LITHIUM BENZOATE, *Merck.*—50% theobromine.—White powd.; decomposes on exposure.—Diuretic; works well with digitalin.—**Dose:** 5—15 grn., in powd. or capsules, followed by water.—MAX. D.: 60 grn. daily.

Uropherin S.

THEOBROMINE AND LITHIUM SALICYLATE, *Merck.*—White powd.—USES, DOSE, ETC., as Uropherin B.

Urotropin,—see FORMIN.

Uva Ursi—U. S. P.

BEARBERRY.—Tonic, Diuretic, Antilithic.—**Dose:** 1-2 drams.—*Preparations:* Ext. (D., 5-15 grn.); F. E. (1:1).

Valerian—U. S. P.

   Dose: 10-30 grn.—*Preparations:* F. E. (1:1); Tr. (1:5); Ammon. Tr. (1:5 arom. spt ammonia).

Validol.

MENTHOL VALERIANATE.—Colorl., syrupy liq.; mild, pleasant odor; cooling, faintly bitter taste.—Nerve Sedative, Carminative.—USES: Hysteria, epilepsy; flatulence, dyspepsia, etc.—**Dose:** 10—20 drops, on sugar.

Vasogen.

OXYGENATED PETROLATUM.—Faintly alkaline, yellowish-brown, syrupy mass, yielding emulsions with water and rendering such active medicaments as creolin, creosote, ichthyol, iodine, pyoktanin, etc., readily absorbable through the skin. Used combined with these, externally as well as internally. *Iodine Vasogen* (80 grn. daily) recommended by inunction in syphilis and glandular swellings, and internally in arterial sclerosis (4—6 grn. twice daily). *Iodoform Vasogen* used in tuberculous processes.

Veratrine Merck.—U. S. P.

White powd.; causes violent sneezing when inhaled; exceedingly irritating to mucous membranes.—SOL. in 2 parts chloroform, 3 parts alcohol, 6 parts ether; slightly in water.—USES: *Intern.,* gout, rheumatism, neuralgia, scrofula, epilepsy; *extern.,* stiff joints, sprains, and chronic swellings.—**Dose:** $\frac{1}{60}$—$\frac{1}{30}$ grn..—MAX. D.: ¾ grn.—EXTERN.: 1—4% in oint.—*Preparations:* Oleate (2%): Oint. (4%).—ANTIDOTES: Tannic acid, emetics, powdered charcoal, stomach pump, stimulants; morphine with atropine hypodermically, heat, recumbent position.

Veratrum Viride—U. S. P.

AMERICAN HELLEBORE.—Cardiac Depressant, Diaphoretic, Diuretic.—*Preparations.* F. E. (D., 1-4 min.); Tr. (D. 3-10 min.).—See also, Veratrine.

Viburnum Opulus—U. S. P.

CRAMP BARK.—Antispasmodic, Sedative—*Preparations:* F. E. (D., 30–60 min.).

Viburnum Prunifolium.

BLACK HAW.—Astringent, Nervine, Oxytocic.—*Preparations:* F. E. (D., 15–60 min.).

Vitogen.

Not completely defined.—(Stated: "Definite, stable compound.—Whitish, odorl., insol. powd.—Surgical Antiseptic, Deodorant.—USED only *extern.*, pure.")

Water, Bitter-Almond, Merck.—U. S. P.

0.1% hydrocyanic acid.—USES: Chiefly as vehicle.—**Dose:** 10—20 ℳ.

Water, Cherry-Laurel, Merck.

0.1% hydrocyanic acid.—Turbid liq.—Anodyne, Sedative, Antispasmodic. —USES: Chiefly as vehicle; also in whooping-cough, asthmatic affections, dyspnea, etc.—**Dose:** 10—20 ℳ.

White Oak—U. S. P.

Astringent.—**Dose:** 30–60 grn., as fl. ext. or decoct.

White Precipitate,—see MERCURY-AMMONIUM CHLORIDE.

Wild-Cherry Bark—U. S. P.

Astringent. Tonic. Sedative.—*Preparations:* F. E. (D., 20–60 min.); Infus. (D., 1–4 oz.); Syr. (D., 1–4 drams).

Xanthoxylum—U. S. P.

PRICKLY ASH.—Diaphoretic, Alterative, Counterirritant.—*Preparation:* F. E. (D., 15–60 min.).

Xeroform.

TRIBROM-PHENOL-BISMUTH.—Yellow-green, alm. odorl. and tastel., insol. powd.; 50% tribromphenol.—Surgical and Intest. Antiseptic.—USES: *Extern.*, infected wounds, buboes, etc.; *intern.*, diarrheas of various kinds. —EXTERN. like iodoform.—**Dose:** 5—15 grn.

Zinc Acetate Merck.—U. S. P.

SOL. in 3 parts water, 36 parts alcohol.—Astringent, Antiseptic, Nervine. —USES: Chiefly *extern.*, collyrium in ophthalmia, injection in urethritis, and gargle in sore mouth or sore throat.—APPLIED: Eye-wash, 1—2 parts to 1000 water; gargle, 5—10 parts to 1000 water; injection, 2—5 parts to 1000 water.—**Dose:** ½—2 grn.

Zinc Bromide Merck.—U. S. P.

SOL. in water, alcohol, ether, ammonia.—USES: Epilepsy, in very diluted solut.—**Dose:** 1—2 grn.—MAX. D.: 10 grn. daily.

Zinc Carbonate Merck.—U. S. P.

USES: Wounds, ulcers, skin diseases, etc.; also face powd.—APPLIED pure or 20% oint. or powd.

Zinc Chloride Merck.—U. S. P.

SOL. in 0.3 part water; in alcohol, ether.—**Dose:** $\frac{1}{16}$—$\frac{1}{3}$ grn.—EXTERN.: Gonorrhea, 1:1000 solut; wounds, 1:100—500; eyes, 1:1000, tuberculous joints, 1:10.—*Preparation:* Solut. (50%).—ANTIDOTES: Alkali carbonates, followed by water or milk; albumen, anodynes, stimulants, tea, etc.

Zinc Cyanide Merck.—Pure.

White, cryst. powd.—Alterative, Antiseptic, Anthelmintic.—USES: Chorea, rheumatism, neuralgia, dysmenorrhea, colic, gastralgia, cardiac palpitation. Small doses at first and gradually increased.—**Dose:** $\frac{1}{10}$–$\frac{1}{4}$ grn.—ANTIDOTES: Stomach siphon, ammonia, mixture of ferrous and ferric sulphates, chlorine inhalation, cold douche, etc.

Zinc Ferro-cyanide Merck.

White powd.—Alterative, Antiseptic.—USES: Dysmenorrhea, rheumatism, chorea, gastralgia, etc.—**Dose:** $\frac{1}{2}$–4 grn.

Zinc Hypophosphite Merck.

SOL. in water.—Antiseptic, Astringent, Antispasmodic.—USES: Gastric and intestinal catarrh, chorea, whooping-cough, epilepsy, skin diseases.—**Dose:** $\frac{1}{2}$–$1\frac{1}{2}$ grn.

Zinc Iodide Merck.—U. S. P.

SOL. in water, alcohol, and ether.—**Dose:** 1–2 grn. ,

Zinc Lactate Merck.

White cryst.—SOL. in 60 parts water.—Anti-epileptic.—**Dose:** $\frac{1}{2}$–1 grn., gradually increased.—MAX. D.: 10 grn. daily.

Zinc Oxide Merck.—U. S. P.

EXTERN: in 5—20% oint. or powd.—USES: *Intern.*, chorea, epilepsy, chronic diarrhea, etc.; *extern.*, wounds, skin diseases, etc.—**Dose:** 1–5 grn.—*Preparation:* Oint. (20%).

Zinc Permanganate Merck.—C. P.

Violet-brown, or almost black, hygroscopic cryst.—SOL. in water.—Antiseptic, non-irritating Antigonorrhoic.—USES: 1:4000 solut. as injection in gonorrhea; and 1 or 2:1000 as eye-wash in conjunctivitis.—INCOMPATIBLES: All easily oxidizable or combustible substances. Explodes when compounded directly with alcohol, glycerin, sugar, dry or fluid vegetable extracts.

Zinc Phosphide Merck.—U. S. P.

INSOL. in the usual solvents—USES: Sexual exhaustion, cerebral affections, melancholia, and chronic skin diseases.—**Dose:** $\frac{1}{20}$–$\frac{1}{4}$ grn., in pill.

Zinc Stearate Merck.

White, agglutinating powd.; turns darker on exposure.—INSOL. in water.—Antiseptic, Astringent.—USES: Gonorrhea, atrophic rhinitis, etc.—APPLIED in substance, or combined with iodole, iodoformogen, etc.

Zinc Sulphate Merck.—U. S. P.—C. P.

WHITE VITRIOL; ZINC VITRIOL.—SOL. in 0.6 part water, 3 parts glycerin.—**Dose:** $\frac{1}{4}$–$\frac{1}{2}$ grn.; *emetic*, 10—30 grn.—ANTIDOTES: Alkali carbonates, tannic acid, albumen, demulcents.

Zinc Sulphocarbolate Merck.

Colorl. cryst.—SOL. in 2 parts water; 5 parts alcohol.—Antiseptic, Astringent.—USES: *Extern.*, gonorrhea, foul ulcers, etc.; *intern.*, typhoid, fermentative diarrhea, etc.—EXTERN. in $\frac{1}{2}$–1% solut.—**Dose:** 2–4 grn.

Zinc Valerianate Merck.—U. S. P.

Decomposes on exposure.—Sol. in 40 parts alcohol, 100 parts water.—USES: Diabetes insipidus, nervous affections, neuralgia, etc.—**Dose:** 1–3 grn.—MAX. D.: 5 grn.

PART II—THERAPEUTIC INDICATIONS

For the Use of the Materia Medica and Other Agents.

Abasia and Astasia.
—*See also, Hysteria.*

Sodium Phosphate: by hypodermic injection once a day for 25 days (Charcot).

Abdominal Plethora.—*See also, Hepatic Congestion, Obesity.*

Aliment: dry diet; avoid much bread, as well as salted or twice cooked meats, rich sauces, etc.

Cathartics, saline and hydragogue: to relieve portal congestion.

Grape cure.

Saline mineral waters.

Abortion.

Acid, Tannic: combined with opium and ipecac.

Cascara Sagrada: as a laxative.

Cimicifuga: as a prophylactic.

Cotton Root.

Creolin: 2 per cent. solution, injected after removal of membranes.

Curettement.

Diet and Hygiene.

Ergot.

Gold Chloride: to avert the tendency to abort.

Iodine: to inner surface of uterus after removal of membranes.

Iron: with potassium chlorate throughout the pregnancy when fatty degeneration present.

Opium or Morphine.

Piscidia.

Potassium Chlorate.

Savin.

Viburnum Prunifolium.

Abrasions.—*See also, Bruises, Burns, etc.*

Benzoin.

Collodion.

Iodoform.

Iodoformogen.

Iodole.

Magnesia.

Solution Gutta-percha.

Sozoiodole salts.

Abscess. — *See also, Suppuration, Boils, Anthrax.*

Acid, Boric: a powerful non-irritating antiseptic dressing.

Acid, Carbolic: as dressing, and as injection after evacuation.

Acid, Tannic.

Aconite: in full dose often aborts.

Alcohol: as a pure stimulant where a large quantity of pus is being poured out, draining the system.

Ammoniac and Mercury Plaster.

Arnica Tincture.

Belladonna: internally, and locally as a liniment or plaster, to abort the preliminary inflammation—e.g. of breast—afterwards to ease pain in addition.

Calcium Phosphate: where abscess is large or chronic, as a tonic.

Calcium Sulphide: small doses, frequently repeated, to hasten maturation or healing, especially in deep-seated suppuration.

Caustic Potassa: for opening abscess in liver, also in chronic abscess where the skin is much undermined, also used to prevent scarring if otherwise opened.

Chlorine Water.

Cod-Liver Oil: in scrofulous cases and in the hectic.

Counter-irritation: to surrounding parts, to check formation or hasten maturation.

Creolin.

Creosote: same as Carbolic Acid, as a stimulant to indolent inflammatory swellings.

Ether: to produce local anesthesia, used as a spray before opening an abscess.

Formaldehyde.

Gaduol: in scrofulous and hectic cases.

Gold Chloride.

Hydrogen Peroxide: to wash out cavity of tubercular or slow abscess.

Ice: after opening.

Iodine: as injection into the sac, and internally to cause absorption of products of inflammation.

Iodoformogen.

Iodoform Gauze: packed into cavity.

Iodole.

Lead Water.

Menthol: in ethereal solution, 10 to 50 per cent., locally applied with camel's hair pencil.

Morphine.

Naphtalin.

Naphtol: 75 grn., alcohol, 10 fl. drs., hot distilled water q. s. to make 3 fl. oz. Inject a few drops.

Oakum: as a stimulating and antiseptic dressing.

Oleate of Mercury and Morphine: relieves the pain, allays the inflammation, and causes the absorption of the products.

Potassium Permanganate: as antiseptic.

Poultices: advantageously medicated, e. g. with belladonna or opium, to allay pain or inflammation.

Quinine.

Resorcin: in syphilitic and other unhealthy sores as an antiseptic.

Salicylic Acid: as antiseptic dressing.

Sarsaparilla: in chronic abscess with profuse discharges.

Sheet Lead: is useful in the chronic abscess of the leg as a dressing.

Silver Nitrate: a strong solution in spirit of nitrous ether, painted around the area of inflammation, will check it in superficial parts.

Sodium Gold and Chloride: in scrofulous abscesses as a tonic.

Sozoiodole salts.

Strontium Iodide.

Sulphides: of potassium, sodium, ammonium, and calcium. They must be used in low doses, and are indicated in scrofulous abscess and in the chronic boils of children. To hasten suppuration.

Tonics.

Veratrum Viride: in full dose often aborts.

Abscess of the Liver.
—*See Hepatic Diseases.*

Acidity of Stomach.

Acids: before meals, or as an acid wine during meals. For acid eructations, especially of sulphuretted hydrogen.

Acid, Carbolic: to stop fermentation or to relieve an irritable condition of the stomach.

Alkalies: after meals, best as bicarbonates; with flatulence give magnesia if there is constipation; l i m e water if there is diarrhea.

Ammonia: in headache from acidity.

Ammonium Bicarbonate.

Atropine: for gastric hypersecretion.

Bismuth: in gastritis due to chronic abscess or chronic alcoholism. Very well combined with arsenic in very chronic cases, with hydrocyanic acid in more acute cases.

Calcium C a r b o n a t e, precipitated.

Cerium Oxalate.

Charcoal: as biscuits.

Creosote: same as carbolic acid.

Ichthalbin.

Ipecacuanha: in small doses in pregnancy where flatulence and acidity are both present.

Kino: useful along with opium.

Lead Acetate: in gastric catarrh and pyrosis.

Lime Water.

Liquor Potassæ: useful for both gastric and urinary acidity.

Magnesium Carbonate.

Magnesium Oxide.

Manganese D i o x i d e: sometimes relieves, probably acting like charcoal.

Mercury: When liver deranged and stools pale.

Nux Vomica: in small doses before meals, especially in pregnancy, or in chronic alcoholism.

Potassium Bitartrate.

Potassium Carbonate.

Pulsatilla: every four hours in hot water.

Silver Nitrate: same as silver oxide.

Silver Oxide: especially u s e f u l when acidity is accompanied by neuralgic pains in stomach.

Sulphurous Acid: if associated with t h e vomiting of a pasty material, presence of sarcinæ.

Tannalbin: when there is abundance of mucus.

Tannic Acid: in acidity associated with chronic catarrh and flatulence. Glycerin 1 minim, tannic acid 4 grn., as pill.

Acne.

Adeps Lanæ: topically.

Alkaline lotions: when skin is greasy and follicles are black and prominent.

Aristol.

Arsenic: in c h r o n i c acne; g e n e r a l l y, though not always, prevents the acne from b r o m i d e or iodide of potassium.

Belladonna: as local application to check a too abundant secretion.

Berberis: for acne of girls at puberty.

Bismuth: as ointment or powder. In acne rosacea, if acute.

Borax: solution very useful.

Cajeput Oil: as stimulant in acne rosacea.

Calcium Sulphide: same as sulphur. For internal use.

Chrysarobin.

Coca.

Cod-Liver Oil.

Copper.

Electricity.

Euresol.

Europhen.

Gaduol: internally, in scrofulous and hectic cases.

Glycerin: both locally and internally.

HydrastineHydrochlorate: as lotion.

Hydrastis.

Ichthalbin: internally.

Ichthyol: externally.

Iodide of Sulphur: in all stages of the disease.

Iodine: is of doubtful value.

Iodole: topically.

Levico Water.

Liquor Hydrarg. Pernitratis: a single drop on an indurated pustule will destroy without a scar.

Magnesium Sulphate.

Mercurials: internally.

Mercury Nitrate: solution topically.

Mercury Bichloride: solution as wash.

Mercury Iodide, red.

Naphtol.

Nitric Acid.

Perosmic Acid.

Phosphorus: in chronic cases in place of arsenic. The phosphates and hypo-phosphites are safer and more valuable. The latter in acne indurata.

Potassium Bromide: sometimes useful in moderate d o s e s in obstinate cases. This salt and the Iodide very often cause acne when t a k e n continuously.

Potassium Chlorate.

Quinine.

Resorcin.

Sand: f r i c t i o n with, useful.

Sodium Bicarbonate.

Strontium Iodide.

Sulphur: internally, and externally as a lotion or ointment, most valuable agent.

Thymol.

Water: Hot sponging several times a day.

Zinc Salts.

Actinomycosis.

Potassium Iodide.

Sodium Salicylate.

Addison's Disease.

Arsenic.

Glycerin: in full doses.

Iron: with anti-emetics and tonics.

Iron Glycerinophosphate.

Levico Water.

Phosphorus.

Skimmed Milk: as diet.

Sozoiodole-Potassium.

Adenitis.—See a l s o, Glandular Affections.

Calcium Phosphate: internally.

Calcium Sulphide: internally.

Carbon Disulphide.

Cod-Liver Oil: internally.

Gaduol: internally.

Ichthalbin: internally.

Ichthyol: topically as antiphlogistic.

Iodole: as cicatrizant.

Sozoiodole - Potassium: as granulator.

Adynamia.—*See also, Anemia, Convalescence, Neurasthenia.*

Acid, Hydriodic.
Acid, Hydrochloric.
Acid, Nitric.
Alcohol.
Arsenic: for swelled feet of old or weakly persons with weak heart.
Calcium Phosphate.
Caffeine.
Camphor.
Cinchona Alkaloids and their salts.
Capsicum.
Digitalis.
Eucalyptol.
Hemogallol.
Hydrastine.
Hydrogen Peroxide.
Iron.
Iron Valerianate.
Levico Water.
Nux Vomica: in dipsomaniacs.
Potassium Chlorate.
Quinine.
Sanguinarine.
Solut. Ammonium Acetate.
Turpentine Oil.
Urethane.
Valerian.

After-Pains.—*See also, Lactation.*

Actæa Racemosa: it restores the lochia in cases of sudden suppression and removes the symptoms.
Amyl Nitrite.
Belladonna: as ointment.
Camphor: 10 grn. with ⅛ grn. morphine.
Chloral: in large doses arrests the pains; contra-indicated in feeble action of the heart.
Chloroform: liniment to abdomen, along with soap liniment.
Cimicifuga: same as ergot.
Copper Arsenite.
Ergot: to keep the uterus constantly contracted and prevent accumulation of clots and the consequent pain.
Gelsemium: stops pains when in doses sufficient to produce its physiological effect.
Morphine: hypodermically very useful, 1-6 to 1-4 grn. with 1-100 grn. atropine.
Opium: the same as morphine.

Pilocarpine: in agalactia.
Poultices: warm, to hypogastrium, relieve.
Quinine: 5 to 10 gr. night and morning, in neuralgic afterpains which do not yield to opiates.
Viburnum.

Ague. — *See Intermittent Fever.*

Albuminuria. — *See also, Bright's Disease, Nephritis.*

Acid, Gallic: lessens albumen and hematuria.
Aconite: to lower a high temperature; and in the onset of acute nephritis in scarlet fever.
Alcohol: hurtful in acute stage; useful when a slight trace of albumen is persistent.
Alkaline Diuretics: to prevent formation of fibrinous plugs in the renal tubules.
Aqua Calcis: in large doses has been found to increase the urine, and decrease the albumen.
Arsenic: beneficial in very chronic cases. Albumen will return if the use of the drug be stopped.
Baths: warm water and hot air and Turkish, to increase action of skin after dropsy or uremic symptoms have appeared.
Belladonna: has been used to diminish the chronic inflammatory condition left by an acute attack.
Broom: as diuretic in chronic renal disease.
Caffeine: to increase secretion of solids especially in cases dependent on cardiac disease. Should be combined with digitalis. Very useful in chronic Bright's disease; should be used with great caution in the acute stage.
Calcium Benzoate.
Cannabis Indica: as diuretic in hematuria.
Cantharides: 1 min. of tincture every three hours, when acute

stage has passed off, to stop hematuria.
Chimaphila: as a diuretic.
Cod-Liver Oil: as a tonic.
Copaiba: to remove ascites and albuminuria dependent on cardiac or chronic Bright's disease, and in some cases of hematuria.
Counter-Irritation: dry cupping most useful when tendency to uremia.
Croton Oil: as liniment to the loins in chronic cases is sometimes useful.
Digitalis: the infusion is the most valuable in acute and tubal nephritis, and in renal disease attended with dropsy due to cardiac disease. Must be given with caution in granular kidney.
Elaterium: as hydragogue cathartic for dropsy; and when uremic symptoms have come on.
Eucalyptus: cautiously for a short time in chronic disease.
Fuchsine: in 1 to 3 grn. doses in the day, in albuminuria of renal origin, in children.
Gaduol: as a tonic.
Glycerinophosphates.
Gold Trichloride: in contracted kidney, in the chronic disease, in doses of 1-20 grn.
Hemo-gallol: in anemia.
Hydrastis: lessens albumen.
Incisions: over the malleoli, to relieve the anasarca of the lower extremities.
Iron: to diminish anemia with a flabby tongue, give the persalts. In dropsy associated with high tension, iron must be cautiously given, and withheld unless improvement is quickly shown. It always does harm if allowed to constipate.
Jaborandi: in uremia and dropsy due either to renal disease or occurring in pregnancy.
Juniper Oil: diuretic.
Lead: lessens albumen and increases the urine.
Levico Water.

Lime Water.

Milk Cure: pure skim-milk diet very useful when tendency to uremia; it also lessens the albumen.

Naphtol.

Nitroglycerin: in acute and chronic albuminuria.

Nitrous Ether: as diuretic.

Oxygen: compressed, will, on inhalation, temporarily diminish albumen.

Pilocarpine.

Potassium salts: especially the iodide and vegetable salts in syphilitic or amyloid disease.

Potassium Bitartrate: as hydragogue cathartic and diuretic.

Potassium Bromide: in uremic convulsions.

Strontium Acetate.

Strontium Lactate: if due to renal atony.

Tannalbin.

Tartrates: as diuretics.

Turpentine: as diuretic, ½ to 1 minim dose every two to four hours.

Water: in large draughts as diuretic when excretion of solids is deficient; and in dropsy.

Alcoholism. — *See also, Delirium Tremens, Vomiting, Neuritis.*

Actæa Racemosa: in irritable dyspepsia.

Ammonia: aromatic spirit of, as substitute for alcohol, to be taken when the craving comes on.

Ammonium Chloride.

Ammonium Acetate.

Arsenic: to lessen vomiting in drunkards, in the morning before food is taken; and also in the irritable stomach of drunkards.

Bismuth: with hydrocyanic acid, to relieve acidity and heartburn.

Bromides: useful during delirium tremens, or to lessen irritability, in 1 dram doses in the wakeful condition which immediately precedes it.

Capsicum: as a substitute for alcohol, and

also to relieve the restlessness and insomnia.

Chloral Hydrate: to quiet nervous system and induce sleep in an acute attack. Must be used with caution in old drunkards.

Cimicifuga.

Cocaine: to remove the craving.

Faradization.

Gelsemium: same as bromides.

Gold and Sodium Chloride.

Hydrastine.

Ichthalbin.

Levico Water: as tonic.

Lupulin: along with capsicum as a substitute for alcohol, also to quiet nervous system in delirium tremens.

Milk: at night.

Nux Vomica: as tonic and stimulant, both to nervous system and generally to aid digestion.

Opium: May be necessary to produce sleep; to relieve the pain of the chronic gastritis and the want of appetite.

Orange: slowly sucked, a substitute for alcohol.

Phosphorus: in chronic cases as nerve tonic.

Picrotoxine: for tremors.

Potassium Bromide.

Quinine: in the "horrors" stage it acts as a sedative to the brain and restores the digestive functions.

Strychnine Nitrate.

Sumbul: in the headache of old drinkers.

Water, cold: a glass taken in small sips at a time as substitute for alcohol

Water, hot: one pint drunk as hot as possible an hour before meals will remove craving.

Zinc Oxide: in chronic alcoholic dyspepsia, and nervous debility. It also allays the craving.

Alopecia. — *See also, Tinea Decalvans.*

Acid, Carbolic: in Alopecia areata.

Acid, Gallic.

Acid, Nitric: with olive oil in sufficient quantity just to make it pugnant.

Alcohol.

Ammonia: very useful; take Ol. amygd. dul., Liq. ammoniæ, each 1 fl. oz., Spt. rosmarini, Aquæ, Mellis, each 3 fl. drams; mix; make lotion (E. Wilson).

Antimonium Tartaratum: as lotion, 1 grn. to 1 fl. oz. water.

Arsenic: internally.

Cantharides Tincture: one part to eight of castor oil rubbed in roots of hair morning and night.

Eucalyptus.

Europhen.

Glycerin: very useful: either alone or in combination appears greatly to assist.

Jaborandi.

Naphtol.

Nutgall.

Pilocarpine: subcutaneous injection has been useful.

Quillaja.

Resorcin.

Savine Oil: Prevents loss of hair in Alopecia pityroides.

Sapo Viridis: very useful as a shampoo night and morning— Take Saponis virid. (German), Alcoholis, each 3 fl. oz. Ol. lavandulæ, 30 drops.

Shaving: sometimes useful after illness.

Sodium Bicarbonate: as a lotion in Alopecia pityroides.

Sulphur Iodide: useful both internally and externally.

Tannin: watery solution or made up into ointment.

Thymol.

Thyraden, and other Thyroid preparations.

Amaurosis and Amblyopia.

Amyl Nitrite: useful in many cases of disease of the optic nerve.

Antipyrine.

Arnica: sometimes useful.

Digitalis: in toxic cases.

Electricity.

Emmenagogues: if due to menstrual disorders.

Mercury: when due to syphilis.

Myotomy: in asthenopia and hysterical amblyopia.

Nitroglycerin.

Nux Vomica.

Phosphorus.

Pilocarpine: in tobacco and alcoholic abuse.

Potassium Bromide.

Potassium Iodide.

Rue: in minute doses in functional dimness of vision, e. g. hysterical amblyopia.

Salicylates.

Santonin: sometimes useful in later stages of iritis and choroiditis, and in loss of power of optic nerve.

Seton: on temple; or blisters, along with iodide of potassium, in amaurosis coming on suddenly, and associated with tenderness of the eyeball on pressure; the disc is sometimes congested.

Silver Nitrate.

Strychnine: very useful in cases of tobacco amaurosis, alcoholic excess, nerve atrophy (without cranial disease), and in traumatic amaurosis.

Veratrine: to eyelids and temples. Care must be taken to keep out of the eye.

Zinc Lactate.

Amenorrhea. — *See also Anemia, Chlorosis.*

Acid, Oxalic.

Aconite: when menses are suddenly checked, as by cold, etc.

Actæa Racemosa: to restore the secretion, and remove the headache, ovarian neuralgia, etc., produced by its sudden stoppage.

Alcohol: in sudden suppression after exposure.

Aloes: alone or with iron. In torpor and anemia; best administered a few days before the expected period.

Ammonium Chloride: in headache.

Apiol: 5–10 min. twice a day for some days before the expected period; if there is a molimen, 15 grn. in a few hours. Useful in

anemia and torpor only.

Arnica.

Arsenic: along with iron in anemia and functional inactivity of the ovaries and uterus.

Asafetida: along with aloes in anemia and torpor of the intestines.

Baptisin.

Berberine Carbonate.

Cantharides: along with iron in torpor of the uterus.

Cimicifuga: at the proper time for a flow.

Cold Sponging: to brace the patient up.

Colocynth: in anemia with constipation.

Croton Oil.

Electricity: locally applied, sometimes useful.

Ergot: in plethoric subjects.

Eupatorium: in hot infusion, if due to cold.

Gold Salts: like asafoetida.

Guaiacum: mild stimulant to the uterus.

Ichthalbin.

Iron: in anemia, q. v.

Iron Iodide.

Iron Phosphate.

Levico Water.

Manganese Dioxide: in amenorrhea of young women; in delayed menstruation, or when a period has been missed through a chill. Perseverance is required, especially in the last case.

Myrrh: a tonic emmenagogue.

Nux Vomica: in combination with iron in anemia.

Polygonum: in torpor; with iron in anemia, aloes in a constipated subject. Contra-indicated in a plethoric condition. Should be given a few days before menses are expected.

Potassium Iodide.

Potassium Permanganate: like manganese dioxide.

Pulsatilla: like aconite.

Quinine.

Rue: in atonic conditions of ovaries or of uterus. Plethora contra-indicates.

Salines: in constipation in plethoric cases.

Sanguinaria: like rue.

Santonin: in two doses of 10 grn. each, one or two days before the expected period.

Savine: like rue.

Senega: a saturated decoction in large doses, a pint daily, about two weeks before period.

Serpentaria: in anemia.

Silver Nitrate: locally, to os uteri at period.

Sitz Baths: hot, alone, or with mustard, for some days before the period; with mustard, if suddenly arrested.

Sodium Borate.

Spinal Ice Bag: to lumbar vertebræ.

Tansy.

Turpentine.

Anemia.

Acids: for a tonic action on the mucous membranes in anemia of young women.

Acid, Gallic: in anemia due to a chronic mucous or other discharge.

Alkalies: potash and soda as gastric and hepatic tonics.

Aloes: as tonic and slight purgative.

Arsenic: in the cases where iron fails of its effect or does not agree with the patient. Also in pernicious anemia.

Bitters.

Bone-marrow.

Bullock's Blood: when iron fails, fresh or dried, by enema.

Cactus Grandiflorus.

Calcium Lactophosphate: during nursing or after exhausting purulent discharge.

Calcium Phosphate: during growth, or where system is enfeebled by drain of any kind.

Calomel.

Cetrarin.

Cold Sponging.

Copper Arsenite.

Diet and Hygiene.

Ferropyrine.

Gaduol.

Galvanization.

Glycerinophosphates.

Gold Salts.

Hemo-gallol.

Hemoglobin.

Hypophosphite of Calcium or Sodium: in cases of nervous debility care must be taken that it does not derange the digestion.

Ichthalbin.

Iron: very useful. When stomach is at all irritable the carbonate is often best. Weak, anemic girls with vomiting after food are best treated with the perchloride. In coated tongue the ammonio-citrate is often best to begin with. The malate has been useful in pernicious anemia. In gastric disturbance and constipation, a combination with rhubarb is often very effectual. Where mucous membrane is very flabby, large doses of the perchloride. Chalybeate waters more often succeed than pharmaceutical preparations; one drop of the solution of perchloride in a tumbler of water is an approximate substitute for them.

Levico Water.

Manganese salts: may be given with iron—not much use alone.

Mercury Bichloride.

Napthol, Beta-.

Nux Vomica: useful sometimes along with iron.

Oxygen: to be inhaled in anemia from loss of blood or suppuration.

Pancreatin: in feeble digestion.

Pepsin: in feeble digestion.

Phosphorus.

Quinine: in malnutrition.

Sea-bathing: good, but not in chlorosis.

Sodium Arsenate.

Sodium Hypophosphite.

Spermine.

Strychnine.

Wine: with the food, to aid digestion.

Aneurism.

Acid, Gallic, and iron.

Aconite: to relieve pain and slow the circulation.

Aliment: low diet; absolute rest.

Barium Chloride: in doses of 1-5 grn. Perhaps raises the arterial tension.

Calcium Chloride.

Chloroform: inhaled to relieve dyspnea.

Digitalis is contra-indicated (Hare.)

Electrolysis: sometimes useful in causing coagulation within the sac.

Ergotin: a local hypodermic injection has been successful.

Eucalyptus.

Iron-Chloride Solution: to cause coagulation on injection into sac.

Lead Acetate: useful, combined with rest.

Morphine: with crotonchloral, for pain.

Potassium Iodide: very useful in doses of 30 grn. Should be combined with the recumbent position.

Strontium Iodide.

Veratrum Viride: along with opium in quieting circulation.

Zinc Chloride.

Angina Catarrhalis.

—See also, Choking, Croup, Laryngitis, Pharyngitis, Throat Tonsilitis, etc.

Acid, Carbolic.

Acid, Gallic.

Alum.

Creolin: by vapor-inhalation.

Iron Chloride: as gargle.

Ichthyol: as gargle.

Potassium Chlorate: as gargle.

Potassium Nitrate.

Silver Nitrate.

Sodium Bicarbonate.

Sozoiodole-Sodium

Angina Diphtheritica. —See Diphtheria.

Angina Pectoris.

Aconite.

Allyl Tribromide.

Antipyrine.

Arsenic: to prevent paroxysms.

Atropine.

Cactus Grandiflorus.

Chamomile: in hysterical symptoms.

Chloral: in full doses.

Chloroform: cautiously inhaled to ease the pain.

Cocaine.

Cold: applied to forehead gives relief.

Convallaria.

Coniine Hydrobromate.

Digitalis.

Ether: to diminish pain, combined with opium in ¼-grn. doses.

Erythrol Tetranitrate.

Morphine: hypodermically.

Nitrite of Amyl: gives great relief during paroxysms; in atheromatous arteries must be used with care.

Nitrites of Sodium and Potassium: less rapid than nitrite of amyl, but have more power to prevent return of symptoms.

Nitroglycerin: like nitrite of sodium.

Phosphorus: during intervals to lessen tendency.

Potassium Bromide: in full doses will relieve the spasm.

Pyridine.

Quinine: when any malarious taint is present.

Spermine.

Spirit Ether.

Strophanthus.

Strychnine: sometimes useful in mild cases in very small doses.

Tonics.

Turpentine Oil: locally to the chest during paroxysms.

Anorexia.—See also, lists of Tonics, Gastric Tonics, etc.

Acid, Nitro-hydrochloric: when following acute disease.

Absinthin.

Berberine Carbonate.

Calomel: when following acute disease; nitro-hydrochloric acid generally preferable, however.

Capsicum: in convalescence.

Chimaphila: in dropsical cases, as a tonic and diuretic.

Cinchonidine.

Cinchonine.

Eupatorium.

Gentian.

Nux Vomica Tincture.

Oleoresin capsicum.

Orexine Tannate: of very wide utility.

Quassia: especially valuable when following malarial fever.

Quassin.

Anthrax.— (Carbuncle.)

Acid, Boric: as dressing.

Acid, Carbolic: as wash and injection after spontaneous discharge, or on lint after opening.
Alcohol: as needed.
Ammonium Acetate.
Ammonium Carbonate: combined with cinchona, after a free purge.
Arnica: fresh extract spread on adhesive plaster and strapped; internal administration is also beneficial.
Belladonna Extract: with glycerin, as local anodyne.
Blister: to cover area, with hole in the center to allow discharge.
Bromine.
Butyl-Chloral Hydrate: to lessen the pain of facial carbuncle.
Calcium Sulphide: one-tenth grn. hourly useful.
Collodion: around base, leaving opening in the center.
Creolin.
Ether: sprayed on for a little time will cause an eschar to separate.
Europhen.
Hydrogen Peroxide.
Ichthalbin: internally.
Ichthyol: topically.
Iodine: locally, to lessen pain and inflammation, should be applied around the base.
Iodoform: useful local antiseptic dressing.
Iodoformogen.
Iodole.
Lead Carbonate.
Menthol.
Mercurial Ointment: early application will abort sometimes.
Opium: locally, mixed with glycerin.
Phosphorus: internally.
Potassium Chlorate and mineral acids: internally administered.
Potassium Permanganate: antiseptic lotion.
Poultices: to relieve pain.
Pyoktanin.
Quinine.
Quinine and Carbolic Acid: internally.
Strapping: concentrically, leaving center free, lessens pain.
Terebene or Oil Turpentine: antiseptic application.

Antrum, Disease of.

Acid, Boric.
Bismuth Subnitrate.
Chloroform.
Iodine.
Zinc Sulphate.

Anus, Fissure of.

Acid, Benzoic: as a local application.
Acid, Carbolic: one drop of 95 per cent. applied to fissure.
Belladonna: locally; relieves spasms.
Bismuth: with glycerin, as a local application.
Calomel: as ointment.
Carron Oil: as a dressing.
Castor Oil: to keep motions soft.
Chloral Hydrate: in dilute solution (2 per cent.) as a dressing.
Chloroform: diluted with half its bulk of alcohol, will aid healing.
Cocaine: in ointment.
Collodion: locally, to protect.
Dilatation, forcible: relieves spasm.
Hydrastis: local application.
Ice: to relieve pain after operation.
Ichthalbin.
Ichthyol.
Iodoform: locally, to heal and relieve pain.
Iodoformogen: very beneficial.
Opium and Gall Ointment: relieves pain.
Potassium Bromide: with five parts of glycerin, locally.
Rhatany: injected after the bowels have been opened by enema.
Silver Nitrate.
Sozoiodole-Potassium.
Sulphur: to keep motions soft.
Tannin: useful as a local application.

Anus, Prolapsus of.
—See Prolapsus Ani.

Aphonia.

Acid, Nitric: in hoarseness from fatigue or indigestion.
Acid, Sulphurous: as spray or inhalation, in clergyman's sore-throat.
Aconite: in the painful contraction of the throat of singers.

Alum: as spray in chronic congestion of throat and larynx, with hoarseness.
Ammonium Chloride: as vapor in laryngeal catarrh.
Argenti Nitras: as local astringent.
Atropine: in hysterical aphonia; must be pushed enough to produce physiological symptoms.
Belladonna.
Benzoin Tincture: by inhalation in laryngeal catarrh.
Borax: a piece the size of a pea slowly sucked in sudden hoarseness.
Chloroform: in hysterical and nervous cases.
Electricity: locally.
Ether: like chloroform.
Glycerite of Tannin: locally to pharynx.
Ignatia: like atropine.
Ipecacuanha: wine as spray in laryngeal catarrh.
Nux Vomica locally applied in impaired nervous power.
Potassium Nitrate: like borax.
Rue Oil: as inhalation in chronic catarrh.
Turkish Bath: in acute laryngeal catarrh.
Uranium Nitrate: as spray in very chronic catarrh.
Zinc Sulphate: local astringent.

Aphthæ. — See also, Cancrum Oris, Gums, Parotitis, Ptyalism, Stomatitis, Odontalgia, Tongue.

Acid, Boric.
Acid, Carbolic.
Acid, Hydrochloric: in small doses and as a local application.
Acids, Mineral: dilute solution as paint.
Acid, Nitric: in small doses.
Acid, Salicylic: as local application.
Acid, Sulphurous: well diluted as solution or spray.
Acid, Tannic.
Alum, Exsiccated: to aphthous ulcers which do not readily heal.
Argenti Nitras: locally.
Bismuth: as local application.

Borax: as honey or as glycerite, either alone or with chlorate of potassium.

Chlorine Water: locally applied.

Copper Sulphate: weak solution painted over the aphthæ.

Coptis Trifolia: infusion is employed in New England.

Creolin.

Glycerin.

Mercury with chalk: to remove the indigestion on which aphthæ frequently depend.

Potassium Chlorate: exceedingly useful as wash, 10 grn. to the oz., alone or with borax, also given internally.

Potassium Iodide: as local application, solution of 1 to 5 grn. to the oz.

Pyoktanin.

Quinine: 1 grn. every two or three hours, in aphthæ consequent on diarrhea in infants.

Rhubarb: as compound rhubarb powder, to remove indigestion.

Saccharin: in 2 or 3 per cent. solut. with sodium bicarbonate.

Sodium Sulphite.

Sozoiodole-Sodium.

Sulphites.

Apoplexy.— *See also, Cerebral Congestion.*

Aconite: to lower blood-pressure and prevent further hemorrhage, where pulse is strong and arterial tension high.

Arsenic: in cerebral congestion proceeding from apoplexy.

Bandaging the limbs,

Belladonna.

Cactus Grandiflorus: when apoplexy is threatened.

Cold Water: to the head when face is congested.

Colocynth: as purgative.

Croton Oil: as purgative, one drop on back of tongue, or part of drop every hour.

Diet and Hygiene, prophylactic: meat and stimulants to be taken very sparingly; exposure to heat, over-exertion, and especially anger, to be avoided.

Elaterium: in suppository, or as enema during attack.

Electricity: to promote absorption, after partial recovery has taken place.

Ice: to head.

Mercurial purge.

Mustard plaster to feet, or mustard foot-bath, and ice to head, keeping head high and feet low.

Nitroglycerin: to lessen cerebral congestion.

Opium and calomel.

Potassium Bromide: in combination with aconite.

Potassium Iodide: to cause absorption of effused blood.

Stimulants: cautiously exhibited, when collapse is present.

Strychnine: hypodermically, if respiration fails.

Venesection or Leeches: to relieve arterial pressure when apoplexy is threatening.

Veratrum Viride.

Appetite, Impaired.—*See Anorexia.*

Appetite, Loss of.— *See Anorexia.*

Arthritis.—(*Gout.*)

Aconite.

Arsenic.

Cimicifugin.

Colchicine.

Colchicum.

Formin.

Gaduol.

Gold.

Ichthyol: topically in 5—10 per cent. oint.

Ichthalbin: internally.

Iodides.

Lithium Salts.

Mercury Bichloride.

Mercury Oleate.

Phenocoll Hydrochlorate.

Potassa Solution.

Potassium Bromide.

Potassium Iodide.

Saliformin.

Sozoiodole-Mercury.

Ascaris.—*See Worms.*

Ascites. — *See also, Dropsy.*

Acidum Nitricum: in cirrhosis of the liver.

Aconite: in scarlatina nephritis at the onset of the attack.

Apocynum Cannabinum: as diuretic.

Arsenic: in old persons with feeble heart.

Asclepias: in dropsy of cardiac origin.

Caffeine: in cardiac dropsy.

Calomel: as diuretic in cardiac dropsy.

Cannabis Indica: as diuretic in acute and chronic Bright's disease with hematuria.

Copaiba: especially useful in hepatic and cardiac dropsy.

Croton Oil: in dropsy, in ½ of a drop doses every morning.

Cytisus Scoparius: in cardiac dropsy and dropsy with chronic Bright's disease.

Diuretics.

Digitalis: best in cardiac dropsy; its action is increased by combination with squill and blue pill.

Elaterium: as hydragogue cathartic.

Gamboge: like elaterium. Large doses tolerated.

Gold.

Jaborandi: in anasarca and uremia.

Jalap: in compound powder as hydragogue cathartic.

Levico Water: as alterant.

Milk Diet: sometimes very useful when kidneys are inadequate.

Pilocarpine.

Podophyllin: in hepatic cirrhosis.

Potassium Bitartrate: in combination with jalap in hepatic cirrhosis.

Saliformin.

Squill: as diuretic in cardiac dropsy.

Stillingia: in hepatic dropsy.

Theobromine Salicylate or its double-salts.

Asphyxia from Chloroform.

Amyl Nitrite.

Artificial respiration.

Cold Douche.

Electricity.

Oxygen.

Astasia.—*See Abasia and Astasia.*

Asthenopia.

Acid, Hydrocyanic: in irritable ophthalmia.
Atropine: to prevent spasms.
Eserine or Pilocarpine: in weak solution, to stimulate ciliary muscle.
Hot Compresses.
Massage.
Myotomy, Intraocular: to relieve spasms.
Physostigma: in the paralysis produced by diphtheria, and in senile asthenopia.
Strychnine.

Asthma.

Acid, Hydriodic.
Acid, Hydrocyanic.
Aconite: in spasmodic cases, also in asthma consequent on nasal catarrh in children.
Alcohol: in combination with amyl nitrite in spasmodic asthma.
Alkalies: in chronic bronchial catarrh.
Allyl Tribromide.
Alum: 10 grn. of dry powdered alum put on the tongue may arrest a spasm.
Ammonia Vapor.
Ammoniacum: 11k Asafetida.
Ammonium Benzoate.
Amyl Nitrite: sometimes checks paroxysm in spasmodic asthma and dysnea due to cardiac hypertrophy. Must not be given in chronic bronchitis and emphysema.
Anemonin.
Anesthetics: as a temporary remedy in severe cases.
Antimony: in asthmatic conditions in children 1-80 grn. of tartar emetic every quarter of an hour.
Antispasmin.
Apomorphine: emetic, in asthma due to a peripheral blocking of the air-tubes.
Arsenic: in small doses in cases associated with bronchitis or simulating hay fever, or in the bronchitis of children, or in the dyspeptic asthma. Inhaled as cigarettes with caution.
Asafetida: as an expectorant where

there is profuse discharge.
Aspidospermine.
Atropine.
Belladonna: internally in large doses to relieve paroxysm. It should only be administered during a paroxysm and then pushed.
Bitter-Almond Water.
Bromides: only available in true spasmodic asthma; soon lose their efficacy.
Caffeine: 1 to 5 grn.
Camphor: 2 grn. combined with 1 grn. of opium, in spasmodic asthma.
Cannabis Indica: sometimes useful in chronic cases.
Chamois-Leather Waistcoat: reaching low down the body and arms, in bronchial asthma.
Chloral Hydrate: during paroxysm.
Chloralamide.
Chloroform: relieves when inhaled from tumbler or with warm water.
Cocaine.
Coffee: very strong, during paroxysm.
Colchicine or Colchicum: in gouty cases.
Compressed or Rarified air.
Coniine Hydrobromate or Conium: palliative in a chronic case.
Counter-irritation: applied for a short time only, at frequent intervals.
Creosote: vapor in bronchitic asthma.
Diet and Hygiene.
Duboisine Sulphate.
Erythrol Tetranitrate.
Ether: in full doses at commencement of attack, or administered by inhalation.
Ethyl Iodide: 15 to 20 drops inhaled may relieve spasm.
Eucalyptus: sometimes along with stramonium, belladonna, and tobacco.
Euphorbia pilulifera.
Galvanism of Pneumogastric Region: positive pole beneath mastoid process, negative pole to epigastrium.

Gelsemium: useful in some cases, but after a time may fail.
Grindelia: to prevent or cut short attack; used as cigarette.
Hyoscine Hydrobromate: in spasmodic asthma.
Ichthalbin.
Iodine: painting the line of the pneumogastric nerve with liniment or tincture in pure spasmodic asthma.
Ipecacuanha: as a spray in bronchial asthma, especially in children; useless in true asthma.
Lobelia: to prevent and cut short paroxysm. Cautiously used in cardiac weakness.
Lobeline Sulphate.
Menthol.
Mercurials: in spasmodic and bronchitic asthma combined.
Morphine: combined with belladonna, very useful.
Nitroglycerin: in bronchitic, nephritic and spasmodic asthma.
Nux Vomica: in dyspeptic asthma.
Oil Eucalyptus.
Oil of Amber.
Opium: hypodermically during paroxysm.
Oxygen: as inhalation during paroxysm.
Pepsin: exceedingly useful in preventing attacks in dyspeptic subjects.
Physostigma.
Pilocarpine Hydrochlorate: in spasmodic asthma, subcutaneously; also in humid asthma if there is no cardiac dilatation.
Potassium Bromide.
Potassium Cyanide.
Potassium Iodide: in large doses when asthma is due to acute bronchial catarrh.
Potassium Nitrate: inhalation of fumes of paper relieves paroxysm. Sometimes advisable to mix a little chlorate with it.
Potassium Nitrite.
Pyridine: in bronchial asthma, vapor to be inhaled.
Quebracho: good in nephritic and spasmodic asthma.

Quinine: during intervals when the attacks are periodical.
Resorcin: relieves dyspnea.
Sandalwood Oil.
Sanguinarine.
Sodium Arsenate: as tonic, acts probably on respiratory centre.
Sodium Iodide.
Sodium Nitrate: like nitroglycerin.
Sodium Phosphate: sometimes efficacious.
Solanine.
Spermine: as tonic.
Stramonium: sometimes very useful. May be made into cigarettes, or 20 grn. of dried leaves may be mixed with nitrate of potassium, and the fumes inhaled. A little powdered ipecacuanha may often be added.
Strontium Iodide.
Strophanthus Tincture.
Strychnine: in weakness of the respiratory center.
Sulfonal.
Sulphurated Potassa.
Sulphur fumes: in bronchitic asthma.
Tobacco: smoking is sometimes beneficial.
Turkish Baths: in bronchial asthma.
Zinc Oxide.

Asthenia.—*See Adynamia, Convalescence.*

Astigmatism.
Suitable Glasses.

Atheroma.—*See also, Aneurism.*
Ammonium Bromide.
Ammonium Iodide: to promote absorption.
Arsenic: often useful, especially where there are cerebral symptoms.
Barium Chloride.
Calcium Lactophosphate.
Cod-Liver Oil.
Digitalis: requires caution; useful in general capillary atheroma.
Hypophosphites.
Phosphates.
Phosphorus: in minute doses along with cod-liver oil, in cases with cerebral symptoms.
Quinine; like arsenic.

Atrophy.
Arsenic: in muscular atrophy.
Electricity.
Massage.
Olive Oil: inunction to atrophied parts.
Strychnine.

Balanitis.—*See also, Phimosis, Gonorrhea.*

Acid, Carbolic.
Acid, Tannic.
Alum.
Alumnol.
Creolin.
Ichthyol.
Lead Water.
Lime Water: as lotion.
Mercury: yellow wash, as lotion.
Silver Nitrate: molded.
Sozoiodole-Potassium: dusting powder.
Sozoiodole-Sodium: lotion.
Tannin or Zinc Oxide: as dusting-powder.
Tannoform.
Zinc Sulphate.

Baldness.—*See Alopecia.*

Barber's Itch.—*See Sycosis.*

Bed-Sores.

Alcohol: as wash to prevent; afterwards dust with powdered starch.
Alum: with white of egg, as local application.
Aristol.
Balsam of Peru and Unguentum Resinæ: equal parts spread on cotton wool.
Bismuth Subnitrate.
Catechu: with lead subacetate, to harden skin.
Charcoal: as poultices, to stop bed-sores.
Galvanic Couplet: of zinc and silver; one element on sore, the other on adjacent part.
Glycerin: prophylactic local application.
Hydrargyri Perchloridum: a solution mixed with diluted alcohol.
Ichthyol.
Incisions: followed by irrigation, if sores tend to burrow.

Iodoform.
Iodoformogen.
Iodole.
Iron Chloride: as tonic.
Medicated Poultices: patient to lie with poultices under the parts likely to be affected; if fetor castaplasma carbonis; if sloughing, addition of Balsam of Peru.
Pyoktanin.
Quinine: local dressing.
Salt and Whisky: topically to harden skin.
Silver Nitrate: dusted over open bed-sores.
Soap Plaster: applied after washing with bichloride solution (1 in 5000) and dusting with iodoform or iodoformogen.
Sozoiodole Potassium.
Styptic Collodion.
Tannate of Lead: at an early stage.
Zinc Oxide: ointment.

Biliousness.—*See also, Dyspepsia, Hepatic Congestion, Duodenal Catarrh.*

Acids, Mineral: nitrohydrochloric acid especially useful in chronic hepatic affections, dysentery and dropsy of hepatic origin.
Aconite: as adjunct to podophyllin.
Alkalies: in indigestion due to obstruction to the flow of bile.
Alkaline Mineral Waters: in catarrh of the bile-duct, early stage of cirrhosis, and obstruction to the hepatic circulation.
Aloes: in constipation, and in deficient secretion of bile.
Ammonium Chloride: in jaundice due to catarrh of the bile-ducts, early stage of cirrhosis; deficient intestinal secretion.
Ammonium Iodide: in catarrh of duodenum and biliary ducts, in the early stage of cirrhosis, in the malarial cachexia; efficacy increased by the addition of arsenic.
Angostura: in bilious fevers.
Argenti Oxidum.
Bromides and Chloral Hydrate.

92

Bryonia: in billious headache.

Calomel: in excessive production with deficient secretion; calomel or blue pill at night and a black draught in the morning.

Calumba: as stomachic tonic.

Carlsbad Water: a tumbler sipped warm on rising very useful.

Chirata.

Colocynth.

Euonymin: at night, followed in the morning by a saline purge.

Friedrichshall Water: a wineglassful in a tumbler of hot water slowly sipped on rising.

Horse Exercise.

Hydrastis: when chronic gastric catarrh is present, in chronic catarrh of the duodenum and bile-ducts, with inspissation of the bile and gall-stones.

Ipecac.

Leptandra.

Manganese: in malarial jaundice.

Mercurial Cathartics: in moderate doses night and morning, or in small doses more frequently repeated. Especially useful when the stools are pale, is the bichloride.

Mercury Iodide, Green.

Mercury Oxide, Yellow.

Milk Cure: in obstinate cases.

Mustard Plaster.

Opium.

Podophyllum: in place of mercury when stools are dark.

Rhubarb: as hepatic stimulant.

Salines.

Salol.

Sodium Phosphate: in bilious sick headache; also in catarrh of the gall-duct in children; dose, 10 grn.

Stillingia: in cirrhosis; torpidity and jaundice following intermittent fever; ascites due to hepatic changes; to be combined with Nux Vomica, in deficient secretion

Bites and Stings.—
    See Stings and Bites.

Bladder Affections.

Acid, Carbolic.
Aseptol.
Berberine Sulphate: for atony.
Codeine.
Formaldehyde.
Gallobromol.
Saliformin.
Sozoiodole-Sodium,

Bladder, Catarrh of.
    —See also, Cystitis

Acid, Benzoic
Ammonium Borate.
Antinosin.
Arbutin.
Betol.
Creolin: by injection.
Ichthyol.
Juniper.
Saliformin.
Salol.
Thymol.

Bladder, Inflammation of.—See Cystitis.

Bladder, Irritable.

See also, Cystitis, Dysuria, Enuresis, Lithiasis, Calculi, Urinary Disorders.

Acid, Benzoic: in large prostate, and alkaline urine.

Alkalies: vegetable salts, especially of potassium when the urine is acid.

Ammonium Benzoate: like benzoic acid.

Aquapuncture.

Arbutin.

Belladonna: in the irritable bladder of children, more especially when causing nocturnal incontinence.

Buchu: in combination with the vegetable salts of potassium, when urine is very acid.

Cannabis Indica.

Cantharides: in women without acute inflammation or uterine displacement; also in irritable bladder produced by chronic enlargement of the prostate.

Copaiba: in chronic irritability.

Cubebs: like copaiba.

Eucalyptol.

Gelseminine.

Hops.

Hyoscyamus.

Indian Corn Silk (Stigmata Maydis): a mild stimulant diuretic; infusion ad lib.

Pareira: in chronic irritable bladder.

Bladder, Paralysis of.

Cannabis Indica: in retention, from spinal disease.

Cantharides: in atonic bladder, painting around the umbilicus with the acetum.

Ergot: in paralysis, either of bladder or sphincter, when bladder is so that urine is retained, and incontinence in sphincter.

Galvanism: in lumbar region.

Nicotine: 1 fl. oz. of a 4 per cent. solution injected by catheter and then withdrawn in a few minutes.

Strychnine.

Blenorrhea. — See Gonorrhea.

Blenorrhea Neonatorum.—See Ophthalmia Neonatorum.

Blepharitis.

Acid, Boric.

Acid, Tannic.

Alkaline Lotions:warm, to remove the secretion.

Alum.

Bismuth.

Borax.

Chloral Hydrate, 5 per cent. solution, to remove scabs and crusts

Copper Sulphate: instil a very dilute solution.

Creolin, 1 or 2 per cent. solution.

Gaduol: as tonic.

Glycerinophosphates: as tonic.

Hydrastis.

Ichthalbin: as alterative.

Ichthyol: topically.

Iron: to remove the anemia usually present.

Mercury-Nitrate Ointment: very useful application. If too strong, dilute with vaselin or simple ointment.

Mercury Oxide, Red.

Pulsatilla: internally and locally.

Pyoktanin: pencil.

Silver Nitrate: pencilling the border of the lid with the solid.
Sodium Bicarbonate.

Blisters.—*See Burns and Scalds.*

Boils.—*See also, Acne, Anthrax.*

Acid, Carbolic: injection.
Acid Nitrate of Mercury: to abort at an early stage.
Acid, Salicylic.
Aluminium Acetate.
Aluminium Aceto-tartrate.
Alumnol.
Arnica: locally as an ointment, and also internally.
Arsenic: to lessen tendency to recurrence.
Belladonna: internally, or as local application.
Boric Acid: as a dressing.
Calcium Chloride.
Calcium Sulphide: to hasten maturation or abort.
Camphorated Alcohol: as local application in early stage.
Camphor, Carbolated.
Caustic.
Chloral Hydrate.
Cocaine: to allay the pain.
Collodion: painted over whole surface to abort papular stage. Over base, leaving centre free, in pustular stage.
Counter-irritation: by plasters surrounding the boil.
Gaduol: as alterative.
Ichthalbin: internally.
Ichthyol: topically.
Lead Subacetate Solution.
Levico Water: as alterative.
Menthol.
Mercury Bichloride.
Mercury Iodide, red.
Mercury Ointment.
Opium: locally to remove pain.
Phosphates: especially of sodium, as a constitutional agent.
Potassium Chlorate: as an alterative.
Poultices: to relieve pain and hasten maturation.
Pyoktanin.

Silver Nitrate: strong solution painted over the skin round boil.
Strapping: properly applied gives great relief.
Subcutaneous Incisions.
Sulphides: in small doses to abort or hasten maturation.
Sulphites.
Sulphur Waters.
Solution Gutta-Percha.
Unguentum Hydrargyri: early applied around will prevent sloughing.

Bone, Diseases of.—*See also, Caries, Exostosis, Nodes, Periostitis, Rachitis, Spina Bifida, etc.*

Calcium Salts: the phosphate in rickets, in delay of union of fractures; the chloride in strumous subjects.
Cod-Liver Oil: in scrofulous conditions.
Gaduol.
Glycerinophosphates.
Hypophosphites.
Iodine: alone, or with cod-liver oil.
Iodoform: as dressing to exposed bone.
Iodoformogen: as dusting-powder.
Iron Iodide.
Mercury Iodide, Red.
Phosphorus.
Pyoktanin.
Strontium Iodide.

Brain, Anemia of.—*See Cerebral Anemia.*

Brain, Fever of.—*See Meningitis, Cerebrospinal Meningitis, Typhoid Fever, Typhus.*

Brain, Inflammation of.—*See Cerebritis.*

Brain, Softening of.—*See Cerebral Softening.*

Breasts, Inflamed or Swollen.—*See Mustitis, Abscess, Lactation, Nipples.*

Breath, Fetid.
Benzoic Acid: in spray.
Camphor.

Carbolic Acid: dilute solution as wash to mouth.
Chlorine: liq. chlori or chlorinated lime as lotion.
Permanganate of Potassium: as wash to mouth.
Thymol.

Bright's Disease, Acute.—*See also, Albuminuria, Hematuria, Scarlet Fever, Uremia.*

Aconite.
Acid, Gallic.
Alkaline salts.
Ammonium Benzoate.
Antipyrine.
Arbutin.
Belladonna.
Bromides.
Caffeine.
Cannabis Indica.
Cantharides.
Digitalis.
Elaterium.
Eucalyptus.
Fuchsine.
Gold Chloride.
Hydrastis.
Hyoscyamus.
Hyoscine Hydrobromate.
Jalap.
Juniper Oil.
Lead.
Mercury Bichloride.
Nitroglycerin.
Oil Turpentine.
Pilocarpine.
Potassium Bitartrate.
Potassium Citrate.
Potassium Iod'de.
Sodium Benzoate.
Sodium Bicarbonate.
Strontium Lactate.
Theobromine salts.

Bright's Disease, Chronic.—*See also, Dropsy, Uremia.*

Acid, Gallic.
Bromides.
Cannabis Indica.
Elaterium.
Eucalyptus.
Fuchsine.
Gold.
Hemo-gallol.
Hydrastis.
Iron.
Jaborandi.
Jalap.
Lead.
Mercury Bichloride.
Nitroglycerin.
Oil Turpentine.
Potassium Bitartrate.
Potassium Iodide.

Bromidrosis. — *See Feet.*

Bronchiectasis.—*See also, Emphysema.*

Chlorine: as inhalation to lessen fetor.
Creosote: as inhalation.
Iodine: as inhalation.
Phosphates and Hypophosphites.
Quinine.
Terebene: as inhalation.

Bronchitis.
Acetanilid.
Acid, Arsenous.
Acid, Benzoic.
Acid, Camphoric.
Acid, Carbolic.
Acid, Hydriodic.
Alum.
Ammonium Benzoate.
Ammonium Chloride.
Ammonium Iodide.
Ammonium Salicylate.
Astringent sprays for excessive secretion.
Anemonin.
Antispasmin.
Antimony Sulphide, Golden.
Antimony and Potassium Tartrate.
Arsenic.
Cetrarin.
Chlorophenol.
Cocaine.
Codeine.
Conium.
Creosote.
Digitalis.
Eserine.
Ethyl Iodide.
Eucalyptol.
Hydrastis.
Iodides.
Iodine.
Mercury Subsulphate.
Myrtol.
Naphtalin.
Nux Vomica Tincture.
Oil Eucalyptus.
Oil Pinus Pumilio.
Oil Pinus Sylvestris.
Oxygen.
Peronin.
Phosphates.
Physostigmine.
Potassium Citrate with Ipecac.
Potassium Cyanide.
Pyridine: an inhalation.
Sodium Benzoate.
Sodium Iodide.
Solanin.
Stramonium.
Sulphur.
Terebene.
Terpine Hydrate.
Thymol.
Zinc Oxide.

Bronchitis, Acute.
—*See also, Cough.*

Acetanilid.
Acid, Carbolic.
Acid, Nitric: when expectoration is free and too copious.
Aconite: one-half to 1 min. every hour at the commencement of an acute catarrhal attack.
Actæa Racemosa: in acute catarrh and bronchitis when the more active symptoms have subsided
Alkalies: to render mucus less viscid.
Amber Oil: counterirritant over spine in children.
Ammoniacum: very useful in old people.
Ammonium Acetate.
Ammonium Carbonate: Where much expectoration and much depression; or where the mucus is very viscid and adherent.
Apomorphine: causes a copious expectoration in the early stage.
Asafetida: like ammoniacum.
Belladonna: in acute bronchitis of children to stimulate respiratory centre.
Benzo and Benzoic Acid: 1 dram inhaled from hot water eases cough and lessens expectoration.
Bleeding: from the superficial jugular veins in severe pulmonary engorgement
Camphor.
Chloral Hydrate: to be used with caution, to allay pain.
Cimicifuga.
Cod Liver Oil: relieves.
Colchicum: in gouty cases.
Copaiba: in advanced stage of disease.
Counter-irritants: dry cupping most efficacious in acute cases; mustard leaves; mustard poultices.
Croton Oil: as liniment; vesication must not be produced.
Cubebs: when secretion is copious.
Demulcents: licorice, linseed.
Eucalyptol.
Eucalyptus: as liniment combined with bella-

donna in the early stage; internally in the late stage.
Garlic, Oil of: in the acute bronchitis of children.
Ipecacuanha: when expectoration is scanty, dryness in chest, ipecacuanha in large doses; also when expectoration has become more abundant but difficult to expel.
Iron.
Jalap: with bitartrate of potassium instead of bleeding in engorgement of the right side of the heart.
Lead: in profuse discharge.
Lobelia: when cough is paroxysmal and there is much expectoration slightly nauseant expectorants are good combined with opium.
Mercury: in some cases useful where there is much congestion and little secretion.
Morphine: one-half grn. combined with Quinine(10grn.)will abort the attack if given early enough.
Muscarine: in doses of ⅙ grn. at the commencement of the attack; well combined with digitalis.
Mustard: poultice in acute bronchitis of children and adults; foot bath.
Opium: as Dover's powder to cut short attack, and along with expectorants to lessen cough.
Pilocarpine : in abundant exudation.
Potassium Chlorate: first increases the fluidity of the expectoration, then diminishes it in quantity, increasing the feeling of relief.
Poultices: in children to encircle the whole chest
Quinine: to reduce temperature.
Sanguinaria: after acute symptoms have subsided.
Senega: in the advanced stage of acute disorder.
Squill Syrup: combined with camphorated tincture of opium

after acute stage is over.

Tartar Emetic: in dry stage to promote secretion; most useful in first stage.

Turpentine Oil: when expectoration profuse: also as inhalation or stupe.

Zinc Oxide.

Bronchitis, Capillary. — *See also, Cough.*

Alum: as a nauseating expectorant and emetic.

Ammonium Carbonate: when much fluid or viscid expectoration and commencing lividity; also as an emetic.

Ammonium Chloride: to promote secretion.

Ammonium Iodide: in small rapid doses relieves much.

Antimony.

Apomorphine: to produce a plentiful fluid secretion; also as nauseant expectorant.

Camphor: as expectorant and stimulant.

Cupping: four to six dry cups over the back often give very great relief, and if the pulmonary congestion appears very great wet cups should be placed instead, and 8 to 10 oz. of blood withdrawn from adult.

Ethyl Iodide: as an inhalation.

Iodides: are very serviceable to diminish viscidity of expectoration if given in very low doses.

Ipecacuanha : as expectorant and emetic.

Mustard: as poultices.

Oil Amber with Olive Oil (1:3): applied to back and chest.

Pilocarpine: in abundant non-purulent exudation; not to be used in dilatation of veins and right side of the heart.

Poultices: over whole chest.

Quinine.

Serpentaria: in children as a stimulant expectorant.

Subsulphate of Mercury: as nauseant, expectorant and emetic.

Turpentine Oil: in languid circulation in the capillaries.

Water: hot and cold dashes if death is imminent from suffocation.

Bronchitis, Chronic. — *See also, Cough, Emphysema.*

Acids: to diminish a chronic copious expectoration.

Acid, Carbolic: as inhalation or spray.

Acid, Gallic: in profuse discharge.

Acid, Nitric: in mixtures, to remedy the effect on digestion produced by sedatives like opium.

Acid, Sulphurous: as inhalation or spray.

Alum: in children with copious expectoration in doses of 3 grn.

Ammonia: when there is difficulty in bringing up expectoration.

Ammoniac: very useful, especially in elderly people.

Ammonium Chloride: to render the secretion less viscid.

Anemonin.

Antimony: when secretion is scanty.

Apocodeine Hydrochlorate.

Apomorphine Hydrochlorate.

Arsenic: in emphysema and asthmatic attack as cigarettes, where there is much wheezing and little bronchitis following the sudden disappearance of eczematous rash

Asafetida: like ammoniacum.

Balsam of Peru: when expectoration is copious.

Balsam of Tolu: the same.

Belladonna: to children choked with secretion give 1 minim of tincture every hour to stimulate respiratory centre. It also lessens the secretion.

Benzoin: as inhalation or as spray.

Burgundy Pitch: emplastrum in chronic bronchitis.

Camphor.

Cannabis Indica: in very chronic cases.

Carbolic Acid Gas: inhaled.

Chamois Waistcoat.

Cheken: the fluid extract renders expectoration easier, and paroxysms less frequent.

Chloral Hydrate: a solution of 10 grn. to the oz. used as a spray to allay cough.

Cimicifuga: sometimes relieves the hacking cough.

Codeine: in place of opium when the latter disagrees.

Cod-Liver Oil: one of the most useful of all remedies.

Colchicine.

Colchicum: in acute cases.

Conium: the vapor to relieve cough.

Copaiba: like Balsam Peru.

Creosote: inhaled to allay cough.

Crude Petroleum: in capsules or pills in chronic bronchitis.

Cubebs: like copaiba.

Digitalis: where heart is feeble, especially in the aged.

Emetics.

Ethyl Iodide.

Eucalyptus: stimulant expectorant.

Euphorbia Pilulifera.

Gaduol: a most useful remedy.

Galbanum: like ammoniac.

Grindelia: expectorant when the cough is troublesome.

Guaiacol.

Guaiacol Vapor.

Hydrastis: in chronic coryza.

Hypnal: for cough.

Iodides and Iodine: as inhalation or liniment to chest, to lessen expectoration in chronic bronchitis; in the hoarse hollow cough of infants after measles.

Iodoform.

Ipecacuanha: the wine as spray in much expectoration; in emetic doses in children where the bronchioles are blocked up with mucus.

Iron: when expectoration is profuse.
Koumys regimen: sometimes very useful.
Levico Water: as tonic.
Lobelia: when there is spasmodic dyspnea.
Mercury: to diminish congestion.
Morphine: to quiet cough, in small doses.
Myrrh.
Myrtol.
Oil Sandalwood.
Opium: to lessen secretion and cough.
Peronin: in place of morphine for the cough.
Phosphates: in very chronic cases.
Physostigma: in chronic cases with great dyspnea.
Physostigmine.
Plumbic Acetate: in profuse secretion.
Potassium Carbonate: in viscid secretion.
Potassium Iodide: in combination with antim. tart. in cases of great dyspnea.
Sanguinaria: with other expectorants.
Senega: when expulsive efforts are feeble.
Serpentaria: like senega
Spinal Ice-bag: in excessive secretion.
Squill: where expectoration is thick.
Steam Inhalations.
Stramonium: in dry cough.
Strychnine: as respiratory stimulant.
Sulphur: where expectoration is copious, bronchitis is severe, and constitutional debility.
Sumbul.
Tar: to lessen secretion and allay chronic winter cough; given in pill or as spray.
Terebene: internally or as inhalation.
Terpin Hydrate.
Turkish Bath: to clear up a slight attack and to render the patient less susceptible to taking cold.
Turpentine Oil: as liniment to chest in children.
Zinc Oxide: to control too profuse a secretion.

Bronchocele. — *See Goiter.*

Bronchorrhea.—*See also, Cough.*

Acid, Carbolic: as spray
Acid, Gallic: remote astringent.
Alcohol: accordingly as it agrees or disagrees with patient.
Alum: a remote astringent.
Ammoniac: in the aged.
Ammonium Carbonate: stimulant expectorant
Ammonium Chloride: stimulant expectorant
Ammonium Iodide: small doses frequently repeated; value increased by the addition of arsenic.
Apomorphine Hydrochlorate.
Asafetida; like ammoniac.
Astringents.
Cod-liver Oil.
Copaiba: stimulant expectorant; to be given in capsules.
Creosote.
Cubebs: like copaiba.
Eucalyptol.
Eucalyptus Oil: sometimes very useful.
Gaduol: efficacious alterant tonic.
Grindelia: respiratory stimulant.
Iodine: as counter-irritant to chest, and as inhalation.
Iodoform.
Iodole.
Lead Acetate: to lessen secretion.
Myrtol: in profuse fetid expectoration.
Oil Pinus Pumilio.
Oil Pinus Sylvestris: as inhalation.
Phosphates: tonic.
Quinine: tonic.
Spinal Ice-bag: to lessen secretion.
Sulphurous Acid: as inhalation or spray.
Terebene.
Terpin Hydrate.
Turpentine Oil: stimulant expectorant, and also as inhalation.

Bruises.

Acid, Sulphurous: as local application constantly applied.
Aconite: liniment locally, to relieve pain.
Alcohol.
Ammonium Chloride.
Arnica: as local application on no more use than alcohol, and sometimes gives rise

to much inflammation; this it will do if the skin is abraded.
Capsicum: to remove discoloration of bruise.
Compressed Sponge.
Convallaria Polygonatum (Solomon's Seal): the juice from the fresh root will take away a "black eye."
Hamamelis: locally.
Ice.
Ichthyol.
Iodoform.
Iodoformogen.
Iodole.
Lead Water: to allay pain.
Oil of Bay: same as Capsicum.
Opium: local application to relieve pain.
Pyoktanin.
Sozoiodole - Potassium: as dusting-powder.
Sozoiodole-Sodium: as wash.

Bubo.—*See also, Chancroid, Syphilis.*

Acid, Carbolic: by injection.
Acid, Nitric: as local application to indolent bubo.
Aristol.
Blisters: followed up by application of tinc. iodi. will often cause absorption.
Calomel.
Chlora Hydrate: 25 per cent. solution, antiseptic and stimulant application.
Copper Sulphate: 4 grn. to the oz.
Creolin.
Diaphtherin.
Europhen.
Hydrargyri Perchloridum epidermis is first removed by a blister and then a saturated solution applied; a poultice is then applied to separate the eschar, leaving a healthy ulcer.
Ice: to relieve pain and lessen inflammation.
Ichthyol.
Iodine: as counter-irritant applied round the bubo.
Iodoform: as local application.
Iodoformogen.
Iodole.
Lead Lotions: compresses soaked in

these will abort, or assist in the healing process.

Mercury: as local application after opening bubo, when syphilitic affection is great.

Peroxide of Hydrogen: wash and dress bubo with lint soaked in it.

Potassa Fusa: to open, instead of the knife.

Potassium Chlorate: applied as fine powder.

Pyoktanin.

Silver Nitrate: lightly applied to surface in indolent bubo.

Sozoiodole - Potassium: incision at first sign of suppuration, followed by washing with antiseptics.

Sulphides: to check suppuration; not so useful as in an ordinary abscess.

Tartar Emetic: when inflammation is acute and fever considerable.

Xeroform.

Bunion. — *See also, Bursitis.*

Iodine: painted on in indolent forms.

Rest: when thickened and painful. Pressure is removed by thick plasters, with a hole in the center.

Burns and Scalds.

Absorbent Dressings.

Acetanilid.

Acid, Boric: useful as ointment or lint dressings, or as boric oil.

Acid, Carbolic: 1 per cent. solution relieves pain and prevents suppuration.

Acid, Picric: dressing.

Acid, Salicylic: 1 in 60 olive oil.

Alkalies: soon remove the pain on exposure to the air after application.

Alum: finely powdered over foul, bleeding granulations.

Antipyrine: in solution or ointment.

Argenti Nitras: wash with a solution of 4 to 8 grn. to the oz. and wrap in cotton wool.

Bismuth Subgallate.

Bismuth Subnitrate: a thick paste with glycerin protective.

Calcium Bisulphite (sol.).

Carron Oil: in recent burns.

Chalk, Oil and Vinegar: applied as a paste of a creamy consistence, relieves pain at once.

Chlorinated Soda: in dilute solution.

Chloroform, Olive Oil and Lime Water: soon relieves the pain.

Cocaine: as lotion to allay the pain.

Cod-Liver Oil.

Cold: Instant application.

Collodion: flexible, to protect from air.

Cotton Wool: to protect from irritation and so lessen pain.

Creolin.

Creosote: like Carbolic Acid.

Diaphtherin.

Digitalis: in shock.

Europhen.

Gallæ Unguentum: 1 part to 8 of lard, to prevent cicatrix.

Ichthalbin.

Ichthyol.

Iodoform: local anesthetic and antiseptic.

Iodoformogen: the same.

Lead Carbonate: i. e. white-lead paint, for small burns; should be applied instantly.

Lead Water.

Linimentum Calcis (lime-water with linseed oil).

Morphine and Atropine: to allay pain.

Naftalan.

Oakum.

Oil and Litharge: applied as a varnish, containing 5 per cent. Salicylic Acid.

Ol. Menthæ Piperitæ: painted on.

Phytolacca: to relieve pain.

Potassium Chlorate: solution 5 grn. to 1 oz.

Pyoktanin.

Resorcin.

Rhubarb Ointment: one part of root to two of lard.

Rhus Toxicodendron.

Soap Suds: instead of alkali, if it is not at hand.

Sodium Bicarbonate: immediate application of a saturated solution.

Sozoiodole - Potassium: as dusting - powder, with starch.

Sozoiodole-Sodium: as wash.

Stimulants, Local: such as Ung. Resinæ, afterwards followed by astringents.

Thymol: one per cent. in olive oil, local anesthetic.

Warm Bath: keep whole body, with exception of head, totally immersed for some days in very extensive burns or scalds. It relieves pain, although it may not save life.

Whiting and Water: mixed to the thickness of cream and smeared over, excluding the air, gives instant relief.

Zinc Ointment and Vaselin: in equal parts for dressing.

Zinc Oxide: as dusting powder.

Bursitis.

Acid, Carbolic: as injection.

Blisters: most useful.

Fomentations: to relieve pain.

Ichthyol.

Iodine: When chronic, Lin. Iodi may be used as a blister, or the liquor, after blistering or aspiration.

Cachexiæ. — *See also, Anemia, Scrofula, Syphilis, etc., and the list of Tonics.*

Acid, Nitric: in debility after acute disease; in combination with the fresh decoction of bark.

Air: fresh.

Aliment: nutritious.

Ammonium Carbonate: with bark; after acute illness.

Arnica: internally, in bad cases.

Arsen-hemol.

Arsenic: in malarial, also in cancerous, cachexia; in chronic malaria, combined with iron.

Baths: Turkish bath. useful.

Calcium Phosphate.
Chalybeate Waters.
Cholagogues: most useful before, or along with other remedies, and especially in malarial cachexia before the administration of quinine.
Cupro-hemol.
Electricity.
Eucalyptus: in general cachectic conditions.
Euonymin: as cholagogue.
Gaduol.
Glycerin: as a food.
Glycerinophosphates.
Gold.
Grape Cure.
Hemo-gallol.
Hemol.
Hydrastine.
Hydrastis: in malaria.
Ichthalbin.
Iodine.
Iron: generally in all anemic conditions.
Levico Water.
Manganese: along with iron and as syrup of double iodide.
Massage: exceedingly useful.
Mercury: in syphilitic cases.
Oils and Fat codliver oil very useful. Cream as an addition to food; oil as inunction.
Phosphates: in scrofula, phthisis and malnutrition.
Podophyllin: as cholagogue, in children of a few months old improperly fed; in alcoholic excess; chronic morning diarrhea.
Potassium Iodide: in syphilitic and resulting conditions.
Purgatives, Saline: as adjuncts to cholagogues.
Quinine: in various forms of cachexia.
Sarsaparilla: in syphilis.

Calculi.— *See also, Gravel.*

Acid, Benzoic.
Acid, Sulphuric, diluted.
Ammonium Borate.
Lithium Benzoate.
Lithium Carbonate.
Lithium Citrate.
Magnesia.
Manganese Dioxide.
Oil Turpentine.
Sodium Benzoate.
Sodium Bicarbonate.
Sodium Phosphate.
Solution Potassa.

Calculi, Biliary.— (*Gall - stones.*)— *See also, Colic, Jaundice.*

Acid, Nitric: hepatic stimulant and alterative.
Acid, Nitro-hydrochloric: same as nitric acid.
Aliment: absence of starch and fat recommended.
Anesthetics: during the passage of the calculus.
Belladonna: relief during spasm.
Carlsbad Waters: prophylactic.
Chloral Hydrate: to relieve pain during paroxysm; good in combination with morphine.
Chloroform: inhalation from tumbler, most useful to relieve paroxysm.
Counter-Irritation: to relieve pain during passage.
Emetics: of doubtful value in aiding the expulsion of the calculus.
Ferri Succinas: as a resolvent for existing stones, and prophylactic.
Ferri Perchlor. Tinctura: like creosote, as an astringent. Useful if renal changes complicate.
Iridin: in doses of 1 grn.for its cholagogue properties.
Mercury: the green iodide, with manna and soap as a pill.
Morphine: 1·5 grn. (repeated if necessary) with 1-120 grn. atropine subcutaneously, to relieve pain and vomiting in paroxysm.
Nitro - hydrochloric Bath: to cause expulsion of calculus and to relieve pain.
Oil: in large doses has been followed by the expulsion of gall-stones.
Salicylate of Sodium: as prophylactic.
Sodium Carbonate: in large quantity of hot water during passage of stone. At first there is usually vomiting, but this soon ceases.

Sodium Phosphate: in 20 or 30 grn. doses before each meal as prophylactic. Should be given in plenty of water.
Turpentine Oil and Ether (Durande's remedy): Equal parts to relieve pain during paroxysm; also occasionally as prophylactic along with a course of Carlsbad or Vichy water.

Calculi, Renal and Vesical.— *See also, Colic, Lithiasis, Oxaluria, etc.*

Acid, Hippuric.
Acid, Nitric: dilute; as injection into the bladder to dissolve phosphatic calculi.
Alkalies, especially Potassa Salts: to resolve calculi, potash and soda to be used.
Alkaline Mineral Waters: especially Vichy and Bethesda.
Ammonium Benzoate: to resolve phosphatic calculi.
Anesthetics: to relieve pain during passage of calculus.
Belladonna: sometimes relieves the pain of the passage of calculus.
Borocitrate of Magnesium: to dissolve uric acid calculus. Formula: Magnesii carb. 1 dram; Acid, citric, 2 drams; Sodii biborat. 2 drams; Aquæ, 8 fl. oz. m. sig.; 2 drams 3 t. p. d.
Calcium Carbonate.
Calumba: to relieve vomiting.
Castor Oil: as purgative.
Chloroform: as in biliary calculi.
Cotton Root: as decoction to relieve strangury.
Counter - Irritants: to lessen pain during passage of calculus.
Formin.
Lead Acetate.
Lithium Salts.
Lycetol.
Lysidine.
Mineral Waters.
Morphine: hypodermically, as in biliary calculi.
Piperazine.

Potassium Boro-Tartrate: more efficient than the magnesium salt; prepared by heating together four parts of cream of tartar, one of boric acid, and ten of water. 20 grn. three times a day well diluted.

Potassium Citrate: in hematuria with uric acid crystals.

Water, Distilled: as drink.

Camp Fever. — *See Typhus.*

Cancer. — *See also, Uterine Cancer.*

Acid, Acetic: as injection into tumors.

Acid, Carbolic: as application or injection into tumor to lessen pain, retard growth and diminish fetor.

Acid, Chromic: as caustic.

Acid, Citric: as lotion to allay pain, 1 in 60.

Acid, Hydrochloric.

Acid, Lactic.

Acid, Salicylic: locally applied as powder or saturated solution.

Acids: internally in cancer of stomach.

Aluminium Sulphate: a caustic and disinfectant application.

Aniline.

Argenti Nitras: a saturated solution injected in several places; to be followed by an injection of table-salt 1 in 1000.

Aristol.

Arsenic: as local application, causes cancer to slough out. Sometimes successful when the knife fails, but is dangerous. Internally, in cancer of stomach, lessens vomiting. Supposed to retard growth of cancer in stomach and other parts.

Arsenic Iodide.

Belladonna: locally relieves pain. Used internally also.

Bismuth Subnitrate: to relieve pain and vomiting in cancer of stomach.

Bromine Chloride: alone or combined with other caustics. To be followed by a poultice.

Bromine, Pure: as caustic to use round cancer.

Calcium Carbonate.

Caustic Alkalies: in strong solution dissolve the cells.

Charcoal Poultices: to lessen pain and fetor.

Chian Turpentine: benefits according to some—acc. to others, it is useless.

Chloral Hydrate: to lessen pain.

Chloroform: vapor as local application to ulcerated cancer.

Codeine: as a sedative in cases of abdominal tumor.

Cod-Liver Oil: in cachexia.

Coffee: disinfectant, applied as fine powder.

Conium: as poultices to relieve pain. Used internally also.

Creolin.

Ferro-Manganous preparations.

Gaduol: in cachexia.

Gas Cautery: a form of actual cautery.

Glycerinophosphates.

Glycerite of Carbolic Acid: same as carbolic acid.

Gold and Sodium Chloride.

Hematoxylin Extract: to a fungating growth.

Hydrastis: as palliative application.

Hydrogen Peroxide.

Hyoscyamus: bruised leaves locally applied.

Ichthyol.

Iodoform: locally to lessen pain and fetor.

Iodoformogen.

Iron and Manganese: internally as tonics.

Levico Water: internally.

Lime: as caustic.

Manganese Iodide.

Mercury Bichloride.

Mercury Nitrate, Acid.

Methylene Blue.

Morphine salts.

Opium: locally and internally, to relieve pain.

Papain: as local application or injection.

Pepsin: as injection into tumor.

Potassium Chlorate: allays the pain and removes the fetor.

Potassium Permanganate.

Potassa Fusa: as escharotic.

Poultices: to relieve pain.

Pyoktanin.

Resorcin.

Sodium Ethylate: a powerful caustic.

Stramonium: ointment to relieve pain.

Terebene: disinfectant dressing.

Vienna Paste.

Warm Enemata: to lessen pain in cancer of rectum.

Zinc Chloride: as caustic.

Zinc Sulphate: as caustic.

Cancrum Oris.—*See also, Aphthæ, Stomatitis.*

Acid, Boric.

Acid, Nitric: undiluted as local caustic.

Arsenic: internally.

Potassium Chlorate: internally in stomatitis; useless in noma.

Quinine: as syrup or enema.

Sodium Borate.

Sozoiodole-Sodium.

Carbuncle.—*See Anthrax.*

Cardiac Affections. —*See Heart.*

Cardialgia.

Antacids.

Bismuth Valerianate.

Charcoal.

Massage.

Caries.—*See also, Necrosis.*

Acid, Carbolic: as a disinfectant lotion; often heals.

Acid, Phosphoric, Diluted: locally.

Aristol.

Calcium Carbonate.

Calcium Chloride.

Cod-Liver Oil.

Gaduol.

Glycerinophosphates.

Gold: in syphiloma of bone.

Iodine: locally and internally.

Iodole.

Iodoform.

Iodoformogen.

Iron.

Phosphates of Calcium and Iron.

Phosphorus.

Potassium Carbonate: concentrated solution, locally applied.

Potassa Fusa: to carious bone to remove disorganized portion.
Potassium Iodide: in syphilitic cases.
Sarsaparilla.
Sozoiodole-Mercury.
Sozoiodole-Potassium.
Sulphuric Acid: injection (one of strong acid to two of water) into carious joints, and locally to carious or necrosed bone. Useful only if disease is superficial.
Villate's Solution: cupri sulph., zinci sulph. 3 parts each, liq. plumb. subacetat. 6 parts, acid acet. 40 parts, as injection into a sinus.

Catalepsy.

Chloroform: inhaled.
Sternutatories.
Turpentine Oil: as enemata and embrocations to spine during paroxysms.

Cataract.

Atropine.
Cineraria Maritima Juice.
Codeine: in diabetic cases.
Diet and Regimen: nutritious in senile cases. Sugar and starch to be avoided in diabetic cases.
Galvanism: in early stage.
Mydriatics: to dilate pupil as a means of diagnosis.
Phosphorated Oil: instilled into the eye will lead to absorption if borne.

Catarrh.—*See also, the various Catarrhs below.*

Acid, Camphoric.
Acid, Hydrocyanic, Dil.
Acid, Sulpho-anilic.
Alantol.
Aluminium Tanno-tartrate.
Antimony Sulphide, Golden.
Antinosin.
Apomorphine Hydrochlorate.
Arsenic Iodide.
Calcium Bisulphite.
Cimicifugin.
Cocaine Carbolate.
Creolin.
Cubeb.

Eucalyptus.
Gaduol.
Ichthalbin.
Ichthyol.
Iodoform.
Iodoformogen.
Menthol.
Naphtalin.
Oil Eucalyptus.
Potassium Cyanide.
Potassium Iodide.
Sodium Bicarbonate.
Sodium Iodide.
Sodium Nitrate.
Sozoiodole-Sodium.
Sozoiodole-Zinc.
Sulphur.
Sulphurated Potassa.
Tannoform.
Terpinol.

Catarrh, Acute Nasal. — *See also, Cough, Hay Fever, Influenza.*

Acid, Carbolic: as inhalation, or much diluted as spray. As gargle, 1 in 100, when catarrh tends to spread from nose into throat and chest, or to ascend from throat into nose.
Acid, Sulphurous: as inhalation, spray or fumigation.
Acid, Tannic: injection of a solution in rectified spirit.
Aconite: internally at commencement, especially in children.
Aconite and Belladonna: in sore-throat and cold with profuse watery secretion, one drop of tinct. of aconite to two of belladonna every hour.
Aconite Liniment: to outside of nose in paroxysmal sneezing and coryza.
Aluminium Aceto-tartrate.
Ammonia: as inhalation in early stage, while discharge is serous.
Ammonium Chloride: in young children.
Ammonium Iodide: one grn. every two hours.
Argenti Nitras: injection of a solution of 10 grn. to the oz.
Arsenic: internally, or as cigarettes, in paroxysmal and chronic cases; valuable in cases which exactly simulate hay fever.
Baths: hot foot-bath

before retiring, Turkish, at commencement; cold bath is prophylactic.
Belladonna: 5 min. of tinct., and afterwards one or two doses every hour until the throat is dry.
Benzoic Acid: in ordinary catarrh, for its stimulant effects.
Bismuth: as Ferrier's snuff. Bismuth subnit., 2 drams; acaciae pulv., 2 drams; morph. hydrochlor., 2 grn.
Camphor: as inhalation.
Chloral.
Chloroform: by inhalation.
Cimicifuga: in coryza accompanied by rheumatic or neuralgic pains in head and face.
Cocaine Hydrochlorate.
Codeine.
Cold Powder: camph. 5 parts dissolved in ether to consistence of cream, add ammon. carbonat. 4 parts, and pulv. opii 1 part. Dose, 3 to 10 grn. To break up or modify cold.
Cubebs: powder as insufflation; also smoked; also the tincture in 2 dram doses with infusion of linseed.
Formaldehyde: by inhalation (2 per cent. solut.).
Hot Sponging: to relieve the headache.
Iodine: as inhalation.
Iodoform and Tannin: as insufflation.
Ipecacuanha: in moderate doses (10 grn.). Dover's powder at night will cut short an attack. The wine as spray to the fauces.
Jaborandi: as tincture. Or hypodermic injection of half a grain of pilocarpine hydrochlorate.
Menthol.
Nux Vomica: in dry cold in the head.
Oil: inunction to whole body to lessen susceptibility; locally to nose; sometimes ointment may be used.
Opium: as Dover's powder at commencement; but not in obstruction to respiration.
Peronin.

Pilocarpine Hydrochlo-
rate (see Jaborandi).
Potassium Bichromate:
solution locally, 1 to
10 grn. in 4 oz.
Potassium Chlorate:
eight or ten lozenges
a day to check.
Potassium Iodide: ten
grn. at bedtime to
avert acute coryza.
Pulsatilla: warm lotion
applied to interior of
nares; or internally
but not in symptoms
of intestinal irrita-
tion.
Quinine: ten grn. with
½ grn. morphine, at
commencement may
abort it.
Resorcin.
Salicylate of Sodium:
two and one-half grn.
every half-hour to re-
lieve headache and
neuralgia associated
with coryza.
Sanguinaria: internally,
and powder locally.
Sea-water Gargle.
Silver Nitrate.
Spray: useful means of
applying solutions
such as ipecacuanha
wine, already men-
tioned.
Sugar: finely powdered
and snuffed up in the
nose in catarrh due to
potassium iodide.
Tartar Emetic: one-
twentieth to one-
twelfth grn. at com-
mencement, especial-
ly in children with
thick and abundant
secretion.
Turkish Bath.
Veratrum Viride: if
arsenic fails.
Zinc Sulphate: as nasal
injection 1 grn. to the
oz.

**Catarrh, Broncho-
Pulmonary.** — See
Bronchitis, Bron-
chorrhea.

Catarrh, Cervical.—
See Uterine Affec-
tions.

**Catarrh, Chronic
Nasal.** — See also,
Ozena.

Acid, Benzoic: inhaled
as vapor.
Acid, Carbolic: one to
100 as spray, or 1 to
200 as douche. One
part with 4 of iodine
tincture as inhalation
or by spray.

Acid, Salicylic.
Acid, Tannic.
Alum: in powder by in-
sufflation, or in solu-
tion by douche.
Ammonia: inhalation.
Ammonium Chloride:
in thick and abundant
secretion.
Asafetida: stimulant
expectorant.
Balsam of Peru: stimu-
lant expectorant.
Bismuth Subnitrate.
Bromine: as vapor, in-
haled with great cau-
tion.
Calomel.
Camphor.
Cocaine.
Cod-Liver Oil.
Cubebs: in powder, by
insufflation, or as
troches.
Ethyl Iodide: as inhala-
tion
Eucalyptol: in chronic
catarrh with profuse
secretion.
Eucalyptus.
Gaduol: as alterative.
Gold chloride.
Hamamelis: snuffed up
nose.
Hydrastis.
Ichthalbin: as altera-
tive.
Ichthyol.
Iodine: vapor inhaled.
Iodole.
Iodoform and Tannin:
insufflated.
Iodoformogen.
Potassium Bichromate.
Potassium Permanga-
nate.
Pulsatilla.
Resorcin.
Sanguinaria: in very
chronic cases.
Silver Nitrate.
Sodium Chloride.
Sodium Phosphate.
Sozoiodole-Potassium.
Turpentine Oil: as lini-
ment to chest.

Catarrh, Duodenal.
— See Duodenal
Catarrh.

Catarrh, Epidemic.
—See Influenza.

Catarrh, Gastric.—
See Gastritis,
Chronic.

**Catarrh, Genito-
Urinary.** — See
Bladder, Catarrh
of; Cystitis; Endo-
metritis; Gonor-
rhea; Leucorrhea,
etc.

Catarrh, Intestinal.
— See Dysentery,
Jaundice.

Catarrh, Vesical.—
See Bladder,
Catarrh of.

Cephalalgia. — See
Headache.

Cerebral Anemia.—
See also, Insomnia.

Ammonia: inhaled is
useful in sudden at-
tacks.
Amyl Nitrite: to act on
vessels.
Arsenic: in hypochon-
driasis of aged people;
best combined with a
minute dose of opium.
Caffeine: in hypochon-
driasis.
Camphor, Monobro-
mated.
Chalybeate Mineral
Water.
Chloral Hydrate: in
small doses, with
stimulants.
Digitalis.
Electricity.
Glycerin.
Gold: melancholic
state.
Guarana: restorative
after acute disease.
Iron.
Levico Water.
Nitroglycerin: to
dilate cerebral ves-
sels. Like amyl
nitrite.
Nux Vomica.
Phosphorus and Phos-
phates: to supply
nutriment.
Quinine.
Strychnine.
Zinc Phosphide.

**Cerebral Concus-
sion.**

Rest: absolute to be en-
joined.
Stimulants to be
avoided.
Warmth: to extremi-
ties.

**Cerebral Conges-
tion.** — See also,
Apoplexy, Coma.

Acid, Hydrocyanic.
Aconite: in acute cases
before effusion has
taken place.
Arsenic: in commenc-
ing atheroma of cere-
bral vessels and ten-
dency to drowsiness
and torpor.

Belladonna: very useful.
Bromides: very useful.
Cathartics: to lessen blood-pressure.
Chloral Hydrate: when temperature is high.
Colchicum: in plethoric cases.
Colocynth: as a purgative.
Croton Oil.
Diet: moderate, animal food sparingly, and stimulants to be avoided.
Digitalis: in alcoholic congestion, and simple congestive hemicrania.
Elaterin.
Elaterium.
Ergot: in want of arterial tone, or miliary aneurisms causing vertigo, etc.
Galvanism of head and cervical sympathetic.
Geiseminum: in great motor excitement, wakefulness, horrors after alcoholic excess.
Potassium Bromide.
Venesection: a suitable remedy in cases of threatening rupture of a vessel.
Veratrum Viride: in acute congestion; the good ceases with exudation.
Water: cold douche to head, and warm to feet, alternately hot and cold to nape of neck.

Cerebral Softening.

Phosphorus.
Potassium Bromide.

Cerebritis.

Ammonium Chloride: locally.
Chloral Hydrate.
Electricity.
Ice.

Cerebro-Spinal Fever.—See Meningitis, Cerebro-Spinal.

Chancre.—See also, Syphilis.

Acetanilid.
Acid, Carbolic: locally.
Alumnol.
Aristol.
Calomel: locally.
Camphor: finely powdered.
Canquoin's Paste: zinc chloride, 1 in 6, made into paste, local.

Caustics: chromic acid, bromine, acid nitrate of mercury, zinc chloride, nitric acid, caustic alkalies.
Copper Sulphate.
Eucalyptus: mixed with iodoform and locally applied.
Europhen.
Formaldehyde.
Hydrogen Peroxide: constantly applied to destroy specific character.
Iodoform.
Iodoformogen: one of the best remedies.
Iodole.
Mercuric Nitrate Solution.
Mercury: internally. Also, locally: black wash; or yellow wash; or corrosive sublimate in solution.
Mercury Salicylate.
Monsel's Solution.
Pyoktanin.
Resorcin.
Sozoiodole-Mercury.
Sozoiodole-Zinc.
Tannoform.

Chancroid.—See also, Bubo.

Acetanilid.
Acid, Carbolic: as injection and local application.
Acid, Nitric: locally as caustic.
Acid, Salicylic.
Acid, Sulphuric: with charcoal.
Acid, Tannic.
Actual Cautery.
Alumnol.
Aristol.
Bismuth Benzoate.
Bismuth Subgallate.
Bismuth and Zinc Oxide: or calomel and bismuth, as substitutes for iodoform.
Camphor: finely powdered.
Caustics: sometimes necessary.
Cocaine.
Eucalyptol: with iodine.
Ferric Iodide: internally in phagedenic cases, or debility.
Ferrum Tartaratum: like ferric iodide.
Hot Sitz-bath.
Hydrogen.
Iodoform.
Iodoformogen: very useful.
Iodole.
Mercury: acid nitrate as local application.

Potassium Chlorate: in fine powder.
Pyoktanin.
Resorcin.
Sozoiodole-Mercury.
Sozoiodole - Potassium, as dusting-powder.
Sozoiodole-Sodium, as wash.
Zinc Chloride.

Chapped Hands and Lips.

Acid, Benzoic.
Acid, Sulphurous: as lotion or as fumigation.
Adeps Lanæ.
Benzoin: compound tincture, 1 part to 4 of glycerin.
Calcium Carbonate, Precipitated.
Collodion.
Camphor Cream.
Glycerin: mixed with half the quantity of eau de cologne; or as glyceritum amyli.
Hydrastis: as lotion.
Lanolin.
Lotio Plumbi.
Lycopodium.
Magnesia.
Solution Gutta-Percha: protective.
Sozoiodole-Sodium.
Starch.
Zinc Carbonate and Oxide.

Chest Pains.—See also, Myalgia, Neuralgia, Pleuritis, Pleurodynia, Pneumonia.

Belladonna: in pleurodynia, as plaster or ointment.
Iodine: in myalgia as ointment.
Strychnine.

Chicken-Pox.

Aconite.
Ammonium Acetate.
Bath: cold in hyperpyrexia; warm as diaphoretic.
Compress, Cold: if sore throat.
Laxatives.

Chilblains.

Acid, Carbolic: with tincture of iodine and tannic acid as ointment.
Acid, Sulphurous: diluted with equal part of glycerin, as spray; or fumes of burning sulphur.

Acid, Tannic.
Aconite.
Alum.
Arnica.
Balsam of Peru: as ointment when broken.
Basilicon Ointment.
Cadmium Iodide: internally.
Cajeput Oil.
Capsicum, Tincture: locally, when unbroken, with solution of gum arabic equal parts on silk.
Chlorinated Lime.
Cod-Liver Oil: internally.
Collodion.
Copper Sulphate: solution of 4 grn. to the oz.
Creolin.
Creosote.
Electricity.
Gaduol: as tonic.
Ichthalbin: as alterant tonic.
Ichthyol: topically.
Iodine: ointment or tincture to unbroken chilblains.
Lead Subacetate.
Sozoiodole-Sodium.
Tincture of Opium: locally to ease itching.
Turpentine Oil.

Chlorosis.—See also, Anemia, Amenorrhea.

Absinthin.
Acid, Gallic.
Arsenic: in place of or along with iron.
Arsen-hemol.
Benzoin.
Berberine Sulphate: inferior to quinine.
Calcium Hypophosphite.
Cetrarin.
Cocculus Indicus: in amenorrhea and leucorrhea.
Cupro-hemol.
Ergot: in chlorotic amenorrhea.
Ferri Iodidum.
Ferro-Manganates.
Ferropyrine.
Gaduol.
Gold.
Glycerinophosphates.
Hemol.
Hemo-gallol: powerful blood-maker.
Hemoglobin.
Hypophosphite of Calcium, or Sodium.
Ichthalbin: effective alterative.
Iron: carbonate, useful form; sometimes best

as chalybeate waters. In irritable stomach the non-astringent preparations; in weak anemic girls, with pain and vomiting after food, the persalts are best.
Levico Water.
Manganese Salts: in general.
Massage: useful, combined with electricity and forced feeding.
Nux Vomica: useful, combined with iron.
Orexine: as appetizer.
Pancreatin: to improve digestion.
Potassium Iodide.
Purgatives: useful, often indispensable.
Sea-Bathing.
Zinc Phosphide.
Zinc Valerianate.

Choking.

Potassium Bromide: in children who choke over drinking, but who swallow solids readily.

Cholera Asiatica.

Acid, Boric.
Acid, Carbolic: 2 min., along with 2 grn. of iodine, every hour.
Acid, Hydrocyanic.
Acid, Lactic.
Acid, Nitric.
Acid, Phosphoric.
Acid, Sulphuric, Diluted: alone, or with opium, is very effective in checking the preliminary diarrhea.
Acid, Tannic. by enemeta.
Alcohol: iced brandy, to stop vomiting, and stimulate the heart.
Ammonia: intravenous injection.
Amyl Nitrite.
Antimony.
Arsenic: in small doses, has been used to stop vomiting.
Atropine: hypodermically in collapse.
Betol.
Cajeput Oil.
Calomel: in minute doses to allay vomiting.
Camphor Spirit: 5 minims with tincture of opium, every ten minutes, while the symptoms are violent; and then every hour.
Cannabis Indica.
Cantharides.

Capsicum.
Castor Oil.
Chloral Hydrate: subcutaneously, alone, or with morphine, in the stage of collapse.
Chloroform: 2 or 3 min., either alone or with opium, every few minutes to allay the vomiting.
Cinnamon.
Copper Arsenite.
Copper salts: sometimes used to stop vomiting.
Corrosive Sublimate.
Counter-Irritation over epigastrium.
Creolin.
Creosote: alone or with opium, to allay vomiting.
Dry Packing.
Enemeta of warm salt solution.
Enteroclysis associated with hot bath.
Ether: subcutaneously.
Guaco.
Hypodermoclysis.
Ice to Spine: for cramps.
Ipecacuanha.
Jaborandi.
Lead Acetate; has been used as an astringent in early stages along with camphor and opium.
Mercury Bichloride.
Morphine: one-eighth to one-fourth of a grain subcutaneously to relieve cramps.
Naphtalin: may be useful.
Naphtol.
Naphtol Benzoate.
Opium: in subcutaneous injection 1-10 to 1-grn. to check the preliminary diarrhea, and arrest the collapse.
Permanganates.
Physostigma.
Podophyllin.
Potassium Bromide.
Quinine.
Resorcin.
Salol.
Strychnine. has been used during the preliminary diarrhea, and also as a stimulant to prevent collapse.
Sulpho-carbolates.
Table Salt Injections into the veins have a marvellous effect during collapse in apparently restoring the patient, but their benefit is generally merely temporary.

Transfusion of Milk: has been used in collapse.

Tribromphenol.

Turpentine Oil: sometimes appears serviceable in doses of 10 to 20 min. every two hours.

Cholera Infantum.

Acid, Carbolic: with bismuth or alone, very effective.

Aliment: milk.

Arsenic: for vomiting in collapse.

Beef Juice.

Bismal.

Bismuth Salicylate.

Bismuth Subgallate.

Bismuth Subnitrate.

Brandy: in full doses.

Caffeine.

Calomel: in minute doses to arrest the vomiting.

Camphor: where there is very great depression.

Castor Oil.

Cold: bath at 75 degrees F. every three or four hours, or cold affusions.

Cold Drinks.

Copper Arsenite.

Creosote.

Creolin.

Cupri Sulphas: in very minute doses up to the one thirty-secondth of a grain.

Diet.

Eudoxine.

Enteroclysis.

Ferri et Ammonii Citras.

Hot drinks, applications, and baths, if temperature becomes subnormal.

Ice to Spine.

Ichthyol.

Iodoform and Oil injections: to relieve tenesmus.

Ipecacuanha: when stools greenish or dysenteric.

Irrigation of Bowels.

Lead Acetate: very useful.

Liquor Calcis.

Mercury: 1-6 grn. of gray powder, hourly. In urgent cases a starch enema should be given, containing a minute quantity of laudanum.

Mustard or Spice plaster to abdomen.

Nux Vomica.

Oleum Ricini.

Opium.

Peptonized Milk.

Podophyllin: if stools are of peculiar pasty color.

Potassium Bromide: in nervous irritability and feverishness.

Potassium Chlorate: as enemata.

Resorcin.

Rhubarb.

Silver Nitrate: after acute symptoms are past.

Sodium Phosphate.

Tannalbin: very useful and harmless.

Tannigen.

Tannin and Glycerin.

Tribromphenol.

Xeroform.

Zinc Oxide: with bismuth and pepsin.

Zinc Sulphocarbolate.

Cholera Morbus.—

See Cholera Simplex.

Cholera Nostras —

See Cholera Simplex.

Cholera Simplex.—

See also, Cholera Asiatica and Infantum.

Acid, Carbolic: with bismuth.

Acid, Sulphuric.

Alcohol: dilute and iced.

Arsenic: to stop vomiting.

Atropine: hypodermically, an efficient remedy.

Borax.

Cajeput Oil: used in India.

Calomel.

Castor Oil with Opium.

Calumba: as anti-emetic.

Camphor: very useful.

Chloroform.

Chloral Hydrate: subcutaneously, very useful.

Chlorine Water.

Copper Arsenite.

Copper salts: as astringent.

Creolin.

Creosote.

Ipecacuanha: very useful.

Lead Acetate: at commencement, after salines, and before administering opium, in order to deplete the vessels.

Morphine: hypodermic.

Mustard: internally as emetic; poultice over chest.

Mustard or Spice plaster to abdomen.

Naphtalin.

Naphtol.

Paraformaldehyde.

Salines: to precede the use of lead acetate.

Salol.

Sumbul.

Tannalbin.

Veratrum Album.

Chordee.

Aconite: 1 min. every hour.

Amyl Nitrite.

Atropine: subcutaneously with morphine.

Belladonna: with camphor and opium, internally, very useful.

Bromides: especially of potassium.

Camphor, Monobromated.

Camphor: internally, useful in full doses.

Cannabis Indica.

Cantharis: one drop of tincture three times a day as prophylactic.

Cocaine Hydrochlorate.

Colchicum: half fl. dr. of tincture at night.

Cubebs.

Digitalis.

Hot Sitz-bath.

Hyoscyamus.

Lupulin: as prophylactic.

Morphine: hypodermically, in perineum at night.

Potassium Bromide.

Tartar Emetic: carried to the extent of producing nausea.

Strychnine.

Tobacco Wine: just short of nauseating, at bedtime.

Chorea.

Acetanilid.

Ammonium Valerianate.

Amyl Nitrite.

Aniline.

Antimony: in gradually increasing doses twice a day, to maintain nauseating effect.

Antipyrine.

Apomorphine.

Arsenic: useful sometimes; must be pushed till eyes red or sickness induced, then discontinued, and then used again.

Belladonna.
Bismuth Valerianate.
Bromalin : agreeable sedative.
Bromo-hemol.
Bromides.
Calcium Chloride: in strumous subjects.
Camphor, Monobromated.
Cannabis Indica: may do good; often increases the choreic movements.
Cerium Oxalate.
Chloralamide.
Chloral Hydrate: sometimes very useful in large doses, carefully watched, also where sleep is prevented by the violence of the movements.
Chloroform: as inhalation in severe cases.
Cimicifuga: often useful, especially when menstrual derangement, and in rheumatic history.
Cocaine Hydrochlorate.
Cocculus : in large doses.
Cod-Liver Oil.
Cold: to spine, or sponging, but not in rheumatism, pain in joints, fever ; best to begin with tepid water.
Conium: the succus is sometimes useful, must be given in large doses.
Copper: the ammonio-sulphate in increasing doses till sickness produced.
Cupro-hemol.
Curare.
Duboisine Sulphate.
Electricity : static.
Ether Spray: instead of cold to spine.
Exalgin.
Gaduol.
Gold Bromide.
Hemol.
Hot Pack.
Hyoscine Hydrobromate.
Hyoscyamus.
Iodides.
Iron: chalybeate waters in anemia and amenorrhea.
Iron Valerianate.
Lactophenin.
Levico Water.
Lobelia : only in nauseating doses.
Mineral Water Baths.
Morphine: subcutaneously in severe cases, until effect is manifested ; by mouth in combination with chloral hydrate best.
Musk.
Nitroglycerin.
Physostigma: three to 6 grains of powder a day for children, 10 to 20 for adult.
Picrotoxin: large doses.
Potassium Arsenite Solution.
Quinine.
Salicin.
Salicylates.
Silver: the oxide and nitrate sometimes do good
Silver Chloride.
Silver Cyanide.
Silver Oxide.
Simulo.
Sodium Arsenate.
Stramonium Tincture.
Strontium Lactate.
Strychnine: useful at puberty or in chorea from fright.
Sulfonal.
Valerian: to control the movements.
Veratrum Viride: has been employed.
Water: cold affusion to spine useful.
Zinc Chloride.
Zinc Cyanide.
Zinc-Hemol: effective hematinic nervine.
Zinc Iodide.
Zinc Oxide.
Zinc Sulphate in small; but very frequent doses, and when the nausea produced is unbearable another emetic to be used.
Zinc Valerianate.

Choroiditis.

Atropine.
Mercury.
Opiates.
Opium.

Chyluria.

Acid, Gallic.
Hypophosphites.
Methylene Blue.
Potassium Iodide.
Sodium Benzoate.
Thymol.

Cicatrices.

Iodine.
Thiosinamine.

Climacteric Disorders.—See also, Metrorrhagia.

Acid, Hydriodic.
Aconite: 1 minim hourly for nervous palpitations and fidgets.
Ammonia: as inhalation. Raspail's Eau Sédative locally in headache: take Sodii chloridum, Liq. ammoniæ, each 2 fl. oz.; Spiritus camphoræ; 3 fl. drs.; Aqua to make 2 pints.
Ammonium Chloride: locally in headache.
Amyl Nitrite.
Belladonna.
Calabar Bean: in flatulence, vertigo, etc.
Camphor: for drowsiness and headache.
Cannabis Indica.
Change: of air and scene useful adjunct.
Cimicifuga: for headache.
Eucalyptol: flushings, flatulence, etc.
Hot Spongings.
Hydrastinine Hydrochlorate.
Iron: for vertical headache, giddiness, and feeling of heat, fluttering of the heart.
Methylene Blue.
Nitrate of Amyl: where much flushed.
Nux Vomica: useful where symptoms are limited to the head.
Opium.
Ovaraden.
Ovarlin.
Physostigma.
Potassium Bromide: very useful.
Potassium Iodide.
Sodium Benzoate.
Stypticin: efficacious, hemostatic and uterine sedative.
Thymol.
Warm Bath.
Zinc Valerianate.

Coccygodynia.

Belladonna: plaster useful.
Chloroform: locally injected.
Counter-irritation.
Electricity.
Surgical Treatment : in obstinate cases.

Coldness.

Atropine.
Chloral Hydrate.
Cocaine Hydrochlorate.
Cold Water: as prophylactic with friction and wrapping up.
Spinal Ice-bag: for cold feet.
Strychnine.

Colic, Biliary.—*See Colic, Renal and Hepatic.*

Colic, Intestinal.

Ammonia: in children.
Anise.
Antacids: in acidity.
Arsenic: when pain is neuralgic in character.
Asafetida: to remove flatulence, especially in children and hysterical patients.
Atropine: in simple spasmodic colic.
Belladonna: especially in children and intestinal spasm.
Caraway.
Chamomile Oil: in hysterical women.
Chloral Hydrate and Bromides: when severe in children.
Chloroform: by inhalation, to remove pain and flatulence.
Cocculus: during pregnancy.
Codeine.
Coriander.
Essential Oils: Aniseed, Cajeput, Camphor, Cardamoms, Cinnamon, Cloves, Peppermint, Rue, Spearmint: all useful.
Ether: internally and by inhalation.
Fennel.
Fomentations.
Ginger: stimulant carminative.
Hyoscine Hydrobromate.
Hyoscyamus.
Lime Water: in children, where due to curdling of milk.
Matricaria: infusion, to prevent, in teething children.
Milk Regimen: in enteralgia.
Morphine: very useful.
Mustard: plaster.
Nux Vomica: useful.
Oil Turpentine.
Opium.
Peppermint.
Potassium Bromide: in local spasm in children, which can be felt through hard abdominal walls.
Poultices: large and warm, of great service.
Rhubarb.
Rue.
Spirit Melissa.
Zinc Cyanide.

Colic, Lead.—*See also, Lead Poisoning.*

Acid, Sulphuric: dilute in lemonade as a prophylactic and curative.
Alum: relieves the pain and constipation.
Atropine.
Belladonna.
Bromides: as solvents alone or with iodides.
Calomel
Castor Oil: given twice a day to eliminate.
Chloroform: internally and externally as liniment.
Croton Oil.
Eggs.
Electro-chemical Baths
Magnesium Sulphate: most useful along with potassium iodide
Milk.
Morphine: subcutaneously to relieve pain.
Opium.
Potassium Iodide: most useful in eliminating lead from the system, and combined with magnesium sulphate to evacuate it.
Potassium Tartrate.
Sodium Chloride.
Strontium Iodide.
Sulphur: to aid elimination.
Sulphurated Potassa.
Sulphur Baths.

Colic, Nephritic.—*See Colic, Renal and Hepatic.*

Colic, Renal and Hepatic.—*See also, Calculi.*

Aliment: abstain from starches and fats.
Alkalies: alkaline waters very useful.
Ammonium Borate.
Amyl Valerianate.
Antipyrine.
Baths: warm, to remove pain.
Belladonna.
Calomel.
Chloroform: inhalation from tumbler during fit.
Collinsonia.
Corn-silk.
Counter-irritation: see list of irritants, etc.
Diet.
Ether: like chloroform.
Formin.
Gelsemium.
Horse-back riding.
Hot application over liver: as a relaxant.

Hydrangea.
Lycetol.
Lysidine.
Olive or Cotton-Seed Oil.
Opium: in small doses frequently repeated, or hypodermically as morphine.
Piperazine.
Sodium Benzoate.
Sodium Salicylate.
Stramonium.
Strophanthus Tincture.
Turpentine Oil.

Collapse.—*See also, Exhaustion, Shock, Syncope.*

Ammonia.
Atropine.
Caffeine.
Digitalin.
Digitoxin.
Ether.
Heat.
Nitroglycerin
Mustard Baths.
Strophanthin.
Strychnine.

Coma.—*See also, Cerebral Congestion, Uremia, Narcotic Poisoning.*

Blisters: on various parts of the body in succession in the critical condition, especially at the end of a long illness.
Cold Douche: in the drunkenness of opium care must be taken not to chill, and it is best to alternate the cold with warm water
Croton Oil: as a purgative in cerebral concussion, etc.
Mustard: to stimulate.
Potassium Bitartrate: purgative where the blood is poisoned.
Turpentine Oil: enema as stimulant.

Condylomata.—*See also, Syphilis, Warts.*

Acid, Carbolic: locally.
Acid, Chromic: with one-fourth of water locally, as caustic.
Acid, Nitric: as caustic, or dilute solutions as a wash.
Arsenic: as caustic.
Europhen.
Ichthyol.
Iodole.
Iodoform: locally.

Iodoformogen.
Mercury: wash with chlorine water, or chlorinated soda, and dust with calomel and oxide of zinc in equal quantities.
Savine.
Silver Nitrate: as caustic.
Sozoiodole-Mercury.
Thuja: strong tincture locally; small doses internally useful.
Zinc Chloride or Nitrate: locally, as a caustic or astringent.
Zinc Sulphate.

Conjunctivitis.

Acid, Boric.
Acid, Carbolic.
Alum: after acute symptoms have subsided; but not if the epithelium is denuded, since perforation may then take place.
Antipyrine.
Argenti Nitras: solution 4 grn. to the fl. dr. in purulent ophthalmia. The solid in gonorrhœal ophthalmia, to be afterwards washed with sodium chloride solution, 4 grn. to the fl. oz.
Atropine.
Belladonna: locally and internally.
Bismuth: locally, in chronic cases.
Blisters: behind ear.
Boroglyceride.
Cadmium: as a wash instead of copper and zinc; the sulphate, 1 grn. to the fl. oz.
Calomel.
Castor Oil: a drop in eye to lessen irritation from foreign body.
Cocaine Hydrochlorate.
Copper Acetate.
Copper Aluminate.
Copper Sulphate: as collyrium.
Creolin Solution, 1 per cent.
Ergot: the fluid extract, undiluted, locally in engorgement of the conjunctival vessels.
Eserine.
Euphrasia: as a mild astringent.
Europhen.
Formaldehyde.
Gallicin.
Hydrastine Hydrochlorate.
Iodole.

Iron Sulphate.
Mercury: as citrine ointment, very useful outside the lids in palpebral conjunctivitis.
Mercury Oxide, Red.
Mercury Oxide, Yellow.
Naphtol.
Opium: fluid extract in eye relieves pain.
Pulsatilla: as wash and internally.
Pyoktanin.
Resorcin.
Retinol.
Silver Nitrate.
Sodium Borate.
Sozoiodole-Sodium.
Tannin: as collyrium.
Zinc Acetate.
Zinc Chloride.
Zinc Sulphate.

Combustiones. — See Burns.

Constipation. — See also, Intestinal Obstruction.

Absinthin.
Aloes, see dinner pill.
Aloin.
Alum.
Ammonium Chloride: in bilious disorders.
Apples: stewed or roast.
Arsenic: in small doses.
Belladonna Extract: one-tenth to ½ grn. in spasmodic contraction of the intestine leading to habitual constipation; best administered along with nux vomica as a pill at bedtime.
Bismuth Formula: take Aluminii sulphas, 1½ grn.; bismuthi subnitratis, 1 grn.; extracti gentianæ, q. s., make pill.
Bisulphate Potassium.
Bryonia.
Calomel.
Carlsbad Waters: tumblerful sipped hot while dressing.
Cascara Sagrada: in habitual constipation, 10 to 20 minims of fluid extract an hour or two after meals.
Castor Oil: 10 to 20 minims in a teaspoonful of brandy and peppermint water before breakfast.
Chloral Hydrate.
Cocculus: When motions are hard and lumpy, and much flatus.

Cod-Liver Oil: in obstinate cases in children.
Coffee: sometimes purges.
Colocynth: compound pill at night.
Croton Oil: when no inflammation is present, very active
Diet.
Dinner Pill: aloes and myrrh; aloes and iron; with nux vomica and belladonna or hyoscyamus, taken just before dinner.
Enemata: soap and water, or castor oil; habitual use tends to increase intestinal torpor; should only be used to unload.
Ergot: to give tone.
Eserine.
Euonymin: cholagogue purgative in hepatic torpor.
Fig: one before breakfast.
Gamboge: in habitual constipation.
Glycerin: suppositories or enemata.
Guaiacum: especially when powerful purgatives fail.
Gymnastics, horseback riding, or massage.
Honey: with breakfast.
Hydrastis: useful in biliousness.
Ipecacuanha: one grn. in the morning before breakfast.
Jalap: along with scammony.
Leptandra.
Lime: saccharated solution after meals.
Licorice Powder, Compound: a teaspoonful at night or in the morning.
Magnesium Bicarbonate: solution useful for children and pregnant women.
Magnesium Oxide.
Magnesium Sulphate.
Manna.
Mercury: in bilious disorders with light stools.
Muscarine: to increase peristalsis.
Nux Vomica: 5 to 10 minims in a glass of cold water before breakfast or before dinner.
Oil Olives.
Opium: when rectum is irritable; also in reflex constipation.
Ox-gall.

Physostigma: 10 minims of tincture along with belladonna and nux vomica in atony of the walls.

Podophyllin or Podophyllum: very useful, especially in biliousness: ten drops of tincture at night alone, or the resin along with other purgatives in pill, especially when stools are dark.

Potassium Bisulphate.

Potassium and Sodium Tartrate.

Prunes: stewed, often efficient; if stewed in infusion of senna they are still more active.

Resin Jalap.

Rhubarb Compound Pill: at night; also for children, mixed with bicarbonate of sodium.

Saline Waters: in morning before breakfast.

Senna: as confection, etc.

Senna: with Cascara Sagrada.

Soap: suppository in children.

Sodium Chlorate.

Sodium Phosphate.

Stillingia: 10 minims of fluid extract.

Strychnine: in atony of the walls.

Sulphates: in purgative natural waters, in small doses.

Sulphur: sometimes very useful as a good addition to compound licorice powder.

Tobacco: 5 minims of the wine at bedtime, or cigarette after breakfast.

Treacle: with porridge, useful for children.

Turpentine Oil: in atonic constipation with much gaseous distention of colon.

Water: draught in the morning before breakfast.

Whole-Meal Bread.

Convalescence.—See also, Adynamia, Anemia.

Acid, Hydriodic.

Alcohol: with meals.

Bebeerine.

Berberine.

Bitters: the simple.

Coca: either extract, or as coca wine for a nervine tonic.

Cod-Liver Oil.

Cream.

Eucalyptus: a tonic after malarial disease.

Glycerinophosphates.

Guarana: same as coca.

Hemo-gallol.

Hydrastine.

Hydrastis: as a substitute for quinine.

Ichthalbin: to promote alimentation.

Iron: as chalybeate waters.

Koumys.

Lime: as lime-water or carbonate of calcium.

Malt Extract, Dry.

Opium: as enema for insomnia.

Orexine: to stimulate appetite, digestion and assimilation.

Pancreatin: to aid digestion.

Pepsin: the same.

Phosphates.

Phosphites.

Quinine.

Sumbul: where great nervous excitability.

Convulsions—See also, Albuminuria, Epilepsy, Hysteria, Puerperal Convulsions, Uremia.

Acid, Phosphoric, Diluted.

Allyl Tribromide.

Amyl Nitrite.

Atropine.

Bromides: in general.

Camphor, Mono-bromated.

Coniine.

Eserine.

Hyoscyamus.

Musk.

Mustard bath.

Nitroglycerin.

Veratrum Viride.

Convulsions, Infantile.

Aconite.

Alcohol: a small dose of wine or brandy arrests convulsions from teething.

Asafetida: a small dose in an enema arrests convulsions from teething.

Baths: warm, with cold affusions to the head.

Belladonna: very useful.

Chloral Hydrate: in large doses—5 grn. by mouth or rectum.

Chloral Hydrate: with Bromide.

Chloroform.

Garlic Poultices: to spine and lower extremities.

Ignatia: when intestinal irritation.

Pilocarpine Hydrochlorate (in uremic).

Spinal Ice-Bag.

Valerian: when due to worms.

Veratrum.

Corneal Opacities.—See also, Keratitis.

Cadmium Sulphate.

Calomel.

Iodine: internally and locally.

Mercury Oxide, Red.

Mercury Oxide, Yellow.

Opium.

Potassium Iodide.

Silver Nitrate: locally.

Sodium Chloride: injected under conjunctiva.

Sodium Sulphate.

Thiosinamine.

Corns.

Acid, Acetic.

Acid, Carbolic.

Acid, Chromic.

Acid, Salicylic: saturated solution in collodion with extract of cannabis indica, ½ dram to 1 fl. oz.

Acid, Trichloracetic.

Copper Oleate.

Iodine.

Mercury Bichloride.

Potassium Bichromate.

Poultices: and plaster with hole in centre to relieve of pressure.

Silver Nitrate.

Sodium Ethylate.

Coryza.—See also, Catarrh.

Acid, Camphoric.

Acid, Sulpho-anilic.

Acid, Tannic.

Aconite: in early stages.

Allium: as a poultice to breast, or in emulsion, or boiled in milk for children.

Amyl Nitrite.

Arsenic: taken for months; for persistent colds.

Antipyrine.

Aristol.

Belladonna.

Bismuth Subnitrate.

Bromides: for associated headache.

Camphor.

Cocaine.

Cubebs.

Formaldehyde: by inhalation (2 per cent. solut.).

Glycerin.
Hamamelis.
Hot Mustard foot-bath.
Ichthyol.
Iodine Fumes.
Iodole.
Iodoformogen.
Losophan.
Menthol.
Pilocarpine Hydrochlorate.
Potassium Iodide.
Quinine.
Salicin.
Sodium Benzoate.
Sozoiodole salts.
Stearates.
Sweet Spirit of Niter.
Tartar Emetic.
Thymol.

Coughs. — *See also,
Bronchitis, Pertussis, Phthisis.*

Acid, Carbolic.
Acid, Hydrobromic.
Acid, Hydrocyanic, Diluted: for irritable cough, and in phthisis, and in reflex cough arising from gastric irritation.
Aconite: in throat-cough and emphysema.
Alcohol: relief by brandy or wine; aggravation by beer or stout.
Alum: as spray or gargle.
Antipyrine.
Antispasmin.
Argenti Nitras: in throat cough, a solution of 8 grn. to the fl. oz. applied to fauces.
Apomorphine: in bronchitis with deficient secretion: and as emetic in children where there is excess of bronchial secretion.
Asafetida: in the after cough from habit, and in the sympathetic whooping-cough of mothers.
Belladonna: in nervous cough and uncomplicated whooping cough.
Blue Pill: in gouty or bilious pharyngeal irritation.
Butyl-Chloral Hydrate: in night coughs of phthisis.
Camphor: internally or locally, painted over the larynx with equal parts of alcohol.
Cannabis Indica.

Carbonic Acid Gas: inhalation in nervous cough.
Cerium: in cough associated with vomiting.
Chloral Hydrate: in respiratory neurosis.
Chloroform: with a low dose of opium and glycerin in violent paroxysmal cough; if very violent to be painted over the throat.
Codeine.
Cod-Liver Oil: one of the most useful of all remedies in cough.
Conium: in whooping cough.
Creosote: in winter cough.
Cubebs: along with linseed in acute catarrh.
Demulcents.
Gaduol: to improve nutrition.
Gelsemium: in convulsive and spasmodic cough, with irritation of the respiratory centre.
Glycerin: along with lemon juice, as an emollient.
Glycerinophosphates.
Glycyrrhizin, Ammoniated.
Grindelia: in habitual or spasmodic cough.
Guaiacol.
Hyoscyamus: in tickling night coughs.
Ichthalbin: as alterative and assimilative.
Iodine: as inhalation in cough after measles, or exposure to cold, associated with much hoarseness and wheezing of the chest.
Iodoform: in the cough of phthisis.
Ipecacuanha: internally and as spray locally; in obstinate winter cough and bronchial asthma.
Ipecacuanha and Squill Pill: in chronic bronchitis at night.
Lactucarium: to relieve cough.
Laurocerasus, Aqua: substitute for hydrocyanic acid.
Linseed: in throat cough.
Lobelia: in whooping-cough and dry bronchitic cough.
Morphine.
Nasal Douche: in nasal cough.
Nux Vomica.

Oil Bitter Almond.
Opiates: morphine locally to the throat and larynx, and generally.
Peronin: admirable sedative, without constipating action.
Potassium Bromide: in reflex coughs.
Potassium Carbonate: in dry cough with little expectoration.
Potassium Cyanide.
Prunus Virginiana.
Pulsatilla: as anemonin ⅙ to 1 grn. dose, in asthma and whooping cough.
Sandalwood Oil.
Sanguinaria: in nervous cough.
Tannin: as glycerite to the fauces in chronic inflammation, especially in children.
Tar Water: in winter cough, especially paroxysmal, bronchial and phthisical.
Theobromine salts.
Thymol.
Valerian: in hysterical cough.
Zinc Sulphate: in nervous hysterical cough.
Zinc Valerianate.

Coxalgia. — *See also, Abscess, Caries, Suppuration, Synovitis.*

Barium Chloride.
Ichthyol.
Iodoform.
Iodoformogen.
Iron Iodide.

Cramp. — *See Spasmodic Affections.*

Cretinism.
Thyroid preparations.

Croup. — *See also, Laryngismus Stridulus, Laryngitis, Diphtheria.*

Acid, Carbolic: spray.
Acid, Lactic: to dissolve membrane (1 in 20); applied as spray or painted over.
Acid, Sulphurous: as spray.
Aconite: in catarrhal croup.
Alum: teaspoonful with honey or syrup every ¼ or ⅙ hour until vomiting is induced; most useful emetic.

Antispasmin.
Apocodeine.
Apomorphine: as an emetic; may cause severe depression.
Aspidospermine.
Calomel: large doses, to allay spasm and check formation of false membrane.
Chloral Hydrate.
Copper Sulphate: 1 to 5 grn., according to age of child, until vomiting is induced.
Creolin Vapors.
Hydrogen Peroxide.
Ichthyol Vapors
Iodine.
Ipecacuanha: must be fresh; if it does not succeed other emetics must be taken.
Jaborandi: beneficial in a few cases.
Lime Water: spray, most useful in adults.
Lobelia: has been used.
Mercury Cyanide.
Mercury Subsulphate: one of the best emetics; 3 to 5 grn., given early.
Papain.
Petroleum.
Pilocarpine Hydrochlorate.
Potassium Chlorate.
Quinine: in spasmodic croup, in large doses.
Sanguinaria: a good emetic; take syrup ipecac, 2 fl. oz.; pulv. sanguin., 20 grn.; pulv. ipecac, 5 grn.; give a teaspoonful every quarter-hour till emesis, then half a teaspoonful every hour.
Senega: as an auxiliary.
Sodium Bicarbonate.
Sozoiodole-Sodium: insufflations.
Sulphurated Potassa.
Tannin: as spray, or glycerite of tannin.
Tartar Emetic: too depressant in young children.
Zinc Sulphate: sometimes used as an emetic.

Croup, Spasmodic. —*See Laryngismus Stridulus.*

Cystitis. — *See also, Bladder, Irritable; Calculus; Dysuria; Enuresis; Hematuria.*

Acid, Benzoic: in catarrh with alkaline urine.

Acid, Boric: as borogly-ceride as injection, in cystitis with an alkaline urine due to fermentation.
Acid, Camphoric.
Acid, Carbolic, or Sulphocarbolates: as antiseptics.
Acid, Gallic.
Acid, Lactic.
Acid, Osmic.
Acid, Oxalic.
Acid, Salicylic: in chronic cystitis with ammoniacal urine.
Aconite: when fever is present.
Alkalies: when urine is acid and the bladder irritable and inflamed.
Ammonium Citrate: in chronic cystitis.
Antipyrine.
Arbutin: diuretic in chronic cystitis.
Belladonna: most useful to allay irritability.
Buchu: especially useful in chronic cases.
Calcium Hippurate.
Cannabis Indica.
Cantharides or Cantharidin: in small doses long continued, where there is a constant desire to micturate, associated with much pain and strain.
Chimaphila: in chronic cases.
Collinsonia.
Copaiba: useful.
Creolin.
Cubebs.
Demulcents.
Eucalyptus: extremely useful in chronic cases.
Gallobromol.
Grindelia.
Guaiacol.
Guethol.
Hot compress over bladder.
Hot Enemata: to relieve the pain.
Hot Sitz Bath.
Hygienic Measures.
Hyoscyamus: to relieve pain and irritability.
Ichthyol Irrigations.
Iodine and Iodides.
Iodoform or Iodoformogen: as suppository.
Kava Kava.
Leeches: to perineum.
Lithium Salts.
Mercury Bichloride: solution to cleanse bladder.
Methylene Blue
Milk Diet.
Myrtol.
Naphtol.
Oil Eucalyptus.

Oil Juniper.
Oil Sandal.
Opium: as enema to relieve pain.
Opium, Belladonna, or Iodoform: suppositories.
areita: in chronic cases.
Picht.
Potassium Bromide: to relieve the pain.
Potassium Chlorate and other Potassium salts, except bitartrate.
Pyoktanin.
Quinine: in acute cases.
Resorcin.
Saliformin.
Salines.
Salol.
Silver Nitrate.
Sodium Benzoate.
Sodium Borate.
Sozoiodole - Sodium: irrigations (1 per cent.).
Sozoiodole-Zinc: irrigations (⅓ per cent.).
Strychnine.
Sulphaminol.
Sulphites: to prevent putrefaction of urine.
Triticum Repens.
Turpentine Oil: in chronic cases.
Uva Ursi: in chronic cases.
Zea Mays: a mild stimulant diuretic.

Cysts. — *See also, Ovaritis.*

Acupuncture.
Chloride of Gold: in ovarian dropsy.
Galvano-puncture.
Iodine: as an injection after tapping.
Silver Nitrate: as an injection.

Cyanosis. — *See also, Asphyxia, Asthma. Dyspnea, Heart Affections.*

Amyl Nitrite.
Oxygen.
Stimulants.

Dandruff.—*See Pityriasis.*

Deafness.

Ammonium Chloride.
Cantharides: as ointment behind the ear.
Colchicum. in gouty persons.
Gargles: in throat-deafness.
Gelseminine.
Glycerin: locally.
Morphine.

Quinine: in Menière's disease.

Tannin: in throat deafness.

Turpentine Oil.

Debility. — *See also, Adynamia, Anemia, Convalescence.*

Acid, Hypophosphorous.

Alcohol: along with food often very useful; liable to abuse—not to be continued too long; effect watched in aged people with dry tongue.

Arsenic: in young anemic persons, alone or with iron, and in elderly persons with feeble circulation.

Berberine.

Bitters: useful as tonic.

Calcium salts: phosphates if from overwork or town life; hypophosphites in nervous debility.

Cholagogue Purgatives: when debility is due to defective elimination of waste.

Cinchona: a fresh infusion along with carbonate of ammonium.

Cod-Liver Oil.

Columbin.

Digitalis: where circulation is feeble.

Eucalyptus: in place of quinine.

Extract Malt, Dry.

Gaduol: in cachexias.

Glycerinophosphates.

Hemo-gallol: as a highly efficacious blood-producer; non-constipating.

Hemol.

Hydrastis: in place of quinine.

Iron: in anemic subjects.

Levico Water.

Magnesium Hypophosphite.

Maltone Wines.

Manganese: alone or with iron.

Morphine: subcutaneously, if due to onanism or hysteria.

Nux Vomica: most powerful general tonic.

Orexine: for building up nutrition when appetite lacking.

Potassium Hypophosphite.

Quinine: general tonic.

Sanguinaria: when gastric digestion is feeble.

Sarsaparilla: if syphilitic taint is present.

Sea-bathing: in chronic illness with debility.

Sodium Arsenate.

Turkish Baths: if due to tropical climate, with caution; in townspeople, when they become stout and flabby.

Decubitus.—*See Bed-Sore.*

Delirium.—*See also, Cerebral Congestion, Fever, Mania.*

Acetanilid.

Alcohol: when delirium is due to exhaustion.

Antimony: along with opium in fever, such as typhus.

Baths, Cold: in fever.

Belladonna: in the delirium of typhus.

Blisters: in delirium due to an irritant poison, and not to exhaustion.

Bromides.

Camphor: in 20 grn. doses every two or three hours in low muttering delirium.

Camphor, Monobrom.

Cannabis Indica: in nocturnal delirium occurring in softening of the brain.

Chloral Hydrate: in violent delirium of fevers.

Cold Douche: place patient in warm bath while administered.

Hyoscyamus.

Morphine: hypodermically.

Musk: in the delirium of low fever, and in ataxic pneumonia of drunkards with severe nervous symptoms.

Opium: with tartar emetic.

Quinine.

Stramonium.

Valerian: in the delirium of adynamic fevers.

Delirium Tremens. —*See also, Alcoholism.*

Acetanilid.

Acid, Succinic.

Alcohol: necessary when the attack is due to a failure of digestion; not when it is the result of a sudden large excess.

Ammonium Carbonate: in debility.

Amylene Hydrate.

Antimony: along with opium, to quiet maniacal excitement and give sleep.

Antispasmin.

Arnica: the tincture when there is great depression.

Beef-tea: most useful.

Belladonna: insomnia when coma-vigil.

Bromoform.

Bromide of Potassium: in large doses, especially when an attack is threatening.

Bromated Camphor: nervine, sedative, and antispasmodic.

Butyl-chloral Hydrate.

Cannabis Indica: useful, and not dangerous.

Capsicum: twenty to thirty grn. doses, repeated after three hours, to induce sleep.

Chloral Hydrate: if the delirium follows a debauch; with caution in old topers and cases of weak heart; instead of sleep sometimes produces violent delirium.

Chloroform: internally by stomach.

Cimicifuga or Cimicifugin: as a tonic.

Coffee.

Cold Douche or Pack: for insomnia.

Conium: as an adjunct to opium.

Croton Oil: purgative.

Digitalis: in large doses has had some success.

Duboisine.

Enemata: nutritive, when stomach does not retain food.

Ethylene Bromide.

Food: nutritious; more to be depended on than anything else.

Gamboge.

Hyoscine Hydrobromate.

Hyoscyamus: useful, like belladonna, probably, in very violent delirium.

Ice to Head: to check vomiting.

Lupulin: as an adjunct to more powerful remedies.

Morphine Valerianate.

Musk.

Nux Vomica.

Opium: to be given with caution.

Paraldehyde.
Potassium Bromide.
Quinine: to aid digestion.
Sodium Bromide.
Stramonium: more powerful than belladonna.
Sumbul: in insomnia and nervous depression and preceding an attack.
Tartar Emetic.
Trional.
Valerian.
Veratrum Viride: very dangerous.
Zinc Oxide.
Zinc Phosphide.

Dementia Paralytica.

Hyoscyamine.
Paraldehyde.
Physostigma.
Thyraden.

Dengue.

Acid, Carbolic.
Acid, Salicylic.
Aconite.
Belladonna.
Emetics.
Opium.
Purgatives.
Quinine.
Strychnine.

Dentition.

Antispasmin.
Belladonna: in convulsions.
Bromide of Potassium: to lessen irritability and to stop convulsions.
Calcium Hippurate.
Camphor, Monobromated.
Calumba: in vomiting and diarrhea.
Cocaine Carbolate.
Hyoscyamus.
Hypophosphites: as tonic.
Phosphate of Calcium: when delayed or defective.
Tropacocaine: weak solution rubbed into gums.

Dermatalgia.

Cocaine.
Menthol.
Tropacocaine.

Dermatitis.

Aluminium Oleate.
Arsenic.
Bismuth Subnitrate.
Cocaine.
Ichthyol.
Lead Water.

Sozoiodole-Sodium.
Tropacocaine.

Diabetes Insipidus.

Acetanilid.
Acid, Gallic.
Acid, Nitric.
Alum.
Antipyrine.
Arsenic.
Atropine.
Belladonna.
Creosote.
Dry Diet.
Ergot: carried to its full extent.
Gold Chloride: in a few cases.
Iron Valerianate.
Jaborandi: in some cases.
Krameria: to lessen the quantity of urine.
Lithium Carbonate or Citrate with Sodium Arsenite: in gouty cases.
Muscarine: in some cases.
Opium: most useful; large doses if necessary.
Pilocarpine.
Potassium Iodide: in syphilitic taint.
Rhus Aromatica.
Strychnine and Sulphate of Iron: as tonics.
Valerian: in large doses.
Zinc Valerianate.

Diabetes Mellitus.

CAUTION: The urine of patients taking salicylic acid gives Trommer's test for sugar.

Acetanilid.
Acid, Arsenous.
Acid, Gallic, with opium
Acid, Lactic.
Acid, Phosphoric, Diluted.
Acid, Phosphoric: to lessen thirst.
Acidulated Water or Non-purgative Alkaline Water: for thirst.
Alkalies: alkaline waters are useful, when of hepatic origin, in obese subjects; and in delirium.
Almond Bread.
Aloin.
Alum.
Ammonium Carbonate.
Ammonium Citrate.
Ammonium Phosphate.
Antipyrine.
Arsenic Bromide.
Arsenic: in thin subjects.

Belladonna: full doses.
Calcium Lactophosphate.
Calcium Sulphide.
Codeine: a most efficient remedy; sometimes requires to be pushed to the extent of 10 grn. or more per diem.
Colchicum and Iodides.
Creosote.
Diabetin.
Diet.
Ergot.
Ether.
Exalgin.
Glycerin: as remedy, and as food and as sweetening agent in place of sugar.
Glycerinophosphates.
Gold Bromide.
Gold Chloride.
Guaiacol.
Hemo-gallol: efficacious hematinic in anemic cases.
Hydrogen Dioxide.
Ichthalbin.
Iodoform.
Iodole.
Iron: most useful along with morphine
Jaborandi.
Jambul.
Krameria.
Levico Water.
Lithium Carbonate or Citrate with Arsenic: if due to gout.
Methylene Blue.
Nux Vomica.
Pancreatin: if due to pancreatic disease.
Pilocarpine Hydrochlorate.
Potassium Bromide.
Purgatives, Restricted Diet and Exercise: if due to high living and sedentary habits.
Quinine.
Rhubarb.
Saccharin: as a harmless sweetener in place of sugar.
Salicylates.
Salines.
Saliformin.
Salol.
Skim-Milk Diet.
Sodium Bicarbonate.
Sodium Carbonate: by intravenous injection in diabetic coma.
Sodium Citrate.
Sodium Phosphate: as purgative.
Sozoiodole-Sodium.
Sulfonal.
Thymol.
Transfusion.
Uranium Nitrate.
Zinc Valerianate.

Diarrhea.—*See also,*
Dysentery, Cholera.

Acid, Boric.
Acid, Camphoric.
Acid, Carbolic.
Acid, Gallic.
Acid, Lactic.
Acids, Mineral; in profuse serous discharges, and in cholera infantum.
Acid, Nitric: with nux vomica to assist mercury, when due to hepatic derangement; combined with pepsin when this is the case with children.
Acid, Nitro-hydrochloric: when there is intestinal dyspepsia.
Acid, Nitrous: in profuse serous diarrhea, and the sudden diarrhea of hot climates.
Acid, Salicylic: in summer diarrhea, and diarrhea of phthisis.
Acid, Sulphuric, diluted in diarrhea of phthisis
Aconite: in high fever and cutting abdominal pains.
Alkalies: in small doses in diarrhea of children, if due to excess of acid in the intestine causing colic and a green stool.
Alum
Aluminium Acetate Solution.
Ammonium Carbonate: in the after-stage, if there is a continuous watery secretion.
Ammonium Chloride: in intestinal catarrh.
Argentic Nitrate: in acute and chronic diarrhea as astringent.
Aristol.
Arnica.
Aromatics: in nervous irritability or relaxation without inflammation.
Arsenic: a few drops of Fowler's solution in diarrhea excited by taking food; in diarrhea with passages of membraneous shreds, associated with uterine derangement; and along with opium in chronic diarrhea of malarial origin.
Belladonna: in colliquative diarrhea.
Betol.
Bismal.
Bismuth Subnitrate: in large doses in chronic

diarrhea; with grey powder in the diarrhea of children.
Bismuth Citrate.
Bismuth Phosphate, Soluble.
Bismuth Salicylate.
Bismuth Subgallate.
Blackberry.
Cajeput Oil: along with camphor, chloroform and opium in serous diarrhea.
Calcium Carbolate.
Calcium Carbonate: the aromatic chalk mixture in the diarrhea of children, and of phthisis and typhus.
Calcium Chloride: in the colliquative diarrhea of strumous children, and in chronic diarrhea with weak digestion.
Calcium Permanganate
Calcium Phosphate: in chronic diarrhea, especially of children.
Calcium Salicylate.
Calcium Sulphate.
Calomel: in minute doses in chronic diarrhea of children with pasty white stools.
Calumba.
Calx Saccharata: in the chronic diarrhea and vomiting of young children.
Camphor: in the early stage of Asiatic cholera, at the commencement of summer diarrhea, acute diarrhea of children, and diarrhea brought on by effluvia.
Camphor, Monobromated.
Cannabis Indica.
Capsicum: in diarrhea from fish; in summer diarrhea; in diarrhea after expulsion of irritant.
Carbon Disulphide.
Cascarilla.
Castor Oil: in the diarrhea of children.
Castor Oil and Opium: to carry away any irritant.
Catechu: astringent.
Chalk Mixture, see Calcium Carbonate.
Charcoal: in foul evacuations.
Chirata.
Chloral Hydrate.
Chloroform: as spirits with opium after a purgative.
Cinnamon.
Cloves.

Cocaine: in serous diarrhea.
Codeine.
Cod-Liver Oil: to children with pale stinking stools.
Cold or Tepid Pack: in summer diarrhea of children.
Copaiba: for its local action in chronic cases.
Copper Arsenite.
Copper Sulphate: one-tenth grn. along with opium in acute and chronic diarrhea, associated with colicky pains and catarrh
Corrosive Sublimate: in small doses in acute and chronic watery diarrhea, marked by slimy or bloody stools of children and adults; and diarrhea of phthisis and typhoid.
Coto Bark: in catarrhal diarrhea.
Cotoin.
Creolin.
Creosote.
Diet: for summer diarrhea.
Dulcamara: in diarrhea of children from teething and exposure
Enterocylsis: when mucous form becomes chronic.
Ergot: in a very chronic diarrhea succeeding to an acute attack.
Erigeron Canadense.
Eucalyptol or Eucalyptus.
Eudoxin.
Flannel Binder: adjunct in children.
Gaduol: as tonic in scrofulous and weakly children.
Galls: in chronic diarrhea.
Geranium.
Ginger.
Guaiacol.
Guaiacol Carbonate.
Guarana: in convalescence.
Hematoxylon: mild astringent, suitable to children from its sweetish taste.
Ice to Spine.
Injection: of starch water, at 100° F., with tinct. opii and acetate of lead or sulphate of copper, in the choleraic diarrhea of children.
Iodine.
Ipecacuanha: drop doses of the wine every hour in the dys-

enteric diarrhea of children, marked by green slimy stools.

Iron Sulphate.

Kino: astringent.

Krameria: astringent.

Lead Acetate: in suppository or by mouth; in summer diarrhea (simple in children, with morphine in adults); with opium in purging due to typhoid or tubercular disease, in profuse serous discharge, and in purging attended with inflammation.

Magnesia: antacid for children.

Magnesium Salicylate.

Menthol.

Mercury: the gray powder in diarrhea of children, marked by derangement of intestinal secretion and stinking stools; to be withheld where masses of undigested milk are passed; in adults, see Corrosive Sublimate.

Monesia Extract.

Morphine Sulphate.

Mustard: plaster.

Naphtalin.

Naphtol.

Naphtol Benzoate.

Nutmeg.

Nux Vomica: in chronic cases.

Oak Bark: infusion, astringent.

Opium: in tubercular and typhoid diarrhea; in acute, after expulsion of offending matter; as an enema, with starch, in the acute fatal diarrhea of children.

Pancreatin.

Paraformaldehyde.

Pepsin: along with nitro-hydrochloric acid.

Podophyllin.

Podophyllum: in chronic diarrhea, with high-colored pale or frothy stools.

Potassium Chlorate: in chronic cases with mucilaginous stools.

Potassium Iodide.

Pulsatilla: in catarrhal.

Quinine.

Resorcin.

Rhubarb: to evacuate intestines.

Rumex Crispus: in morning diarrhea.

Salicin: in catarrh and chronic diarrhea of children.

Saline Purgatives.

Salol.

Silver Chloride.

Silver Nitrate.

Silver Oxide.

Sodium Borate.

Sodium Carbolate.

Sodium Paracresotate.

Sodium Phosphate.

Sodium Thiosulphate.

Starch, Iodized.

Tannalbin: has a very wide range of indications.

Tannigen.

Tannin with Opium: in acute and chronic internally, or as enema.

Thymol.

Tribromphenol.

Veratrum Album: in summer diarrhea.

Zinc Sulphate.

Diphtheria.

Acid, Benzoic: in large doses.

Acid, Boric; or Borax: glycerin solution locally.

Acid, Carbolic: as spray or painted on throat; internally with iron.

Acid, Carbolic, Glycerite of: painted over twice a day.

Acid, Hydrochloric: dilute as gargle, or strong as caustic.

Acid, Lactic: a spray or local application of a solution of 1 dram to the oz. of water, to dissolve the false membrane.

Acid, Salicylic: locally as gargle, or internally.

Acid, Sulphurous.

Acid, Tartaric.

Aconite.

Alcohol: freely given, very useful.

Alum.

Ammonium Chloride.

Antidiphtherin.

Antitoxin.

Apomorphine: as an emetic.

Argentic Nitrate: of doubtful value.

Arsenic: internally.

Asaprol.

Aseptol.

Belladonna: at commencement, especially useful when tonsils are much swollen and there is little exudation; later on, to support the heart.

Bromine: as inhalation.

Calcium Bisulphite: solution, as paint.

Calomel.

Chloral Hydrate.

Chlorinated-Soda Solution: as gargle or wash.

Chlorine Water: internally; locally in sloughing of the throat.

Cold: externally.

Copper Sulphate: as emetic.

Creolin.

Creosote.

Cubeb.

Eucalyptol.

Ferropyrine.

Guaiacum: internally.

Hydrogen Peroxide.

Ichthyol: paint.

Ice: to neck, and in mouth; with iron chloride internally if suppuration threatens.

Iodine: as inhalation.

Iron: the perchloride in full doses by the mouth, and locally painted over the throat.

Lemon Juice: gargle.

Lime Water: most serviceable in adults, as a spray.

Mercury: internally as calomel or cyanide, 1-20 to 1-40 of a grn.

Mercury Bichloride.

Mercury Oxycyanide.

Methylene Blue.

Milk Diet.

Oil Turpentine.

Oxygen: inhalations, with strychnine and atropine hypodermically. If suffocation is imminent, intubation or tracheotomy may be necessary.

Papain: as solvent of false membrane.

Pepsin: as membrane solvent, locally.

Pilocarpine Hydrochlorate: internally; sometimes aids in loosening the false membrane.

Potassa Solution: internally.

Potassium Bichromate: as emetic.

Potassium Chlorate: internally, frequently repeated, and locally as a gargle.

Potassium Permanganate: as gargle.

Pyoktanin: topically.

Quinine: strong solution or spray.

Resorcin: spray.

Sanguinaria: as emetic. See under Croup.

Sassafras Oil: locally.

Sodium Benzoate: in large doses, and powder insufflated.

Sodium Borate.

Sodium Hyposulphite, or Sulphites: internally and locally.

Sodium Sulphocarbolate.

Sozoiodole - Potassium: as dusting - powder with sulphur.

Sozoiodole-Sodium: as preceding; or as solution.

Strychnine: subcutaneously for paralysis.

Sulphocarbolates.

Sulphur.

Tannin: five per cent. solution as a spray.

Thymol.

Tolu Balsam.

Tonics.

Tribromphenol.

Dipsomania. — *See Alcoholism.*

Dropsy. — *See also, Ascites, Hydrocele, Hydrocephalus, Hydrothorax, etc.*

Aconite: at once in dropsy of scarlet fever if temperature should rise.

Acupuncture: in œdemas about the ankles, to be followed up by hot bathing; not much use in tricuspid disease.

Ammonium Benzoate: in hepatic dropsy.

Ammonium Chloride: in hepatic dropsy.

Antihydropin: a crystalline principle extracted from cockroaches; is a powerful diuretic in scarlatinal dropsy; 15 grn. as a dose for an adult; the insect is used in Russia.

Apocynum.

Arbutin.

Arsenic: in dropsy of feet from fatty heart, debility, or old age.

Asclepias Syriaca: may be combined with apocynum.

Broom: one of the most useful diuretics, especially in scarlatinal, renal, and hepatic dropsy.

Bryonia: as drastic purgative, and diuretic.

Cactus Grandiflorus: tincture.

Caffeine: in cardiac and chronic renal dropsy.

Calomel.

Cannabis Indica: as a diuretic.

Chenopodium Anthelminticum: in scarlatinal dropsy.

Chimaphila: in renal dropsy.

Cimicifugin.

Colchicum: in hepatic, cardiac, and scarlatinal dropsy.

Colocynth.

Convallaria: used by the Russian peasantry.

Copaiba: especially in hepatic and cardiac dropsy: not certain in renal.

Digitalin.

Digitalis: in all dropsies, but especially cardiac dropsies. Infusion is best form.

Digitoxin.

Diet: dry.

Elaterium or Elaterin: useful hydragogues cathartics, especially in chronic renal disease; should not be given in exhaustion.

Erythrophleum: in cardiac dropsy instead of digitalis.

Ferropyrine.

Fuchsine.

Gamboge never to be used !

Gold.

Hellebore: in post-scarlatinal dropsy.

Hemo-gallol: when marked anemia present.

Iron: to correct anemia: along with saline purgatives.

Jaborandi: in renal dropsy with suppression of renal function.

Jalap: in some cases.

Juniper: exceedingly useful in cardiac, and chronic, not acute renal trouble.

Levico Water.

Magnesium Sulphate.

Mercury.

Milk Diet.

Nitrous Ether: useful alone, or with other diuretics.

Oil Croton.

Oil Juniper.

Parsley: a stimulant diuretic.

Paracentesis Abdominis

Pilocarpine Hydrochlorate.

Potassium Bicarbonate.

Potassium Bitartrate and Acetate with Compound Jalap Powder: most useful of the hydragogue cathartics.

Potassium Carbonate.

Potassium Iodide: in large doses, sometimes a diuretic in renal dropsy.

Potassium Nitrate: as diuretic.

Potassium and Sodium Tartrate.

Resin Jalap.

Resorcin.

Rhus Toxicodendron.

Saliformin.

Saline Purgatives.

Scoparin.

Scoparius Infusion.

Senega: in renal dropsy.

Squill: in cardiac dropsy.

Strophanthus: in cardiac dropsy.

Sulphate of Magnesium: a concentrated solution before food is taken.

Taraxacum.

Theobromine and salts.

Turpentine Oil: in albuminuria.

Duodenal Catarrh. —*See also, Jaundice, Biliousness.*

Acid, Citric.

Acid, Nitro-hydrochloric.

Arsenic: in catarrh of bile-ducts as a sequela.

Bismuth.

Calomel.

Gold and Sodium Chloride.

Hydrastis: in catarrh associated with gall stones.

Ipecacuanha.

Podophyllum.

Potassium Bichromate.

Rhubarb.

Salol.

Sodium Phosphate.

Dysentery.— *See also, Diarrhea, Enteritis.*

Acid, Boric: continuous irrigation with a two-way tube.

Acid, Carbolic.

Acid, Gallic.

Acid, Nitro-hydrochloric.

Acid, Nitrous: in the chronic dysentery of hot climates.

Acid, Tannic.

Aconite: when much fever.

Alum: to control the diarrhea.
Aluminium Acetate: solution.
Ammonium Chloride.
Aristol.
Arnica: where much depression.
Arsenic: Fowler's solution along with opium if due to malaria.
Baptisin.
Belladonna.
Benzoin: in chronic cases.
Berberine Carbonate: in chronic intestinal catarrh.
Bismal.
Bismuth.
Bismuth Subgallate.
Bismuth Subnitrate.
Calomel: in acute sthenic type.
Castor Oil: in small doses, with opium.
Cathartics: to cause local depletion.
Cold: Enemata of ice cold water to relieve pain and tenesmus.
Copaiba: in some cases.
Copper Arsenite.
Copper Sulphate.
Corrosive Sublimate: in small doses, when stools are slimy and bloody.
Creolin.
Creosote.
Enemata.
Ergotin: in very chronic type.
Glycerin: with linseed tea, to lessen tenesmus.
Grape Diet.
Hamamelis: where much blood in motions.
Hydrogen Peroxide.
Ice Water: injections.
Injections: in early stages, emollient; in later, astringent.
Iodine.
Ipecacuanha: in 30 grn. doses on empty stomach, with complete rest; or as enema, with small quantity of fluid; milk is a good vehicle.
Iron: internally, or as enemata.
Lead Acetate, by mouth, or as enema or suppository, along with opium.
Lemon Juice.
Magnesium Salicylate.
Magnesium Sulphate: in acute cases in early stage.
Mercury Bichloride.

Morphine Sulphate.
Naphtalin.
Naphtol, Alpha.
Nux Vomica: in epidemic cases, and where prune juice stools and much depression.
Oil Eucalyptus.
Opium: to check the diarrhea; given after the action of a saline.
Potassium Bitartrate: in advanced stages where much mucus.
Potassium Chlorate: as enema.
Quinine Sulphate: in large doses in malarial cases, followed by Ipecacuanha.
Saline Purgatives.
Salol.
Silver Chloride.
Silver Nitrate: as injection.
Silver Oxide.
Soda Chlorinata: as enema.
Sodium Carbolate.
Sodium Nitrate.
Strychnine
Sulphur: in chronic cases.
Tannalbin.
Tannin: conjoined with milk diet in chronic disease.
Tribromphenol.
Turpentine Oil: with opium when the acute symptoms have passed off; also in epidemic of a low type.
Zinc Oxide.
Zinc Sulphate: by mouth or enema.

Dysmenorrhea.

Acetanilid.
Acid, Salicylic.
Aconite: in congestive form in plethorics; or sequent to sudden arrest.
Aloes.
Ammonium Acetate.
Ammonium Chloride.
Amyl Nitrite: in neuralgic form.
Anemonin.
Antipyrine.
Apiol (Oil of Parsley): as emmenagogue in neuralgic form; to be given just before the expected period.
Arsenic: when membranous discharge from uterus.
Atropine.
Belladonna: in neuralgic form; along with synergists.

Borax: in membranous form.
Butyl-Chloral Hydrate: in neuralgic form.
Cajeput Oil.
Camphor: frequently repeated in nervous subjects.
Cannabis Indica: very useful.
Cerium Oxalate.
Cetrarin.
Chloralamide.
Chloral Hydrate.
Chloroform: vapor locally.
Cimicifuga: in congestive cases at commencement.
Cimicifugin.
Codeine.
Conium.
Copper Arsenite.
Electricity: the galvanic current in neuralgic; an inverse current in congestive.
Ergot: in congestive cases at commencement, especially if following sudden arrest.
Ether.
Ethyl Bromide.
Ferropyrine.
Gelseminine.
Gelsemium.
Ginger: if menses are suddenly suppressed.
Gold and Sodium Chloride.
Gossypium.
Guaiacum: in rheumatic cases.
Hamamelis: often relieves.
Hemogallol.
Hemol.
Hot Sitz-bath.
Hydrastinine Hydrochlorate.
Ipecacuanha; as an emetic.
Iron: in anemia.
Magnesium Sulphate.
Manganese Dioxide.
Morphine: like opium.
Nux Vomica: in neuralgic form.
Opium: exceedingly useful in small doses of 3 to 5 min. of tincture alone, or along with 3 or 4 grn. of chloral hydrate.
Picrotoxin.
Piscidia Erythrina.
Pulsatilla: like aconite.
Quinine.
Rue.
Silver Oxide.
Sodium Borate.
Strychnine.
Stypticin: useful uterine sedative.

Sumbul.
Triphenin.
Viburnum.
Water: cold and hot, alternately dashed over loins in atonic cases.
Zinc Cyanide.

Dyspepsia.—*See also, Acidity, Biliousness, Flatulence, Gastralgia, Pyrosis.*

Absinthin.
Acids: before or after meals, especially nitro-hydrochloric acid.
Acid, Carbolic.
Acid, Gallic: in pyrosis.
Acid, Hydrochloric, Dilute: after a meal, especially if there is diarrhea.
Acid. Hydrocyanic: in irritable cases.
Acid, Lactic: in imperfect digestion.
Acid, Nitric: with bitter tonics.
Acid, Nitro-hydrochloric.
Acid, Sulphurous; in acid pyrosis and vomiting.
Acid, Tannic: in irritable dyspepsia.
Alcohol: along with food when digestion is impaired by fatigue, etc.
Alkalies: very useful before meals in atonic dyspepsia, or two hours after.
Aloes: as dinner pill, along with nux vomica, in habitual constipation.
Arsenic: 1 min. of liquor before meals in neuralgia of the stomach, or diarrhea excited by food.
Asafetida.
Belladonna: to lessen pain and constipation.
Berberine.
Bismuth Citrate.
Bismuth Subgallate.
Bismuth Subnitrate: when stomach is irritable; and in flatulence.
Bitters: given with acids or alkalies, to stimulate digestion.
Bryonia: in bilious headache.
Calabar Bean: in the phantom tumor sometimes accompanying.
Calcium Saccharate.
Calcium Sulphite.

Calomel.
Calumba: very useful.
Cannabis Indica.
Capsicum: in atonic dyspepsia.
Cardamoms.
Castor Oil.
Cerium Nitrate.
Cerium Oxalate.
Cetrarin.
Chamomile.
Charcoal: for flatulence
Chloral Hydrate.
Chloroform.
Cholagogues: often very useful.
Cinchona.
Cocaine: in nervous dyspepsia, $\frac{1}{4}$ grn. twice or three times a day.
Cod-Liver Oil: in the sinking at the epigastrium in the aged without intestinal irritation.
Colchicum: in gouty subjects.
Cold Water: half a tumbler half an hour before breakfast.
Columbin.
Creosote: if due to fermentative changes.
Diastase of Malt.
Eucalyptus: in atonic dyspepsia due to the presence of sarcinæ.
Gentian: in atony and flatulence.
Ginger: an adjunct.
Glycerin.
Glycerinophosphates.
Gold: the chloride in nervous indigestion.
Hops: a substitute for alcohol.
Hot Water: a tumbler twice or three times between meals, in acid dyspepsia, flatulence, and to repress craving for alcohol.
Hydrastis or Hydrastine Hydrochlorate: in chronic dyspepsia or chronic alcoholism.
Hydrogen Peroxide.
Ichthalbin.
Ipecacuanha: useful adjunct to dinner pill, in chronic irritable dyspepsia.
Iron and Bismuth Citrate.
Iron Phosphates.
Kino: in pyrosis.
Lime Water.
Magnesia: in acid dyspepsia.
Malt Extract, Dry.
Manganese: in gastrodynia and pyrosis.
Mercury: as cholagogue.

Morphine: subcutaneously in irritable subjects.
Naphtol.
Naphtol Benzoate.
Nux Vomica: exceedingly useful in most forms along with mineral acids.
Opium: in sinking at the stomach partially relieved by food which, at the same time, produces diarrhea, a few drops of tincture before meals; with nux vomica in palpitation, etc.
Orexine Tannate: very potent.
Pancreatin: $1\frac{1}{2}$ or 2 hours after meals, very useful.
Papain.
Pepper: in atonic indigestion
Pepsin: sometimes very useful with meals; and in apepsia of infants.
Picrotoxin.
Podophyllin: a cholagogue, used instead of mercury; useful along with nux vomica and mineral acids.
Potassa, Solution of.
Potassium Bicarbonate.
Potassium Carbonate.
Potassium Iodide.
Potassium Permanganate: like manganese.
Potassium Sulphide.
Ptyalin.
Pulsatilla.
Quassia.
Quinine: in elderly people, and to check flatulence.
Resorcin.
Rhubarb.
Saccharin.
Salol.
Sanguinaria: in atonic dyspepsia.
Serpentaria.
Silver Nitrate: in neuralgic cases.
Silver Oxide.
Sodium Sulphocarbolate: in flatulence and spasm after a meal.
Sodium Thiosulphate.
Sozoiodole-Sodium.
Strontium Bromide.
Strychnine.
Taraxacum.
Terebene.
Turkish Bath: in malaise after dining out.
Wahoo (Euonymin): as a cholagogue.
Xanthoxylum: as stomachic tonic.

Dysphagia.

Acid, Hydrocyanic: as gargle.
Bromide of Potassium: in hysterical dysphagia; or dysphagia of liquids in children.
Cajeput Oil: in nervous dysphagia.
Cocaine: in tonsillitis, etc., as cause, 4 per cent. solution painted over.
Iced Fluids: slowly swallowed in spasmodic dysphagia.
Iron.
Quinine.
Strychnine.

Dyspnea. — *See also, Angina Pectoris, Asthma, Bronchitis, Croup, Emphysema, Phthisis.*

Acid, Hydrocyanic, Diluted.
Adonis Aestivalis: tincture.
Adonidin.
Ammonium Carbonate.
Amyl Nitrite.
Arsenic.
Aspidospermine.
Bitter Almond Water.
Cherry Laurel Water.
Chloroform.
Cimicifugin.
Dry Cupping over back: when due to cardiac or pulmonary trouble.
Ether.
Ethyl Iodide.
Grindelia.
Hyoscyamus.
Lobeline.
Morphine.
Opium.
Oxygen.
Pilocarpine Hydrochlorate.
Potassium Cyanide.
Potassium Iodide.
Pyridine.
Spermine.
Stramonium: tincture.
Strophanthin.
Strychnine.
Terebene.
Terpin Hydrate.
Theobromine and Sodium Salicylate.
Thoracentesis: if there is pleural effusion.
Valerian.

Dysuria. — *See also, Vesical Sedatives; Bladder, Irritable; and Cystitis.*

Alkalies: when urine very acid.
Arbutin.

Belladonna.
Camphor: in strangury.
Cannabis Indica: in hematuria.
Cantharides: tincture.
Chimaphila.
Conium.
Digitalis.
Ergot: in paralysis, when bladder feels imperfectly emptied.
Gelsemium.
Gelseminine.
Hyoscyamus.
Nitrous Ether.
Opium.

Ear-ache.

Almond Oil.
Atropine: along with opium.
Blisters: behind the ear.
Brucine.
Cardiac Sedatives: internally.
Chloroform: on swab, behind and in front of ear.
Cocaine: as spray.
Ether Vapor: to tympanum.
Glycerin.
Heat, Dry: locally.
Hop Poultice.
Illicium.
Inflation of Eustachian tube with Politzer's air bag.
Lead Acetate and Opium: as wash.
Leeching: behind ear.
Menthol and Liquid Petrolatum as spray.
Opium.
Pulsatilla.
Puncturing of tympanum if it bulge, followed by careful cleansing and insufflation of boric acid.
Water: hot as it can be borne, dropped into the ear.

Ear Affections. — *See also, Ear-ache, Deafness, Myringitis, Otalgia, Otitis, Otorrhea, Vertigo.*

Acid, Boric.
Bismuth Subgallate.
Cocaine Hydrochlorate (ringing).
Electricity.
Iodoformogen.
Iodole.
Pyoktanin.
Sodium Bromide.
Sodium Borate, Neutral.
Sozoiodole-Zinc.
Tropacocaine.

Ecchymosis. — *See also, Bruises, Purpura.*

Alcohol: externally.
Ammonia.
Arnica: internally and externally.
Compressed Sponge: bound over.
Ice.
Massage.
Solomon's Seal (Convallaria): the juice of the root, especially in a "black eye."

Eclampsia. — *See Puerperal Convulsions.*

Ecthyma.

Borax.
Cod-Liver Oil: internally and locally.
Chrysarobin.
Copper Salts.
Gaduol: internally, as resolvent tonic.
Grape regimen.
Ichthalbin: internally, as assimilative and regulator of nutritive processes.
Ichthyol: topically.
Lead: locally.
Quinine: for the malnutrition.
Zinc Oxide: locally.

Ectropium and Entropium.

Collodion.
Silver Nitrate.

Eczema.

Acetanilid.
Acid, Carbolic: internally and locally.
Acid, Salicylic: locally, if there is much weeping.
Acid, Picric.
Alkalies: weak solutions as a constant dressing.
Alum: to check a profuse discharge; not curative.
Alumnol.
Ammonium Carbonate: along with fresh infusion of cinchona.
Ammonium Urate.
Anacardium Orientale.
Argentic Nitrate: simple solution, or solution in nitric ether, painted over, in chronic form.
Aristol.
Arsenic: applicable only in squamous and

chronic form, not in acute.

Belladonna: internally, or atropine subcutaneously, in acute stage.

Benzoin: compound tincture painted on to relieve itching.

Bismuth: where there is much exudation, the powder, or ointment, either of subnitrate or carbonate.

Bismuth Subgallate.

Bismuth Subnitrate.

Black Wash.

Blisters: in chronic cases, especially of hand.

Borax: the glycerite in eczema of the scalp and ears.

Boric Acid Ointment: topically, especially in eczema of the vulva.

Calcium Lithio-carbonate.

Calcium Sulphide.

Camphor: powder to allay heat and itching.

Cantharides.

Cashew Nut Oil: ointment in chronic cases.

Chloral Hydrate: as ointment half dram in oz. of petrolatum; or as lotion.

Chrysarobin.

Cinchona: powdered bark locally as an astringent.

Citrine Ointment: locally, alone or with tar ointment, in eczema of the eyelids.

Cocaine: to allay itching in scrotal eczema.

Cocoa Nut Oil: in eczema narium.

Cod-Liver Oil: in eczema of children due to malnutrition; and locally to skin to prevent cracking.

Collodion.

Conium.

Copper Sulphate: as-tringent.

Croton Seeds: tincture of, as ointment.

Creolin.

Diaphtherin.

Electricity: central galvanization in very obstinate cases.

Eucalyptol: with iodoform and adeps lanæ in dry eczema.

Eugenol.

Gaduol: internally in scrofula or malnutrition.

Gallicin.

Gallanol.

Gallobromol.

Gelanthum.

Glycerin: as local emollient after an attack.

Glycerite of Aloes: in eczema aurium.

Hamamelis: locally to allay itching.

Hygienic measures and Diet.

Ichthalbin: internally, as assimilative and tonic.

Ichthyol: locally.

Iodole.

Iodoformogen.

Iris Versicolor: in chronic gouty cases.

Iron Arsenate.

Iron Sulphate.

Jaborandi.

Lead Carbonate.

Lead Salts: where there is much inflammation and weeping, a lotion containing a glycerin preparation; if dry and itching, a strong solution or an ointment.

Levico Water.

Lime Water: a sedative and astringent; in later stages with glycerin.

Lithia: in gouty subjects.

Losophan.

Menthol.

Mercury, Ammoniated.

Mercury Oleate.

Methylene Blue: in eczema of the lids.

Naphtol.

Nutgall.

Oil Croton.

Oil of Cade: with adeps lanæ.

Phosphorus.

Phytolacca: in obstinate cases.

Plumbago: ointment in eczema aurium.

Potassium Acetate: internally.

Potassium Cyanide: to allay itching.

Potassium Iodide.

Potato Poultice: cold, sprinkled with zinc oxide, to allay itching.

Pyoktanin.

Resorcin.

Rhus Toxicodendron: internally and externally; where there is much burning and itching, and in chronic eczema of rheumatism worse at night-time.

Salol.

Soap: a glycerin soap to wash with, night

and morning, will allay itching; green soap.

Sodium Arsenate.

Sozoiodole-Potassium.

Starch Poultice.

Sulphides or Sulphur: internally, and as baths; but not in acute stage.

Sulphur Iodide.

Tannin Glycerite: after removal of the scales; or tar, or other ointment, may be required to complete cure.

Tannoform.

Tar: ointment; and internally as pill or capsule in very chronic form.

Thymol.

Thyraden.

Turkish Bath.

Viola Tricolor: infusion along with senna; externally as ointment.

Warm Baths: in acute stages.

Yolk of Egg: with water locally.

Zinc: the oxide and carbonate as dusting powders; the oxide as ointment if the raw surface is indolent after inflammation has subsided.

Zinc Oleate.

Elephantiasis.

Anacardium Orientale.

Arsenic: along with five or six times as much black pepper.

Cashew Nut Oil.

Gurjun Oil.

Iodine: internally and externally.

Oil Chaulmoogra.

Sarsaparilla.

Emissions and Erections. — See also, Chordee, Spermatorrhea, and the list of Anaphrodisiacs.

Acetanilid.

Antispasmin.

Belladonna.

Bromalin.

Bromides.

Bromo-hemol.

Camphor, Monobromated.

Chloral Hydrate.

Cimicifuga.

Cocaine Hydrochlorate.

Hygienic Measures.

Hyoscine.

Iron.

Potassium Citrate.

Strychnine and Arsenic: in full dose.
Warm Bath: before retiring.

Emphysema.—*See also, Asthma, Bronchitis, Dyspnea.*

Apomorphine: when secretion is scanty.
Asafetida.
Arsenic: in subjects who are affected with dyspnea on catching a very slight cold. Especially valuable if following on retrocession of rash.
Aspidospermine.
Belladonna: if bronchitis and dyspnea are severe.
Bleeding: when right side of heart engorged.
Chloral Hydrate: in acute if sudden, a single large dose; if long continued, small doses.
Cod-Liver Oil: one of the best remedies.
Coniine.
Compressed Air: inhaled.
Cubebs: the tincture sometimes relieves like a charm.
Digitalis.
Ether: internally, as inhalation.
Euphorbia Pilulifera.
Ethyl Iodide: as inhalation.
Gaduol.
Grindelia: in most respiratory neuroses.
Hemogallol.
Hemol.
Hypophosphites.
Iron.
Lobelia: where there is severe dyspnea, or capillary bronchitis.
Morphine.
Oxygen: in paroxysmal dyspnea.
Potassium Iodide.
Purging: instead of bleeding.
Physostigma.
Quebracho.
Resorcin.
Senega.
Stramonium.
Strychnine: as a respiratory stimulant.
Terebene.
Turpentine Oil.

Empyema.

Ammonium Acetate.
Aspiration, or free Incisions.

Acid, Carbolic: as injection to wash out cavity.
Acid, Salicylic: same as above.
Carbolate of Iodine: same as above.
Chlorine Water: same as above.
Creosote.
Gaduol: as tonic.
Ichthalbin: as assimilative and alterative.
Iodine: same as carbolic acid.
Iodoform.
Iodoformogen.
Pyoktanin.
Styrone.
Quinine: same as carbolic acid.

Endocarditis.—*See also, Pericarditis.*

Acid, Salicylic: in the rheumatic form.
Aconite: in small doses frequently at commencement.
Alkalies.
Antirheumatics
Blisters.
Bryonia.
Calomel.
Chloral Hydrate: in moderate doses.
Digitalis.
Ice-bag over precordium.
Iron.
Leeches or Wet Cups: in early stages, to abort.
Lithium Citrate or Acetate.
Mercury: to prevent fibrinous deposits; conjointly with alkalies if of rheumatic origin.
Opium: in full doses.
Potassium Iodide.
Potassium Salts: to liquefy exudation.
Quinine: in full doses at commencement.
Veratrum Viride.

Endometritis.—*See also, Uterine Congestion and Hypertrophy.*

Acid, Carbolic: locally applied, undiluted, on cotton wool probe, in chronic form.
Acid, Chromic: strong solution, 15 grn. in 1 fl. dram of hot water in catarrh.
Acid, Nitric.
Alumnol.
Aristol.
Calcium Bisulphite: solution.

Ergot: subcutaneously.
Europhen.
Formaldehyde.
Glycerin: locally.
Gold and Sodium Chloride.
Hot Water Injections.
Hydrargyri Bichloridum: injection.
Hydrastinine Hydrochlorate.
Ichthyol.
Iodine.
Iodoform.
Iodoformogen.
Iodole.
Iodo-tannin: solution of iodine in tannic acid, on cotton-wool.
Methylene Blue.
Sozoiodole-Zinc.
Stypticin.

Enteric Fever.—*See Typhoid Fever.*

Enteritis.—*See also, Diarrhea, Dysentery, Cholera, Peritonitis, Typhlitis.*

Aconite: in acute cases
Argentic Nitrate: in chronic form.
Arsenic: in small doses along with opium.
Bismuth and Ammonium Citrate.
Bismuth Subgallate.
Bismuth Subnitrate.
Bismuth-Cerium Salicylate.
Calcium Salicylate.
Calomel: in obstructive enteritis with constipation, pushed to salivate.
Castor Oil: especially in the chronic enteritis of children. Very useful along with opium.
Chlorine Water.
Copper Arsenite.
Copper Sulphate: in minute doses.
Eudoxine.
Extract Monesia.
Ichthalbin.
Iron.
Lead Acetate: sedative astringent.
Linseed: infusion as drink
Magnesium Sulphate: the most valuable purgative.
Naphtalin.
Naphtol Benzoate.
Opium.
Podophyllum.
Poultice, Hot.
Resorcin.
Skim Milk: as diet, alone or with limewater.

Sodium Nitrate.
Tannalbin.
Tannigen.
Turpentine Oil.
Ulmus: infusion as drink, or leaves as poultice.

Enuresis.

Acid, Camphoric.
Antipyrine.
Atropine.
Belladonna: very useful for children, but the dose must be large.
Buchu: in chronic cases.
Cantharides: internally; very useful in middle-aged women or the aged.
Chloral Hydrate: in children.
Collodion: to form a cap over prepuce.
Ergot: in paralytic cases.
Iodide of Iron: in some cases.
Iodine.
Lupuline.
Pichi.
Potassium Bromide.
Potassium Nitrate: in children.
Quinine.
Rhus Aromatica.
Rhus Toxicodendron.
Santonin: when worms present.
Strychnine: very useful in the paralysis of the aged, and incontinence of children.
Turpentine Oil.

Epididymitis. — See also, Orchitis.

Aconite: in small doses frequently repeated.
Belladonna.
Collodion.
Guaiacol: locally.
Heat, Moisture, and Pressure: in later stages, to relieve induration.
Ice-bags.
Ichthyol.
Iodine: grn. 4 to adeps lanæ oz. 1, locally. to relieve induration.
Mercury and Belladonna: as ointment.
Mercury and Morphine: locally as oleate if persistent.
Naftalan.
Potassium Iodide.
Pulsatilla: in very small doses along with aconite.
Punctures: to relieve tension and pain.

Rest in bed: elevation of pelvis and testicles, suspension of any local gonorrheal treatment.
Silver Nitrate: strong solution locally applied to abort.
Strapping and suspending testicle.

Epilepsy. — See also, Hystero-Epilepsy, Convulsions.

Acetanilid.
Acid, Boric.
Acid, Camphoric.
Acid, Hydrobromic.
Acid, Perosmic.
Adonis Vernalis.
Ammonium Bromide.
Ammonium Valerianate.
Ammonium or Sodium Nitrite.
Amyl Nitrite.
Amylene Hydrate.
Aniline Sulphate.
Antipyrine.
Apomorphine: to prevent; in emetic doses.
Argentic Nitrate: sometimes useful, but objectionable from risk of discoloring the skin
Arsenic: in epileptiform vertigo.
Asafetida.
Atropine.
Anesthetics: rarely.
Belladonna: in petit mal, in nocturnal epilepsy and anemic subjects; perseverance in its use is required.
Bismuth Valerianate.
Blisters: over seat of aura.
Borax.
Bromides of Potassium, Sodium, Strontium, Lithium, and Iron: most generally useful; dose should be large; in cases occurring in the day-time, in grand mal, reflex epilepsy, and cerebral hyperemia.
Bromalin: mild yet very efficacious.
Bromo-hemol.
Bryonia.
Caesium and Ammonium Bromide.
Calabar Bean.
Calcium Bromide.
Calcium Bromo-iodide.
Camphor: has been, but is not now, much used.
Camphor, Monobromated.

Cannabis Indica.
Cautery: frequently and lightly repeated.
Cerium Oxalate.
Chloral Hydrate: full dose at bed-time in nocturnal attacks.
Chloroform: inhalation in hystero-epilepsy.
Chloralamide.
Cod-Liver Oil.
Conium.
Copper Acetate.
Copper Ammonio-sulphate: sometimes useful.
Copper Sulphate.
Cupro-hemol.
Digitalis.
Diet.
Duboisine.
Electricity.
Ethylene Bromide.
Fluorides.
Gaduol.
Gold Bromide.
Hydrargyri Biniodidum: in syphilitic history.
Hydrastinine Hydrochlorate.
Hyoscyamine.
Ignatia.
Iron: in uterine obstruction, in cerebral and genital anemia; alone, or the bromide along with the bromide of potassium.
Iron Valerianate.
Lithium Bromide.
Lobelia: has been used as a nauseant to relieve the spasms.
Mercury.
Musk: has been tried.
Nickel.
Nitrite of Amyl: inhaled will cut short a fit; if there is appreciable time between aura and fit will prevent it. and cut short status epilepticus.
Nitrite of Sodium: in petit mal in 1 grn. dose thrice daily.
Nitroglycerin: like nitrite of amyl, but slightly slower in action.
Opium.
Paraldehyde: instead of bromides.
Phosphorus.
Physostigma.
Picrotoxin: weak and anemic type: or nocturnal attacks: must be persisted in.
Potassium Bromate.
Potassium Bromide.
Potassium Iodide: with bromide; alone in syphilitic history.

Potassium Nitrite.
Quassia: injections when due to worms.
Quinine.
Rubidium - Ammonium Bromide.
Rue: when seminal emissions also are present.
Santonin: has been tried.
Seton: in the back of the neck.
Silver Salts.
Simulo Tincture.
Sodium Fluoride.
Solanum Carolinense: in epilepsy of childhood.
Spermine.
Stramonium Tincture.
Strontium Bromide.
Strychnine: in idiopathic epilepsy and especially in pale anemic subjects; not if there is any organic lesion.
Sulfonal.
Sumbul.
Tartar Emetic.
Turpentine Oil: if due to worms.
Valerian: sometimes does good, especially if due to worms.
Zinc Salts: the oxide, or sulphate; epileptiform vertigo due to gastric disturbance is often relieved by the oxide.

Epistaxis. — See also, Hemorrhage.

Acetanilid.
Acid, Acetic.
Acid, Gallic: along with ergot and digitalis.
Acid, Trichloracetic.
Aconite: in small and frequent doses to children, and in plethora.
Alum: powder snuffed or blown up the nostrils.
Antipyrine.
Aristol.
Arnica: in traumatic cases.
Barium Chloride: to lower arterial tension.
Belladonna.
Blister over Liver.
Cocaine: locally in hemorrhage from the nasal mucous membrane.
Compression of Facial Artery.
Digitalis: the infusion is best.
Ergot: subcutaneously, or by stomach.

Erigeron Oil.
Europhen.
Ferropyrine.
Hamamelis.
Hot Foot-bath, or Hot- or Cold-water Bags applied to dorsal vertebræ.
Ice: over nose and head.
Iodole.
Iodoformogen.
Ipecacuanha: until it nauseates or produces actual vomiting.
Iron: as spray the subsulphate or perchloride.
Krameria.
Lead Acetate.
Plugging anterior and posterior nares necessary, if epistaxis is obstinate.
Tannin: locally applied.
Transfusion: if death threatens from loss.
Turpentine Oil: internally in passive hemorrhage.
Warm Baths: to feet and hands, with or without mustard.
Warm Water Bags: to spine.

Epithelioma.

Acid, Lactic.
Acid, Picric.
Aniline.
Arsenic.
Aristol.
Calcium Carbide.
Diaphtherin.
Europhen.
Iodoform.
Iodoformogen.
Iodole.
Levico Water.
Mercury, Acid Nitrate: applied to part with glass rod.
Methylene Blue.
Papain.
Pyoktanin.
Resorcin.

Erysipelas. — See also, Phlegmon.

Acid, Benzoic: the soda salt 2 to 3 drams in the twenty-four hours.
Acid, Boric: lotion in phlegmonous erysipelas.
Acid, Carbolic: lint soaked in two per cent. solution relieves pain; subcutaneously ½ dram, alcohol ½ dram, water 2 oz.
Acid, Salicylic: as ointment, or dissolved in collodion as paint.

Acid, Sulphurous: equal parts with glycerin locally.
Acid, Picric.
Acid, Tannic.
Aconite: at commencement may cut it short; valuable when skin is hot and pungent and pulse firm; also in erysipelatous inflammation following vaccination.
Alcoholic Stimulants: if patient passes into typhoid state.
Alumnol.
Ammonium Carbonate: when tendency to collapse, and in typhoid condition; internally and locally; more adapted to idiopathic, especially facial erysipelas.
Antipyrine.
Belladonna.
Bismuth Subgallate.
Bismuth Subnitrate.
Bitters and Iron.
Borax.
Calomel.
Chloral Hydrate.
Collodion: locally in superficial erysipelas, useless when cracked.
Creolin.
Creosote.
Digitalis: infusion locally.
Europhen.
Fuchsine.
Hamamelis.
Hot Fomentations.
Ichthyol.
Iodine: solution not too strong painted over.
Iodole.
Iron: large doses frequently, and local application.
Lactophenin.
Lead Acetate.
Lead Carbonate.
Manganese Dioxide.
Mercury Oxycyanide.
Naphtol.
Neurodin.
Pilocarpine.
Potassium Iodide.
Potassium Permanganate: solution locally, and internally.
Potassium Silicate.
Quinine: in large doses.
Resin Jalap.
Resorcin: antipyretic and antiseptic.
Rhus Toxicodendron.
Salol.
Silver Nitrate: strong solution locally applied for an inch or

two beyond inflamed area.
Sodium Salicylate: antipyretic.
Tartar Emetic: small doses frequently.
Thermodin.
Thiol.
Tinct. Ferric Chloride.
Traumaticin.
Trichlorphenol.
Triphenin.
Turpentine.
Veratrum Viride.
White Lead: paint locally.
Zinc Oxide.

Erythema.

Acids: in cases of indigestion.
Acid, Picric.
Aconite.
Adeps Lanæ.
Alum: lotion.
Belladonna: in simple erythema.
Bismuth Subgallate.
Bismuth Subnitrate.
Cold Cream.
Gelanthum.
Ichthyol.
Lead: the glycerite of the carbonate.
Quinine: in erythema nodosum.
Rhus Toxicodendron.
Sozoiodole-Sodium.
Tannoform.
Zinc: locally, as ointments or lotions.

Excoriations. — See also, Intertrigo.

Bismuth Subgallate.
Bismuth Subnitrate.
Ichthyol.
Iodoformogen.
Iodole.
Lead Acetate.
Lead Carbonate.
Lead Cerate.
Lead Nitrate.
Lead Subacetate.
Lead Tannate.
Sozoiodole-Potassium.
Tannoform.
Traumaticin.
Zinc Carbonate.
Zinc Oxide.

Exhaustion. — See also, Adynamia, Convalescence, Insomnia, Myalgia, Neurasthenia.

Acetanilid.
Calcium Carbonate.
Calcium Phosphate.
Cimicifuga.
Coca.
Cocaine.
Coffee.

Hemol-gallol.
Iron Phosphate.
Kola.
Opium.
Phosphorus.
Potassium Bromide.
Stimulants.

Exhaustion, Nervous.

Acid, Hypophosphorous.
Arsenic.
Bromo-hemol.
Coca.
Cupro-hemol.
Iron Valerianate.
Kola.
Levico Water.
Sodium Hypophosphite
Spirit Ammonia.

Exhaustion, Sexual.

Cocaine.
Cornutine Citrate.
Muira Puama.
Phosphorus.
Solanin.
Zinc Phosphide.

Exophthalmos.

Acid, Carbolic.
Acid Boric.
Acid, Hydriodic.
Acid, Picric.
Acid, Salicylic.
Arsenic.
Barium Chloride: to raise arterial tension.
Belladonna.
Bromides.
Cactus Grandiflorus.
Cannabis Indica.
Chalybeate Waters: for the anemia.
Convallaria.
Coto.
Digitalis: if functional in young subjects; often relieves in other cases.
Digitoxin.
Duboisine.
Galvanism of the cervical sympathetic, and pneumogastric nerves.
Glycerinophosphates.
Gold Bromide.
Iodothyrine.
Iron: for the anemia.
Mercury Oleate.
Myrtol.
Resorcin.
Sparteine Sulphate.
Strophanthus.
Thyraden.
Veratrum Viride.
Zinc Valerianate.

Exostosis.

Aconite.
Iodine.
Mercury.
Potassium Iodide.

Eye Diseases. — See also, Amaurosis, Amblyopia, Asthenopia, Cataract, Conjunctivitis, Corneal Opacities, Glaucoma, Iritis, Keratitis, Myopia, Opthalmia, Photophobia, Retina, Strabismus, etc.—See also lists of Mydriatics, Myotics and other agents acting on the eye.

Acetanilid.
Acid, Boric.
Ammonium Acetate: solution.
Arecoline Hydrobromate.
Atropine.
Belladonna.
Bismuth Subgallate.
Cadmium Sulphate.
Calomel.
Chloroform.
Cineraria Juice.
Cocaine.
Copper Salts.
Erythrophleine Hydrochlorate.
Eserine.
Formaldehyde.
Homatropine.
Hydrastine Hydrochlorate.
Ichthalbin: internally.
Ichthyol.
Iodole.
Iodoformogen.
Iron Sulphate.
Lead Acetate.
Mercury Bichloride.
Mercury Nitrate.
Mercury Oleate.
Mercury Oxide, Red.
Mercury Oxide, Yellow.
Morphine.
Phenol, Monochloro-, Para-.
Phyostigmine (Eserine).
Pilocarpine Hydrochlorate.
Pulsatilla.
Pyoktanin.
Resorcin.
Rhus Toxicodendron: tincture.
Rubidium Iodide.
Santonin.
Scoparin Hydrobromate.
Silver Nitrate.
Sozoiodole salts.
Strychnine.
Tropacocaine.
Zinc Acetate.
Zinc Permanganate.
Zinc Sulphate.

Eye-Lids, Affections of.—*See also, Blepharitis, Conjunctivitis, Ecchymosis, Ectropion, Ptosis, etc.*

Acid, Tannic.
Ammonium Chloride.
Cadmium Sulphate.
Calomel.
Coniine : for spasm.
Copper Sulphate.
Mercury and Morphine: for stye.
Pulsatilla.
Pyoktanin.
Sozoiodole-Sodium.
Zinc Sulphate.

False Pains.

Acetanilid.
Neurodin.
Opium.
Tartar Emetic.
Triphenin.

Fauces, Inflammation of.—*See also, Throat, Sore.*

Acid, Tannic.
Silver Nitrate.

Favus.

Acid, Boric : locally in ethereal solution.
Acid, Carbolic : as a local parasiticide.
Acid, Salicylic : like above.
Acid, Sulphurous : like above.
Alumnol.
Cod-Liver Oil : in a debilitated subject.
Copper Oleate.
Gaduol.
Gallanol.
Mercury : the oleate as a parasiticide ; also lotion of bichloride 2 grn. to the oz. of water.
Myrtol : parasiticide.
Naftalan.
Naphtol.
Oil Cade.
Oils : to get rid of scabs and prevent spread.
Potassium Bichromate.
Resorcin : parasiticide.
Sulphurated Potassa.
Sozoiodole-Sodium.

Feet.—Perspiring, Fetid, Tender, Swelled, etc.—*See also, Bromidrosis, Chilblains.*

Acid, Boric.
Acid, Chromic.
Acid, Salicylic.

Acid, Tannic.
Alum.
Arsenic : grn. 1-60 to 1-40 in swelling of old persons.
Belladonna.
Borax : stocking soaked in saturated solution each day and allowed to dry while on.
Chloral Hydrate.
Calcium Carbonate, Precipitated.
Cotton, instead of woolen, stockings.
Formaldehyde.
Hamamelis.
Hydrastine Hydrochlorate.
Lead.
Lead Plaster and Linseed Oil : equal parts, applied on linen to feet, every third day, for sweating.
Potassium Bichromate.
Potassium Permanganate.
Rest : absolutely for swollen feet may be necessary.
Salicylic Acid and Borax : equal parts, in water and glycerin, for sweating and tender feet.
Sodium Bicarbonate.
Sodium Chloride.
Tannoform : with starch or talcum, as dusting-powder in stocking ; very efficacious.

Felon.—*See Onychia.*

Fermentation, Gastro-Intestinal.—*See Flatulence.*

Fever. — *See also, the titles of the fevers in their alphabetical order.*

Acetanilid.
Acids or Acid Drinks : to allay thirst and aid digestion.
Acid, Carbolic.
Acid, Carbonate.
Acid, Citric.
Acid, Hydrochloric.
Acid, Phosphoric.
Acid, Salicylic: in rheumatic fevers, or in hyperpyrexia.
Acid, Sulphurous.
Acid, Tartaric.
Aconite · small doses frequently in all sympathetic fevers.
Alcohol : often useful, but effect watched carefully,—quickly

discontinued if it does not relieve symptoms.
Alkalies : febrifuges, and increase urinary solids.
Ammonia : in sudden collapse.
Ammonium Acetate : very useful as diaphoretic, chiefly in milder forms.
Ammonium Carbonate: in scarlet fever and measles, and in any typhoid condition.
Ammonium Picrate: in malarial fever.
Antipyrine : to reduce temperature.
Arnica : full doses of the infusion in sthenic reaction; low doses of the tincture in asthenia.
Arsenic : in malarious fevers ; and in prostrating acute fevers to raise the patient's tone.
Belladonna: in eruptive fevers and in delirium.
Bitters : with acid drinks to quell thirst, e.g. cascarilla, orange peel, etc.
Blisters: flying blisters in various parts of the body in the semicomatose state.
Bromides.
Calomel : in the early stages of typhoid.
Camphor : in adynamic fevers, and in delirium, in 20 grn. doses every two or three hours, and effects watched.
Carbolate of Iodine : in the later stages of typhoid; and in chronic malarial poisoning.
Castor Oil : as purgative.
Chloral Hydrate: in the violent delirium and wakefulness of typhus, etc., and to reduce fever.
Cimicifuga : when cardiac action is quick and tension low.
Cinchonine.
Coca : as a supportive and stimulant in low fevers.
Cocculus : in typhoid, to lessen tympanitis.
Coffee : in place of alcohol.
Cold Applications: affusions, packs and baths, to lessen hyper-

pyrexia, and an excellent stimulant, tonic and sedative; the pack in acute fevers, especially on retrocession of a rash.

Digitalis: in inflammatory eruptive fevers, especially scarlet fever, as an antipyretic; much used also in typhoid.

Elaterium: hydragogue cathartic.

Eucalyptus: in intermittent fevers.

Gallanol.

Gelsemium: in malarial and sthenic fevers, especially in pneumonia and pleurisy.

Guaiacol: topically.

Hot Afusions: for headache sometimes better than cold.

Hydrastis: inferior to quinine in intermittent fever.

Ice: to suck; bag to forehead.

Lactophenin.

Lemon Juice: an agreeable refrigerant drink.

Menthol.

Mercury: small doses at the commencement of typhoid or scarlet fever.

Musk: a stimulant in collapse; along with opium in an acute specific fever.

Neurodin.

Opium: in typhoid delirium; with tartar emetic if furious; at the crisis aids action of alcohol.

Phenacetin.

Phenocoll Hydrochlorate.

Phosphate of Calcium: in hectic.

Potassium Bitartrate.

Potassium Citrate.

Potassium Nitrate.

Potassium Tartrate.

Potassium and Sodium Tartrate.

Quinine: in malarial, typhoid, and septic fevers; the most generally applicable antipyretic.

Resorcin: antipyretic and antiseptic.

Rhus Toxicodendron: in rheumatic fever, and scarlet fever with typhoid symptoms.

Salicin: in rheumatic fevers, or in hyperpyrexia.

Salicylate of Sodium: in rheumatic fevers, or in hyperpyrexia.

Salol.

Sodium Benzoate: in infectious and eruptive fevers; antiseptic and antipyretic.

Strychnine: subcutaneously for muscular paralysis as a sequela.

Sulphate of Magnesium: as a depletive and purgative.

Tartar Emetic: in small doses, with opium, if delirium is not greater than wakefulness; if greater, in full doses, with small doses of opium; diaphoretic; in ague aids quinine, also in acute.

Thermodin.

Triphenin.

Turpentine Oil: stimulant in typhoid, puerperal, and yellow, and to stop hemorrhage in typhoid.

Valerian.

Veratrum Viride: in delirium ferox.

Warm Sponging: in the simple fevers of children.

Fibroids.—See Tumors.

Fissures. — See also, Rhagades.

Bismuth Subnitrate.
Collodion.
Creolin.
Ichthyol.
Iodoformogen.
Iodole.
Papain.
Pyoktanin.
Traumaticin.

Fissured Nipples.—See also, Rhagades.

Bismuth Oleate.
Cacao Butter.
Ichthyol.
Sozoiodole-Potassium.
Traumaticin.

Fistula.

Bismuth Oxyiodide.
Capsicum: as weak infusion locally.
Chlorine Water.
Creolin.
Diaphtherin.
Ichthyol.
Potassa.
Sanguinaria: as injection.

Flatulence.—See also, Colic, Dyspepsia.

Abstention from sugar, starchy food, tea.

Acid, Carbolic: if without acidity, etc.

Acid, Sulphurous: if due to fermentation.

Alkalies: before meals.

Ammonia: in alkaline mixture a palliative.

Asafetida: in children; simple hysterical or hypochondriacal.

Belladonna: if due to paresis of intestinal walls.

Benzo-napthol.

Bismuth: with charcoal, in flatulent dyspepsia.

Calcium Saccharate.

Calumba: with aromatics.

Camphor: in hysterical flatulence, especially at climacteric.

Carbolated Camphor.

Carlsbad Waters: if due to hepatic derangement.

Carminatives.

Charcoal.

Chloroform: pure, in drop doses in gastric flatulence.

Creosote.

Essential Oils.

Ether: in nervousness and hypochondriasis.

Eucalyptol: at climacteric, if associated with heat flushings, etc.

Galvanism.

Hot Water: between meals.

Ichthalbin.

Ipecacuanha: in constipation, oppression at epigastrium, and in pregnancy.

Manganese Dioxide.

Mercury: when liver is sluggish.

Muscarine: in intestinal paresis.

Nux Vomica: in constipation, pain at top of head.

Oleoresin Capsicum.

Pepper.

Physostigma: in women at change of life.

Picrotoxin.

Podophyllin with Euonymin, Leptandra, Chirata and Creosote.

Potassium Permanganate: in fat people.

Rue: most efficient.

Sodium Sulphocarbolate.

Strontium Bromide.

126

Sulphocarbolates: when no acidity, and simple spasms.
Terebene.
Turpentine Oil: few drops internally, or as enema in fevers, peritonitis, etc.
Valerian.
Xanthoxylum.

Fluor Albus. — *See Leucorrhea.*

Flushing and Heat. — *See also, Climacteric Disorders.*

Eucalyptol: at climacteric.
Iron: most useful.
Nitrite of Amyl: if associated with menstrual irregularity (accompanying symptoms, cold in the extremities, giddiness, fluttering of the heart); inhalation, or internally in one-third of a drop doses; effects sometimes disagreeable.
Nux Vomica: with tinct. opii in the hysteria of middle-aged women.
Ovaraden or Ovariin: at menopause.
Potassium Bromide.
Valerian.
Zinc Valerianate: at climacteric.

Fractures and Dislocations. — *See also, Wounds.*

Acid, Carbolic.
Arnica: internally and locally.
Calcium Glycerinophosphate: internally, to hasten union.
Chloroform.
Iodine: antiseptic dressing.
Iodoformogen.
Iodole.
Opium.
Phosphate of Calcium: internally; quickens union.
Sozoiodole-Sodium.

Freckles.

Acid, Boric.
Acid, Lactic.
Alkaline Lotions.
Benzoin.
Borax.
Copper Oleate.
Iodine.
Lime-Water.
Mercuric Chloride: locally, with glycerin, alcohol, and rose

water. Three-fourths of grn. to the oz.
Olive Oil.
Potassium Carbonate.
Resorcin.

Frost-Bite. — *See also, Chilblains.*

Acid, Carbolic.
Acid, Tannic.
Adeps Lanæ.
Aluminium Acetotartrate.
Camphor Cream.
Creosote.
Ichthyol.
Sozoiodole-Potassium.
Sozoiodole-Zinc.
Styrax.

Furunculus. — *See Boils.*

Gall Stones. — *See Calculi, Biliary.*

Gangrene. — *See also, Wounds, Gangrenous.*

Acid, Carbolic: locally in strong solution to act as caustic; as a dressing to promote healthy action.
Acid, Chromic: local escharotic.
Acid, Citric.
Acid, Nitric: next to bromine the most useful escharotic.
Acid, Pyroligneous.
Acid, Salicylic: locally.
Ammonium Chloride.
Balsam of Peru.
Bromal.
Bromine: escharotic in hospital gangrene.
Charcoal: as poultice.
Chlorine Water: to destroy fetor.
Cinchona.
Creosote.
Eucalyptol: along with camphor in gangrene of lungs, to prevent spread and lessen the fetor.
Lime Juice and Chlorine Water: in hospital gangrene.
Myrtol: to destroy fetor and promote healthy action.
Oakum: dressing.
Opium.
Oxygen: as a bath.
Potassa: as caustic.
Potassium Chlorate.
Potassium Permanganate.
Quinine.
Resorcin: antiseptic, antipyretic.

Sanguinaria.
Sodium Sulphate.
Tannoform.
Terebene.
Turpentine Oil: internally, and by inhalation.
Zinc Chloride.

Gastralgia. — *See also, Acidity, Dyspepsia, Gastrodynia, Neuralgia.*

Acetanilid.
Acid, Carbolic.
Acid, Hydrocyanic: if purely nervous.
Acid, Salicylic: used in paroxysmal form; like quinine.
Acupuncture: sometimes gives great relief.
Alkalies.
Alum: if pyrosis.
Arsenic: in small doses.
Arsenic with Iron.
Atropine: in gastric ulcer.
Belladonna.
Bismuth: in irritable gastralgia.
Bismuth and Pepsin.
Bismuth Subnitrate.
Bromides.
Cannabis Indica.
Cerium Oxalate.
Charcoal: in neuralgia.
Chloral Hydrate: to relieve pain.
Chloroform: two or three drops on sugar.
Codeine.
Cod-Liver Oil.
Counter-irritation and a vigorous revulsive, especially useful in hysteria.
Creosote.
Diet and Hygiene.
Emesis and Purgation: when due to indigestible food.
Enemata.
Ergot.
Ether: a few drops.
Ferropyrine.
Galvanism: of pneumogastric and sympathetic.
Hot Applications.
Magnesium Oxide.
Manganese Dioxide.
Massage.
Menthol.
Methylene.
Milk Diet.
Morphine: subcutaneously, in epigastrium, very useful; or with bismuth and milk before each meal.
Nitroglycerin: quickly eases.

Nux Vomica: to re-move morbid condition on which it depends.
Opium.
Pancreatin.
Papain.
Pepsin.
Potassium Cyanide.
Potassium Nitrite.
Pulsatilla.
Quinine: if periodic in character.
Resorcin.
Silver Chloride.
Silver Iodide.
Silver Nitrate: nervine tonic.
Silver Oxide.
Sodium Salicylate.
Strontium Bromide.
Strychnine.
Suppository of Gluten, Glycerin and Soap: to overcome constipation.
Triphenin.
Valerian.
Zinc Oxide.

Gastric Dilatation.

Acid, Carbolic.
Bismuth Salicylate.
Bismuth Subnitrate.
Calcium Lactophosphate.
Charcoal.
Cod-Liver Oil or Gaduol, if due to rachitis.
Diet.
Enemas Nutrient.
Faridization of Gastric walls.
Gentian and Columba.
Ichthyol.
Iron Iodide.
Lavage.
Naphtol.
Nux Vomica.
Physostigma.
Sodium Phosphate.
Strontium Bromide.
Strychnine.

Gastric Pain. — See Gastralgia.

Gastric Ulcer. — See also, Hematemesis.

Acid, Carbolic.
Acid, Gallic.
Arsenic: in chronic ulcer it eases pain and vomiting, and improves the appetite.
Atropine: arrests pain and vomiting.
Bismuth Oxyiodide.
Bismuth Subgallate.
Bismuth Subnitrate: in very large doses.
Cannabis Indica.
Carlsbad Salts: before meals.
Castor Oil.

Charcoal: in chronic ulcer to allay pain.
Chloroform.
Cocaine.
Codeine.
Cold Compresses.
Counter-irritation.
Creosote.
Diet and Hygiene.
Hydrogen Peroxide.
Ice-bag: to epigastrium.
Iron.
Lead Acetate: to check hematemesis.
Lime Water with Milk: and diet.
Magnesium Sulphate.
Massage and Electricity
Mercuric Chloride: small dose before meals.
Mercury Iodide, Red.
Mercury Oxide, Red.
Methylene Blue.
Milk.
Monsel's Solution.
Morphine: like atropine.
Nutritive Enemata
Opium.
Pepsin.
Peptonized Milk.
Potassium Iodide: with bicarbonate, to lessen flatulent dyspepsia.
Potassium Sulphite.
Resorcin.
Silver Nitrate: to relieve pain and vomiting.
Silver Oxide.
Sodium Phosphate.
Sodium Tellurate.
Spice Plaster.
Stimulants: guardedly.
Tannin.
Turpentine Oil: frequently repeated, to check hemorrhage.
Zinc Carbonate.
Zinc Oxide.
Zinc Sulphocarbolate.

Gastritis.

Acid, Hydrocyanic: to allay pain.
Acid, Tannic.
Alum: when vomiting of glairy mucus.
Ammonium Chloride: in gastric catarrh.
Arsenic: in drunkards.
Atropine: in chronic cases.
Bismuth: in catarrh.
Caffeine: especially when associated with migraine.
Calumba.
Cinchona.
Eucalyptus: in chronic catarrh.
Hydrastis.

Ice: to suck; and to epigastrium.
Ipecacuanha: in catarrh.
Lead Acetate: along with opium.
Nutrient Enemata.
Nux Vomica.
Opium.
Silver Nitrate: in chronic gastritis.
Silver Oxide.
Veratrum Viride should never be used.

Gastritis, Acute.

Belladonna.
Bismuth Subnitrate.
Calomel.
Demulcents.
Mercury.
Morphine.
Oils.
Opium.
Sodium Paracresotate.
Warm Water, internally, or Stomach Pump: to unload stomach at onset.

Gastritis, Chronic.
—See also, Dyspepsia, Gastralgia.

Alkalies.
Bismuth Salicylate.
Bismuth Subnitrate.
Bismuth and Ammonium Citrate.
Caffeine.
Calcium Salicylate.
Cinchona.
Ichthalbin: internally, as regulator and tonic.
Mercury.
Morphine.
Orexine Tannate.
Papain.
Pepsin.
Podophyllum.
Pulsatilla.
Resorcin.
Silver Nitrate: by irrigation.
Sodium Paracresotate.
Strontium Bromide.
Thymol.
Zinc Oxide.
Zinc Sulphate.

Gastrodynia. — See Gastralgia.

Gastrorrhea. — See Pyrosis.

Gingivitis.

Alum.
Aseptol
Borax.
Myrrh.
Potassium Chlorate.
Pyoktanin.
Sozoiodole-Potassium.

Glanders and Farcy.

Acid, Carbolic.
Ammonium Carbonate.
Arsenic.
Creosote.
Escharotics.
Iodine.
Iron.
Potassium Bichromate.
Potassium Iodide.
Quinine.
Strychnine.
Sulphur Iodide.
Sulphites.

Glandular Enlargement. — See also, Bubo, Wen, Goiter, Tabes Mesenterica, Parotitis, Tonsillitis, etc.

Acid, Carbolic: injections of a two per cent. solution.
Ammoniacum Plaster: as counter-irritant on scrofulous glands.
Ammonium Chloride.
Antimony Sulphide.
Arsenic.
Barium Chloride.
Belladonna.
Blisters: to scrofulous glands.
Cadmium Chloride.
Calcium Chloride: in enlarged and breaking-down scrofulous glands.
Calcium Sulphide: for glands behind jaw with deep-seated suppuration.
Cod-Liver Oil.
Conium: in chronic enlargements.
Creosote.
Gaduol.
Gold Chloride: in scrofula.
Guaiacum.
Hydrastis.
Ichthalbin: internally.
Ichthyol: topically.
Iodides.
Iodine: internally; and painted around, not over the gland.
Iodoform: as a dressing to breaking-down glands.
Iodoformogen: equable and persistent in action on open glands.
Iodole: internally.
Lead Iodide: ointment.
Mercury: internally; locally the oleate of mercury and morphine.
Pilocarpine: in acute affections of parotid and submaxillary.

Potassium Iodide: ointment over enlarged thyroid and chronically inflamed glands.
Sozoiodole-Mercury.
Sulphides.
Thiosinamine.
Valerian.

Glaucoma.

Atropine has caused this disease.
Duboisine like atropine.
Eserine: lowers intraocular tension.
Iridectomy: the only cure.
Quinine.

Glottis, Spasm of.— See Laryngismus Stridulus.

Gleet.—See also, Gonorrhea.

Acid, Tannic.
Acid, Trichloracetic.
Airol.
Aloes.
Argentamine.
Argonin.
Aristol.
Betol.
Bismuth Oxyiodide or Subnitrate: suspended in glycerin or mucilage.
Blisters: to perineum useful in obstinate gleet.
Cantharides: minim doses of tincture frequently repeated.
Copaiba: internally, and locally smeared on a bougie and introduced; best used in chronic form.
Copper Sulphate: as injection.
Creosote.
Eucalyptol: in very chronic gleet.
Gallobromol.
Hydrastine Hydrochlorate.
Iodoform.
Iodoformogen.
Iodole.
Iron: either perchloride or sulphate as injection, along with opium.
Juniper Oil: like copaiba.
Kino.
Lead Acetate: injection is sometimes used.
Lime Water.
Mercury: half a grn. of bichloride in six ounces of water.

Naphtol.
Oil Juniper.
Oil Turpentine.
Peru, Balsam of
Piper Methysticum.
Potassium Permanganate.
Protargol.
Salol.
Sandalwood Oil: useful both locally and generally.
Silver Citrate.
Sozoiodole-Sodium.
Tannin, Glycerite of: as injection.
Terebene.
Thalline Sulphate.
Tolu, Balsam of.
Turpentine Oil: in a condition of relaxation.
Uva Ursi.
Zinc Acetate.
Zinc Sulphate: as injection.

Glossitis.

Alum.
Bismuth: locally.
Electrolysis: in simple hypertrophy, and cystic.
Iron.
Leeches.
Purgatives.
Quinine.

Glottis, Œdema of. —See also, Croup, Laryngitis.

Acid, Tannic.
Alum.
Ammonium Carbonate: as emetic.
Conium.
Emetics.
Ethyl Iodide.
Inhalations.
Scarification.
Tracheotomy.

Glycosuria.—See Diabetes.

Goiter.—See also, Exophthalmos.

Ammonium Chloride.
Ammonium Fluoride.
Cadmium Oleate.
Electricity.
Ferric Chloride.
Iodides.
Iodine: internally, and locally as ointment or tincture, and as injection.
Iodoform.
Iodoformogen.
Iodothyrine.
Mercuric Biniodide: as ointment, to be used

in front of hot fire, or in hot sun.
Potassium Bromide.
Potassium Iodide.
Strophanthus.
Strychnine.
Thyraden.

Gonorrhea.—*See also, Chordee, Gleet. Orchitis; Rheumatism; Gonorrheal; Urethritis, Urethral Stricture, Vaginitis.*

Acid, Benzoic: internally.
Acid, Boric.
Acid, Camphoric.
Acid, Chromic.
Acid, Cubebic.
Acid, Gallic.
Acid, Tannic.
Acid, Trichloracetic.
Aconite: in acute stage.
Airol.
Alcohol not to be touched.
Alkalines: salts, or waters, as citrates or bicarbonates, to make urine alkaline.
Alum : as an injection.
Aluminium Tannate.
Alumnol.
Antimony: if acute stage is severe.
Aristol.
Argentamine.
Argonin.
Belladonna.
Bismuth Oxyiodide.
Bismuth Subgallate.
Bismuth Subnitrate.
Buchu: more useful after acute stage.
Cadmium Sulphate: astringent injection.
Cannabis Indica: to relieve pain and lessen discharge.
Cantharides in small doses where there is pain along urethra and constant desire to micturate. The tincture in minim doses three times daily in chordee.
Chloral Hydrate.
Cinnamon Oil.
Cocaine: injection to relieve the pain.
Colchicum: in acute stage.
Collinsonia.
Copaiba: after acute stage.
Copper Acetate.
Copper Sulphate.
Creolin.
Cubebs : either alone or mixed with copaiba.
Diet and Hygiene.

Ergotin.
Erigeron, Oil of.
Eucalyptus, Oil of.
Europhen.
Ferropyrine.
Formaldehyde.
Gallobromol.
Gelsemium.
Glycerite of Tannin: injection in later stage.
Hamamelis.
Hot Sitz-bath.
Hydrastine Hydrochlorate.
Hydrastis: an injection.
Hydrogen Peroxide.
Ichthyol.
Iodole.
Iron : astringent injection in later stage.
Kaolin.
Kava Kava.
Largin : very effective.
Lead Acetate.
Lead Nitrate.
Lead Subacetate, solution of.
Lead Water and Laudanum.
Mercury Benzoate.
Mercury Bichloride : weak solution, locally.
Mercury Salicylate.
Methylene Blue.
Methyl Salicylate.
Naphtol.
Opium.
Potassium Citrate.
Potassium Permanganate.
Protargol.
Pulsatilla.
Pyoktanin.
Pyridine.
Quinine : stimulant in later stage.
Quinoline Tartrate.
Resorcin.
Salol.
Sandalwood Oil: internally and locally.
Silver Nitrate: as injection, said to cut short at commencement.
Silver Oxide.
Sodium Bicarbonate.
Sodium Dithio-Salicylate.
Sodium Salicylate.
Sozoiodole-Sodium.
Sozoiodole-Zinc.
Terpin Hydrate.
Thalline Sulphate.
Turpentine Oil.
Urinating: with penis in hot water, to relieve ardor urinæ.
Veratrum Viride: in early stage of acute fever.
Warm Baths: lasting ½ to 2 hours, in early stage.
Zinc Permanganate.
Zinc salts in general.

Gout.—*See also, Arthritis, Lithemia.*

Acid, Arsenous.
Acid, Carbonic.
Acid, Di-iodo-salicylate
Acid, Salicylic.
Aconite.
Alkalies.
Alkaline Mineral Waters.
Alkaline Poultice.
Ammonia Water.
Ammonium Benzoate.
Ammonium Phosphate.
Ammonium Tartrate.
Antipyrine.
Argentic Nitrate.
Arnica.
Arsenic.
Asaprol.
Asparagin.
Belladonna.
Blisters.
Calcium Sulphate.
Chicory.
Chloral Hydrate.
Chloroform.
Cod-Liver Oil.
Colchicine.
Colchicum.
Cold Water.
Collodion.
Colocynth with Hyoscyamus: to unload bowels.
Diet.
Diuretics and Alkaline drinks.
Ether : hypodermically.
Formin.
Fraxinus.
Gaduol.
Gentian.
Glycerinophosphates.
Guaco.
Horse Chestnut Oil.
Hydrogen Sulphide.
Ichthalbin : internally, as resolvent and alterative.
Ichthyol: topically.
Iodide of Potassium.
Iodine.
Iodoform.
Iron Iodide.
Levico Water.
Lithium Salts.
Lycetol.
Lysidine.
Magnesia.
Manganese.
Morphine.
Oil of Peppermint.
Piperazine.
Piper Methysticum.
Potassæ Liquor.
Potassium Acetate.
Potassium Bromide.
Potassium Permanganate.
Potassium Silicate.
Prunus Virginiana.
Quinine.
Rubefacients.

Salicylates: large doses.
Saliformin.
Sodium Arsenate.
Sodium Benzoate.
Sodium Bicarbonate.
Sodium Carbonate.
Sodium Chloride.
Sodium Salicylate.
Stimulants.
Strawberries.
Strontium Bromide.
Strontium Lactate.
Strontium Salicylate.
Strychnine.
Sulphides: in chronic cases.
Sulphur.
Sulphur Baths.
Sulphurated Potassa.
Tetraethyl-ammonium Hydroxide.
Trimethylamine.
Turkish Baths.
Veratrine: as ointment.
Vichy Water.
Water: distilled.

Granulations, Exuberant.

Acid, Chromic.
Alum, Dried.
Cadmium Oleate.
Copper Sulphate.
Potassium Chlorate.
Silver Nitrate.
Zinc Chloride.

Griping.—See *Colic.*

Growths, Morbid.—
See *Tumors.*

Gums, Diseases of.
—See also, *Mouth,
Sore; Scurvy, Teeth.*

Acid, Carbolic.
Acid, Salicylic.
Alum.
Areca.
Catechu: as a mouth wash.
Cocaine: locally.
Ferric Chloride.
Ferropyrine.
Formaldehyde.
Hamamelis.
Iodine Tincture: locally.
Krameria.
Myrrh.
Pomegranate Bark.
Potassium Chlorate.
Potassium Iodide.
Salol.
Tannin.

Hay Fever.—See also,
*Asthma, Catarrh,
Conjunctivitis, Influenza.*

Acid, Boric.
Acid, Carbolic.
Acid, Salicylic.
Acid, Sulphurous.

Aconite.
Ammonia.
Argentic Nitrate.
Arsenic: as cigarette
Atropine.
Brandy Vapor.
Bromine.
Camphor.
Cantharides: tincture.
Chlorate of Potassium.
Cocaine.
Coffee, strong.
Formaldehyde.
Grindelia.
Hamamelis.
Ichthyol: as spray.
Iodides.
Ipecacuanha.
Lobelia.
Morphine.
Muscarine.
Menthol.
Opium.
Pilocarpine.
Potassium Chlorate.
Potassium Iodide: internally and locally.
Quinine: locally as injection or douche.
Resorcin.
Sozoiodole salts.
Stearates.
Strychnine.
Terpin Hydrate.
Tobacco.
Turkish Baths.
Veratrum Viride.

Headache.—See also,
Hemicrania.

Acetanilid.
Acid, Acetic.
Acid, Hydrobromic.
Acid, Nitrohydrochloric: for pain just
above eyeballs without constipation, also
for pain at back of neck.
Acid, Phosphoric.
Acid, Salicylic.
Aconite: when circulation excited.
Actæa Racemosa.
Aloin.
Ammonia: aromatic
spirits, ½ to 2 drams.
Ammonium Carbonate.
Ammonium Chloride:
10 to 15 grn. doses in
hemicrania.
Ammonium Valerianate.
Antacids.
Antipyrine.
Arsenic: in brow ague.
Atropine: locally to
eye in migraine.
Belladonna: frequently
given in frontal headache, especially at
menstrual period, or
if from fatigue.
Berberine.
Bismuth Valerianate.

Bleeding.
Bromides: in large doses.
Bryonia: in bilious headache.
Butyl-chloral Hydrate.
Caffeine, with antipyrine or sodium bromide.
Cajeput Oil: locally.
Camphor: internally,
and saturated solution externally.
Camphor with acetanilid or antipyrine,
in nervous headache.
Cannabis Indica: in
neuralgic headache.
Capsicum: plaster to
nape of neck.
Carbon Disulphide.
Carbon Tetrachloride.
Chamomile.
Chloralamide.
Chloroform, Spirit of:
in nervous headache.
Cimicifuga: in nervous
and rheumatic headache, especially at
menstrual period.
Coffee and Morphine.
Colchicum.
Cold Affusion.
Croton Oil.
Cup, to nape of neck,
in congestion.
Digitalin: (German)
1-16 grn. twice a day
for congestive hemicrania.
Electricity.
Ergot.
Ergotin.
Ethylene Bromide.
Erythrol Tetranitrate.
Ether Spray: locally,
for frontal headache
after illness or fatigue.
Eucalyptol.
Ferropyrine.
Friedrichshall Water.
Galvanism.
Gelsemium.
Guarana.
Heat: as hot water-bag
or poultice to nape of neck.
Hot Sponging.
Hot Water.
Hydrastis: in congestive headache with
constipation.
Hyoscyamus.
Ice-bag: applied to
head, or leeches back
of ears, in severe cases.
Ichthalbin: to improve
digestion and nutrition.
Ignatia: in hysterical
headache.
Iodide of Potassium:
in rheumatic head-

ache with tenderness of scalp.
Iris: in supra-orbital headache with nausea.
Kola.
Lithium Bromide.
Magnesium Carbonate.
Magnesium Citrate.
Magnesium Oxide.
Magnesium Sulphate: for frontal headache with constipation.
Menthol: as local application.
Mercury: in bilious headache.
Methylene Blue.
Morphine.
Mustard: as foot-bath, or poultice to nape of neck.
Neurodin.
Nitrite of Amyl: as inhalation when face pale.
Nitroglycerin.
Nux Vomica: frequently repeated in nervous or bilious headache.
Oxygen Water.
Paraldehyde.
Phenacetin.
Phosphorus.
Picrotoxine: in periodical headache.
Podophyllum: when constipation.
Potassium Cyanide: as local application.
Pulsatilla.
Quinine.
Salicylate of Sodium: three grn. dose every half hour exceedingly useful.
Sanguinaria: in gastric derangement.
Sitz-bath.
Skull-cap: as prophylactic.
Sodium Bicarbonate: with bitters before meals in frontal headache at the junction of hairy scalp and forehead, or pain in upper part of forehead without constipation. As wash to the mouth when headache depends on decayed teeth.
Sodium Bromide.
Sodium Chloride.
Sodium Phosphate: as laxative in bilious headache.
Spectacles: where the headache depends on inequality of focal length or astigmatism.
Strontium Bromide.
Strychnine.

Tea: strong black or green, often relieves nervous headache.
Thermodin.
Triphenin.
Valerian: in nervous and hysterical cases.
Veratrum Viride.
Zinc Oxide.

Headache, Bilious.
—*See Biliousness.*

Heart Affections.—
See also, Angina Pectoris, Dropsy, Endocarditis, Pericarditis, Syncope.

Aconite.
Adonidin.
Adonis Æstivalis.
Ammonia and Ether, followed by Digitalis and Alcohol: in heart failure.
Ammonium Carbonate: in heart failure.
Amyl Nitrite.
Arsenic.
Barium Chloride: in heart failure.
Butyl-Chloral Hydrate.
Cactus Grandiflorus.
Caffeine.
Camphor.
Chloral Hydrate: in neurotic palpitation and pseudo-angina pectoris.
Cimicifuga.
Convallaria.
Convallamarin.
Diet and Exercise.
Digestives.
Digitalis.
Digitoxin.
Erythrol Tetranitrate.
Hoffmann's Anodyne.
Hydragogue Cathartics
Hyoscyamus.
Iron.
Iron with Arsenic and simple Bitters.
Kola.
Morphine.
Nicotine: for functional disturbance.
Nitroglycerin.
Nux Vomica.
Oleander.
Opium.
Potassium Iodide.
Sparteine Sulphate.
Strontium Bromide.
Strontium Iodide.
Strophanthus.
Strychnine.
Suprarenal Gland.
Theobromine and Sodium Salicylate.
Uropherin.
Venesection.
Veratrine Ointment
Veratrum Viride.

Heartburn.—*See Pyrosis.*

Heart, Dilated.

Amyl Nitrite.
Cocaine.
Digitalis.
Ergot.
Iron.
Mercury.
Morphine.
Nitroglycerin.
Purgatives.
Sodium Nitrite.
Sparteine.

Heart, Fatty.

Arsenic.
Belladonna.
Cimicifuga.
Cod-Liver Oil.
Digitoxin.
Ergot.
Iron.
Nitrite of Amyl.
Strychnine.

Heart, Hypertrophied.

Aconite: to be used with care when valvular disease is present.
Bromides.
Camphor: in palpitation and dyspnea.
Cimicifuga.
Digitalis: in small doses.
Ergot.
Galvanism.
Iron.
Lead Acetate: in palpitation.
Nitrite of Amyl.
Potassium Iodide.
Veratrum Viride.

Heart, Palpitation of.

Acid, Hydrocyanic.
Aconite: internally.
Amy Nitrite.
Belladonna: internally useful in cardiac strain.
Bromides: in fluttering heart.
Camphor.
Cimicifuga.
Cocaine.
Digitalis.
Eucalyptus.
Hot Bath.
Hyoscyamus: in nervous palpitation.
Lead.
Milk Cure: in gouty persons.
Nux Vomica.
Posture: head hung forward, body bent, arms by the sides, and breath held for a few seconds.
Potassium Bromide.

132

Potassium Iodide.
Senega.
Spirit Ether.
Valerian: in nervous cases with dyspnea.
Veratrine: as ointment to chest.

Heart, Valvular Disease of.— *See also, Endocarditis.*

Aconite: to quiet action; to be used with caution.
Adonidin.
Arsenic.
Barium Chloride.
Cactus Grandiflorus.
Caffeine.
Cimicifuga.
Comp. Sp. of Ether.
Digitalis: in mitral disease; to be avoided in purely aortic disease, but useful when this is complicated with mitral.
Iron.
Jalap Resin.
Morphine: to relieve pain and dyspnea.
Nitrites: to lessen vascular tension.
Nitroglycerin.
Nux Vomica.
Purgatives: to lessen tension and remove fluid.
Salicin.
Sodium Salicylate.
Strophanthus.
Strychnine: as cardiac tonic.
Veratrum Viride.

Hectic Fever.—*See Perspiration, Night-Sweats, Phthisis.*

Hematemesis.

Acid, Acetic.
Acid, Gallic.
Acid, Sulphuric.
Alum.
Ammonium Chloride.
Ergot: hypodermically.
Hamamelis.
Ice: exceedingly useful.
Ipecacuanha.
Iron Perchloride, or Subsulphate.
Krameria.
Lead Acetate.
Logwood.
Magnesium Sulphate.
Silver Nitrate.
Tannin.
Turpentine Oil.

Hematocele, Pelvic.

Acid, Carbolic.
Bromides.
Hemostatics.
Iodides.

Iron.
Mercury Bichloride.
Opium.
Potassium Iodide.
Tonics.

Hematuria.

Acid, Acetic.
Acid, Gallic.
Acid, Tannic.
Alum: internally, or as injection into the bladder.
Ammonia.
Ammonium Benzoate.
Bursa Pastoris.
Camphor.
Cannabis Indica.
Chimaphila.
Copaiba.
Creosote.
Digitalis.
Ergot.
Erigeron.
Hamamelis.
Ipecacuanha.
Iron Perchloride.
Krameria: extract in large dose.
Lead Acetate.
Matico.
Myrtol.
Potassium Bitartrate.
Quinine.
Rhus Aromatica.
Sodium Hyposulphite.
Turpentine Oil.

Hemeralopia and Nyctalopia.

Acetanilid.
Amyl Nitrite.
Blisters: small, to external canthus of the eye.
Calcium Chloride.
Calcium Phosphate.
Electricity.
Mercury: locally.
Quinine: in large doses internally.
Strychnine.

Hemicrania. — *See also, Migraine.*

Acetanilid.
Aconite.
Ammonium Chloride.
Amyl Nitrite.
Antipyrine.
Arsenic.
Belladonna.
Bromides.
Caffeine.
Camphor.
Cannabis Indica.
Cimicifuga.
Digitalis.
Euphorin.
Exalgin.
Menthol.
Mercury.
Neurodin.
Nux Vomica.

Podophyllum.
Potassium Bromide.
Potassium Nitrite.
Quinine Valerianate.
Sanguinaria.
Sodium Chloride.
Thermodin.
Triphenin.
Valerian.

Hemiopia.

Glycerinophosphates.
Iodides.
Iodipin.
Iron.
Phosphates.
Potassium Bromide.
Quinine.
Strychnine.

Hemiplegia.—*See also, Paralysis, Facial.*

Glycerinophosphates.
Physostigma.
Picrotoxin.
Potassium Iodide.
Spermine.
Strychnine.

Hemoptysis.—*See also, Hematemesis.*

Acetanilid.
Acid, Acetic.
Acid, Gallic: very useful.
Acid, Phosphoric.
Acid, Pyrogallic.
Acid, Sulphuric.
Acid, Tannic.
Aconite.
Alum.
Ammonium Chloride.
Apocodeine.
Arnica.
Astringent Inhalations.
Atropine.
Barium Chloride.
Bromides.
Bursa Pastoris.
Cactus Grandiflorus.
Calcium Chloride.
Chloral Hydrate.
Chlorodyne.
Chloroform: to outside of chest.
Copaiba.
Copper Sulphate.
Digitalis.
Dry Cups: to chest.
Ergot or Ergotinin.
Ferric Acetate: very weak solution, constantly sipped.
Ferri Persulphas.
Hamamelis: very useful.
Hot Water Bag: to spine.
Hydrastinine Hydrochlorate.
Ice.
Ipecacuanha.
Iron: and absolute rest.

133

Larix: tincture.
Lead Acetate: very useful.
Matico.
Morphine.
Oil Turpentine.
Opium.
Potassium Bromide.
Potassium Chlorate.
Potassium Nitrate: when fever is present, along with digitalis or antimony.
Silver Oxide.
Sodium Chloride: in dram doses.
Subsulphate of Iron.
Tannin.
Veratrum Viride.

Hemorrhage and Hemorrhagic Diathesis.—*See also, Dysentery, Ecchymosis, Epistaxis, Hematemesis, Hemoptysis; Hemorrhage Post-Partum, Intestinal; Menorrhagia, Metrorrhagia, Purpura, Wounds, etc.*

Acid, Chromic.
Acid, Citric.
Acid, Gallic.
Acid, Tannic.
Aconite.
Alum.
Antipyrine.
Belladonna.
Copper Sulphate.
Creolin.
Creosote.
Digitalis.
Gaduol.
Geranium.
Hamamelis.
Iron.
Iron Subsulphate.
Iron Sulphate.
Hydrastinine Hydrochlorate.
Hydrastis Tincture.
Iodoform or Iodoformogen.
Lead Acetate.
Manganese Sulphate.
Nux Vomica.
Stypticin.
Turpentine Oil.

Hemorrhage, Intestinal.—*See also, Hemorrhoids, Dysentery, Typhoid.*

Acid, Gallic.
Acid, Sulphuric.
Acid, Tannic.
Belladonna: for rectal ulcers.
Camphor.
Castor Oil.
Ergotin.

Enemas, Styptic.
Ferric Chloride.
Hamamelis: very useful.
Ice.
Iodine.
Iron.
Lead Acetate.
Opium.
Potassium Bitartrate.
Turpentine Oil.

Hemorrhage, Post-partum.

Acid, Acetic.
Acid, Gallic.
Achillea.
Amyl Nitrite.
Atropine.
Capsicum.
Cimicifuga.
Compression of Aorta.
Digitalis.
Enemata, Hot.
Ergot: most efficient.
Ether Spray.
Hamamelis: for persistent oozing.
Hot Water: injection into uterus.
Hydrastinine Hydrochlorate.
Ice: to abdomen, uterus, or rectum.
Iodine.
Ipecacuanha: as emetic dose; good.
Iron Perchloride Solution: 1 in 4, injected into the uterus.
Mechanical Excitation of Vomiting.
Nux Vomica: along with ergot.
Opium: one-dram dose of tincture, with brandy, in profuse bleeding.
Pressure over uterus.
Quinine.

Hemorrhage, Uterine and Vesical.

Cornutine.
Creosote.
Hydrastis.
Hydrastinine Hydrochlorate.
Stypticin.

Hemorrhoids.

Acid, Carbolic: injection into piles.
Acid, Chromic.
Acid, Gallic.
Acid, Nitric: as caustic; dilute as lotion.
Acid, Salicylic.
Acid, Tannic.
Alkaline Mineral Waters: useful.
Aloes: as purgative.
Alum: in bleeding piles; powder, crystal or ointment.

Argentic Nitrate.
Belladonna.
Bismuth.
Bromide of Potassium
Calomel.
Castor Oil.
Chalybeate Waters.
Chlorate of Potassium.
Cocaine.
Chrysarobin.
Cold Water Injection: in the morning.
Cubebs.
Ergot.
Ferri Perchloridum.
Ferri Protosulphas: as lotion.
Ferropyrine.
Galls Ointment with Opium: very useful.
Grapes.
Glycerin.
Hamamelis: internally; and locally as lotion, injection, enema, or suppository.
Hydrastine.
Hydrastis: as lotion and internally.
Hyoscyamus: bruised leaves or ointment locally.
Ice.
Ichthyol: topically.
Ichthalbin: internally.
Iodoform or Iodoformogen: as ointment or suppository.
Iodole.
Iron.
Leeches.
Lead.
Liquor Potassæ.
Magnesia.
Malt Extract, Dry: as nutrient.
Nux Vomica: very useful.
Ol. Lini.
Ol. Terebinthinæ.
Opium.
Pitch Ointment.
Podophyllum.
Potassium Bitartrate.
Potassium Chlorate, with Laudanum: as injection.
Potassium and Sodium Tartrate.
Poultices: to effect reduction.
Rheum.
Saline Purgatives.
Senna: as confection; or better, compound liquorice powder.
Sodium Chlorate.
Sozoiodole-Potassium.
Sozoiodole-Sodium.
Stillingia: in constipation and hepatic disease.
Stramonium.
Sulphides.

Sulphur: as confection, to produce soft passages.
Sulphurous Waters.
Tannoform.
Tobacco.
Turpentine Oil.

Hepatalgia.

Ammonium Chloride.
Nux Vomica.
Quinine.

Hepatic Cirrhosis.—
See also, Ascites.

Acid, Nitrohydrochloric.
Ammonium Chloride.
Arsenic.
Diuretin.
Gold and Sodium Chloride.
Iodides.
Iodoform.
Iodole.
Mercurials.
Sodium Phosphate.

Hepatic Diseases.—
See also, Biliousness, Calculi, Jaundice, Cancer, Hepatalgia, Hepatic Congestion, Hepatic Cirrhosis, Hepatitis, Jaundice.

Acids, Mineral.
Ammonium Chloride: for congestion, torpor and enlargement.
Calomel.
Cholagogues.
Euonymin.
Glycerinophosphates: for hypersecretion.
Iodine or Iodides.
Iron.
Levico Water.
Mercurials: as cholagogues.
Nux Vomica.
Ox-gall.
Phosphorus.
Podophyllum.
Potassium salts.
Quinine: for congestion.
Sanguinaria.
Sodium Phosphate.
Sulphur.
Taraxacum.
Turpentine Oil.

Hepatitis.

Acid, Nitro-Hydrochloric.
Aconite.
Alkaline Mineral Waters.
Ammonium Chloride.
Bryonia.
Chelidonium.
Colchicum.

Iodine: as enema.
Leeches.
Mercury.
Nitre and Antimony.
Rhubarb.
Sulphurous Waters.
Tartar Emetic.

Hepatitis and Hepatic Abscess.—See also, Jaundice.

Acid, Nitric.
Acid, Nitrohydrochloric.
Aconite: in early stages.
Active treatment for dysentery if present.
Alkalies and Colchicine.
Ammonium Chloride.
Antimony with Nitre.
Aspiration when pus forms.
Blister or Mustard-plaster.
Calomel.
Colchicine.
Diet.
Hot clothes or counter-irritation.
Iodine.
Mercury.
Potassium Iodide.
Quinine and Iron: after abscess develops.
Saline Purgatives: preceded by calomel.
Sweet Spirit of Niter: with potassium citrate, or diuretics, to regulate kidneys.
Tartar Emetic.
Veratrum Viride.

Hernia.

Chloral Hydrate: as enema.
Chloroform.
Ether and Belladonna.
Ether Spray.
Forced Enemata.
Iodine.
Morphine.
Oil.
Opium.
Sternutatories.
Thyroid preparations.

Herpes.

Acetanilid.
Acid, Tannic.
Alum.
Ammoniated Mercury.
Anthrarobin.
Arsenic.
Bismuth Subgallate.
Bismuth Subnitrate.
Calomel.
Europhen.
Glycerin.
Hydroxylamine Hydrochlorate.
Ichthalbin: internally.
Ichthyol: locally.

Iodole.
Iron Arsenate.
Lenirobin.
Levico Water.
Magnesium Citrate.
Myrtol.
Naphtol.
Potassium Carbonate.
Rhus Toxicodendron.
Silver Nitrate.
Sozoiodole salts.
Zinc Sulphate.

Herpes Circinatus.
—See Tinea Circinata.

Herpes Tonsurans
(Pityriasis Rosea).
—See also, Seborrhea

Acid, Carbolic: 2 parts with 3 parts each glycerin and water, applied twice daily.
Alkalies: internally, often control mild cases.
Baths: followed by shampooing and brisk friction.
Borax: saturated solution, to cleanse scalp; or glycerite, as paint.
Chrysarobin.
Cod-Liver Oil or Linseed Oil: as lotion.
Gaduol: as tonic.
Ichthalbin: internally, as alterative tonic and regulator of digestive functions.
Lead-Subacetate Solution: with equal part glycerin and 2 parts water, as lotion when inflammation high.
Mercury: internally in obstinate cases; Donovan's solution highly successful.
Mercury-Ammonium Chloride: as 1 per cent. ointment.
Mercury Oleate, 5 per cent.: as paint.
Mercury Iodide: as 2 per cent. ointment.
Pyrogallol.
Sozoiodole-Mercury.
Sozoiodole-Potassium.
Sulphur: as 1 to 8 ointment every morning; with almond-oil inunction at night.
Sulphurated Potassa: ½ oz. to pint lime water, as lotion.
Thyraden: as stimulant of cutaneous circulation.

Herpes Zoster.

Acid, Carbolic.
Aconite and Opium: locally.

Alcohol : locally.
Atropine.
Belladonna.
Calomel.
Celandine.
Chloroform.
Collodion.
Copper Acetate.
Dulcamara.
Europhen.
Ferri Perchloridum.
Galvanism.
Ichthalbin : internally.
Ichthyol : locally.
Iodole.
Levico Water.
Menthol.
Mercury.
Methylene Blue.
Morphine.
Myrtol.
Phosphorus.
Rhus Toxicodendron.
Silver Nitrate : strong
   solution locally.
Spirits of Wine.
Tar.
Traumaticin.
Veratrine : as ointment.
Zinc Ointment.
Zinc Oxide.
Zinc Phosphide.

Hiccough.

Amber, Oil of.
Amyl Nitrite.
Antispasmin.
Apomorphine.
Belladonna.
Bismuth.
Camphor.
Cannabis Indica.
Capsicum.
Chloral.
Chloroform.
Cocaine.
Ether.
Iodoform.
Jaborandi.
Laurel Water.
Morphine : hypodermi-
   cally.
Musk.
Mustard and Hot Water.
Nitroglycerin.
Nux Vomica.
Pepper.
Potassium Bromide.
Pressure over phrenic
   nerve, hyoid bone,
   or epigastrium.
Quinine : in full doses.
Spirit Ether.
Sugar and Vinegar.
Sulfonal.
Tobacco-smoking.
Valerian.
Zinc Valerianate.

Hordeolum (Stye).—
See also, Eyelids.

Iodine Tincture.
Mercury Oleate with
   Morphine.

Pulsatilla : internally,
   and externally as
   wash, often aborts.
Silver Nitrate.

Hydrocele.—See also,
Dropsy, Orchitis.

Acid, Carbolic.
Ammonium Chloride.
Chloroform.
Iodine.
Silver Nitrate.

Hydrocephalus,
Acute.— See also,
Dropsy.

Blisters : to the nape of
   neck useful.
Bromide of Potassium.
Croton Oil : liniment.
Elaterium.
Ergot.
Iodide of Potassium.
Iodoform or Iodofor-
   mogen : dissolved in
   collodion, or as oint-
   ment to neck and
   head ; along with
   small doses of calo-
   mel as enemata.
Leeches.
Mercuric Chloride :
   small doses internally.
Tartar Emetic : oint-
   ment.
Turpentine Oil : by
   mouth or as enema at
   commencement.

Hydrocephalus,
Chronic.—See also,
Meningitis, Tuber-
cular ; Dropsy.

Blisters.
Cod-Liver Oil.
Iodide of Iron.
Iodide of Potassium.
Iodine.
Mercury.
Potassium Bromide.

Hydropericardium.
—See Dropsy.

Hydrophobia.

Acid, Acetic or Hy-
   drochloric.
Acid, Carbolic.
Actual Cautery.
Acupuncture.
Amyl Nitrite.
Arsenic.
Asparagus.
Atropine.
Belladonna.
Bromide of Potassium.
Calabar Bean.
Cannabis Indica.
Chloral Hydrate.
Chloride of Potassium.
Chloroform : to control
   spasms.
Coniine.

Curare.
Escharotics.
Ether.
Euphorbia.
Excision of Bitten Part.
Gelsemium.
Hoang-nan.
Hyoscine Hydrobrom-
   ate.
Hyoscyamine.
Iodine.
Jaborandi.
Mercury.
Morphine.
Nicotine.
Nitroglycerin.
Pilocarpine.
Potassium Chlorate.
Potassium Permangan-
   ate : as lotion to
   wound.
Potassium Iodide.
Quinine.
Sabadilla.
Silver Nitrate to wound,
   is of no use, even
   though applied im-
   mediately.
Stramonium.

Hydrothorax. — See
also, Dropsy.

Blisters.
Broom.
Digitalis : as diuretic.
Diuretin.
Dry Diet.
Elaterium.
Iodine : injections after
   tapping.
Iron Chloride : tincture.
Jaborandi.
Mercury.
Morphine.
Pilocarpine.
Resin of Copaiba.
Sanguinaria.
Veratrum Viride.

Hyperidrosis. — See
Perspiration.

Hypochondriasis. —
See also, Melan-
cholia.

Alcohol : as temporary
   stimulant.
Arsenic : in the aged.
Asafetida.
Bromo-hemol.
Bromide of Potassium.
Caffeine.
Cimicifuga : in puer-
   peral, and spermator-
   rhea.
Cocaine Hydrochlorate.
Codeine.
Colchicum.
Creosote.
Electricity.
Gold Chloride : when
   giddiness and cere-
   bral anemia.

Hyoscyamus: in syphilophobia.
Ignatia.
Musk.
Opium: in small doses.
Ox-Gall.
Peronin.
Spermine.
Sumbul.
Valerian.

Hysteria.

Acetanilid.
Acid, Camphoric.
Acid, Valerianic.
Aconite.
Actæa Racemosa.
Alcohol.
Aloes: in constipation.
Allyl Tribromide.
Ammonia, Aromatic Spirits of.
Ammoniated Copper.
Ammonium Carbonate.
Ammonium Valerianate.
Amyl Nitrite.
Amyl Valerianate.
Anesthetics.
Antipyrine.
Antispasmin.
Antispasmodics.
Apomorphine.
Arsenic.
Asafetida.
Atropine: in hysterical aphonia.
Belladonna.
Bromalin.
Bromide of Calcium.
Bromide of Potassium.
Bromide of Sodium.
Bromide of Strontium.
Bromo-hemol: as nervine and hematinic.
Camphor: in hysterical excitement.
Camphor, Monobromated.
Cannabine Tannate.
Cannabis Indica.
Cerium Oxalate.
Chloral Hydrate.
Chloralamide.
Chloroform.
Cimicifuga: in hysterical chorea.
Cimicifugin.
Cocaine Hydrochlorate.
Codeine.
Cod-Liver Oil.
Cold Water: poured over mouth to cut short attack.
Conium.
Creosote.
Electricity: to cut short attack.
Ether.
Ethyl Bromide.
Eucalyptus.
Faradism.
Gaduol.
Galbanum: internally, and as plaster to sacrum.

Galvanism.
Garlic: to smell during the paroxysm.
Glycerinophosphates.
Gold and Sodium Chloride.
Hyoscyamus.
Ignatia.
Ipecacuanha: as emetic.
Iron Bromide.
Iron Valerianate.
Levico Water.
Lupulin: when sleepless.
Massage.
Morphine Valerianate.
Musk.
Neurodin.
Nux Vomica.
Oil Amber.
Oil Wormseed.
Opium: in small doses.
Orexine: as appetizer and digestant.
Paraldehyde.
Pellitory: for "globus."
Phosphates.
Phosphorus: in hysterical paralysis.
Pulsatilla.
Santonin: if worms present.
Simulo.
Spirit Nitrous Ether: to relieve spasm.
Sumbul.
Sulfonal.
Tartar Emetic.
Trional.
Valerian.
Volatile Oils.
Zinc Iodide.
Zinc Oxide.
Zinc Sulphate.
Zinc Valerianate.

Hystero-Epilepsy.

Electricity.
Nitroglycerin.
Picrotoxin.
Spermine.

Ichthyosis.

Baths.
Cod-Liver Oil.
Copper Sulphate.
Elm Bark: decoction useful.
Glycerin.
Ichthyol.
Naphtol.
Sodium Bicarbonate.
Thyroid preparations.
Zinc Oxide.
Zinc Sulphate

Impetigo.—See also, Eczema.

Acetate of Lead.
Acid, Boric.
Acid, Chrysophanic: locally.
Acid, Hydrocyanic: to relieve itching.

Acids, Mineral: internally.
Acid, Nitric.
Adeps Lanæ.
Arsenic.
Calcium Chloride.
Cod-Liver Oil.
Gaduol: internally as alterative tonic.
Glycerite of Tannin
Grape Cure.
Gutta-Percha.
Ichthalbin: internally, as a regulator of digestive functions and as alterative.
Ichthyol: locally.
Iron Arsenate.
Laurel Water: to relieve itching.
Lead Nitrate.
Levico Water.
Mercuric Nitrate.
Mercury: locally.
Oil Cade.
Potassium Chloride.
Poultices.
Quinine.
Salol.
Solution Arsenic and Mercury Iodide.
Sozoiodole-Potassium.
Sozoiodole-Zinc.
Sulphate of Copper.
Sulphur: internally.
Tannin: locally.
Tannoform.
Tar.
Zinc Ointment.
Zinc Oxide.

Impetigo Syphilitica.

Iodipin.
Mercuro-iodo-hemol.
Sozoiodole-Mercury.

Impotence.—See also, Emissions, Spermatorrhea.

Acid, Phosphoric.
Arseniate of Iron.
Cannabis Indica.
Cantharides.
Cimicifuga.
Cold Douche: to perineum and testicles, in atonic types.
Cubebs.
Damiana.
Ergotin: hypodermically about dorsal vein of penis, when it empties too rapidly.
Glycerinophosphates.
Gold Chloride: to prevent decline of sexual power.
Muira Puama.
Nux Vomica: very useful.
Phosphorus.
Potassium Bromide.
Sanguinaria.

137

Serpentaria.
Spermine.
Strychnine.
Testaden.
Turpentine Oil.
Zinc Phosphate: very
useful.

Indolent Swellings.

Ichthalbin: internally.
Ichthyol: topically.
Potassium Iodide.

Induration.

Ichthalbin: internally.
Ichthyol: locally.
Iodipin.
Potassium Iodide.

Infantile Diarrhea.
—See Diarrhea.

Inflammation. — *See
also, Bronchitis,
Pleuritis, etc. Also
list of Antiphlogis-
tics.*

Acetanilid.
Acid, Salicylic: most
valuable.
Aconite: at the com-
mencement of all in-
flammations, superfi-
cial or deep-seated:
best given in small
doses frequently re-
peated until pulse and
temperature are re-
duced.
Alcohol: as antipyretic
and stimulant, espec-
ially useful in blood-
poisoning.
Alkalies.
Ammonium Chloride.
Ammonium Tartrate.
Antimony: 10 to 15 min.
of vinum antimonii
frequently repeated
at commencement.
Arnica.
Arsenic.
Astringents.
Atropine.
Barium Chloride.
Belladonna: in gouty
and rheumatic inflam-
mation and cystitis.
Blisters.
Borax
Bryonia: in serous in-
flammations, after
heart or pulse lowered
by aconite.
Cannabis Indica: in
chronic types.
Chloral Hydrate: when
temperature is high
and much delirium.
Cocaine Hydrochlorate:
in acute types.
Cod-Liver Oil: in chron-
ic inflammation.
Colchicine.
Cold.

Copaiba.
Digitalis.
Electricity.
Ergot.
Exalgin.
Flaxseed: for inflamed
mucous membranes.
Fomentations.
Gelsemium.
Hop Poultice.
Ice: locally applied.
Ichthalbin: internally.
Ichthyol: locally.
Iodine: locally.
Lead.
Leeches.
Magnesium Sulphate.
Mercury: in deep-seated
inflammations, espec-
ially those of serous
membranes,and iritis,
and syphilitic cases.
Mercury inunctions.
Neurodin.
Nitrates.
Opium: exceedingly
useful to check it at
commencement, and
relieve pain after-
wards.
Phosphorus.
Pilocarpine.
Poultices.
Pulsatilla: when puru-
lent discharge from
eyes, ears or nose:
and in epididymitis
Purgatives.
Pyoktanin.
Quinine: in peritonitis
and in acute inflam-
mations, along with
morphine.
Salicin.
Sodium Salicylate:
most useful,especially
in rheumatic affec-
tions.
Saline Cathartics.
Silver Nitrate.
Sozoiodole-Sodium.
Stramonium.
Sulphides: to abort or
to hasten maturation.
Tartar Emetic.
Triphenin.
Turpentine Oil: as
stupe.
Veratrum Viride.
Water: cold, as com-
presses.

Inflammation, In-
testinal.—*See En-
teritis.*

Influenza.

Acetanilid.
Acid, Agaric.
Acid, Boric.
Acid, Camphoric.
Acid, Carbolic: as spray
and gargle.
Acid, Sulphurous: by
fumigation or inhala-
tion.

Aconite, Sweet Spirit of
Nitre, and Citrate of
Potassium, in com-
bination: valuable in
early stage.
Actæa Racemosa.
Alcohol.
Ammonium Acetate,
with Nitrous or Chlor-
ic Ether.
Ammonium Salicylate.
Antispasmin.
Antipyrine.
Belladonna.
Benzene.
Bismuth Salicylate.
Bromides.
Camphor.
Camphor, Monobrom-
ated.
Cannabis Indica.
Chloralamide.
Chloral Hydrate.
Cimicifuga.
Cocaine Hydrochlorate.
Cold Baths as Antipyr-
etic.
Cubebs.
Digitalin.
Ergot, Cannabis Indica,
with Bromides: often
relieve vertigo.
Eucalyptus.
Glycerinophosphates.
Hot Sponging.
Ichthyol.
Menthol.
Naphtol.
Opium with Ipecacu-
anha: useful for
cough.
Phenacetin.
Potassium Bicarbonate.
Potassium Nitrate:
freely diluted, as lem-
onade.
Quinine: useful, especi-
ally in later stages.
Salipyrine.
Salol.
Salol with Phenacetin.
Sandalwood Oil.
Sanguinaria: some-
times very useful.
Sodium Benzoate.
Sodium Salicylate.
Spirit Nitrous Ether.
Steam, Medicated: in-
halations.
Strychnine.
Tartar Emetic.
Thermodin.
Thymol.
Triphenin.
Turkish Baths: useful.

Insanity and De-
mentia.—*See also,
Delirium, Hypo-
chondriasis, Mania,
Melancholia.*

Chloral Hydrate.
Codeine.
Colchicine.

138

Coniine.
Duboisine.
Hyoscine Hydrobro-
mate.
Hyoscyamine.
Opium.
Potassium Bromide.
Scopolamine Hydrobro-
mate.
Spermine.
Sulphonal.
Thyraden.
Zinc Phosphate.

Insomnia. — See also,
Nervousness; also,
list of Hypnotics.

Acetanilid.
Aconite: one min. of
tinct. every quarter
hour when skin is dry
and harsh.
Alcohol: sometimes
very useful.
Ammonium Valeriana-
ate.
Amylene Hydrate.
Atropine with Mor-
phine: 1-120 to 1-100
grn. atropine to ¼ or
½ grn. morphine.
Bath: cold in cerebral
anemia, hot in ner-
vous irritability.
Belladonna.
Bleeding.
Bromo-hemol.
Butyl-Chloral Hydrate:
if heart is weak.
Camphor, Monobroma-
mated.
Cannabis Indica: alone
or with hyoscyamus.
Cannabine Tannate.
Chloralamide.
Chloral-Ammonia.
Chloral Hydrate: very
useful, alone or with
bromide of potas-
sium; the addition of
a small quantity of
opium to the combi-
nation assists its ac-
tion.
Chioralimide.
Chloralose.
Chlorobrom.
Chloroform.
Cocaine Hydrochlorate.
Codeine.
Coffee: causes insom-
nia, but has been
recommended in
insomnia from de-
ficient nervous pow-
er, or chronic alcohol-
ism.
Cold Douche.
Digitalis: when defi-
cient tone of vaso-mo-
tor system.
Duboisine.
Ether: in full dose.
Ethylene Bromide.

Galvanization.
Gelsemium: in simple
wakefulness.
Glycerinophosphates.
Hot-water bags to feet
and cold to head if
due to cerebral hyper-
emia.
Humulus: a hop-pillow
sometimes useful in
the aged.
Hyoscine Hydrobro-
mate.
Hyoscyamus: alone or
with cannabis indica;
useful to combine
with quinine.
Hypnone.
Ignatia: in nervous ir-
ritability.
Methylene Blue.
Morphine.
Musk: in irritable and
nervous cases.
Narceine.
Narcotine.
Opium: most powerful
hypnotic: given
alone or in combina-
tion.
Paraldehyde.
Pellotine Hydrochlor-
ate.
Phosphorus: in the
aged.
Potassium Bromide: in
full doses, alone or
with other hypnotics.
Removal Inland.
Scopolamine Hydrobro-
mate.
Sitz Bath.
Sodium Bromide.
Sodium Lactate.
Spermine.
Strychnine.
Sulfonal.
Sumbul: in nervous ir-
ritability and chronic
alcoholism.
Tannate of Cannabin.
Tartar Emetic: along
with opium when
there is a tendency to
congestion of the
brain, which opium
alone would increase.
Tetronal.
Trional.
Urethane.
Valerian.
Warm Bath.
Warmth: internally and
externally.
Water.
Wet Compress.
Wet Pack.

**Intercostal Neural-
gia.** — See Neural-
gia.

Intermittent Fever.
—See also, Malaria;
also list of Anti-
periodics.

Acetanilid.
Acid, Carbolic.
Acid, Nitric: in obstin-
ate cases.
Acid, Salicylic.
Aconite.
Alcohol.
Alum.
Ammonium Carbazo-
tate: one-half to one
grn. in pill.
Ammonium Chloride.
Amyl Nitrite.
Antipyrine.
Apiol: in mild cases, 15
grns. during an hour,
in divided doses, four
hours before the par-
oxysm.
Arsenic: exceedingly
useful, especially in
irregular malaria.
Atropine: subcutan-
eously, to arrest or cut
short cold stage.
Berberine: in chronic
cases.
Bleeding.
Brucine.
Calomel.
Camphor: taken before
the fit to prevent it.
Capsicum: along with
quinine as adjuvant.
Chamomile.
Chloral Hydrate: as an-
tipyretic when fever
is high; and to check
vomiting or convul-
sions in adults and
children during mal-
arious fever.
Chloroform: to prevent
or cut short cold
stage.
Cimicifuga: in brow
ague.
Cinchonidine or Cin-
chonine: useful and
cheap.
Coffee.
Cold Compress.
Cool drinks and spong-
ing.
Cornus Florida: a sub-
stitute for quinine.
Digitalis.
Elaterium.
Emetics: if chill fol-
lows full meal.
Eucalyptus Globulus:
during convalescence.
Eupatorium.
Ferric Sulphate.
Ferrous Iodide.
Gelsemium: pushed un-
til it produces dilated
pupils or double
vision.

Grindelia Squarrosa: in hypertrophied spleen.
Guaiacol.
Hot Bath.
Hydrargyri Bichloridum.
Hydrastis: in obstinate cases.
Hydroquinone.
Hyoscyamine.
Ice Pack: if fever is long continued and excessive.
Iodine Tincture: to prevent recurrence of ague.
Ipecacuanha: most useful as emetic.
Iron.
Leptandra Virginica: after disease is lessened by quinine.
Mercury.
Methylene Blue.
Morphine: along with quinine as an adjuvant.
Mustard: to soles of feet.
Narcotine: two to five grn. three times a day sometimes very useful.
Nitrite of Amyl: by inhalation to relieve or shorten cold stage.
Nitrite of Sodium.
Nitroglycerin.
Nux Vomica.
Ol. Terebinthinæ.
Opium: in full doses to prevent chill.
Pepper: along with quinine.
Phenocoll Hydrochlorate.
Phosphates.
Phosphorus.
Pilocarpine Hydrochlorate.
Piperin.
Podophyllin.
Potassium Arsenite: solution.
Potassium Bromide.
Potassium Chloride.
Potassium Nitrate: ten grn. in brandy and water, or dry on tongue, to prevent fit.
Purgatives.
Quassia.
Quinetum.
Quinine: as prophylactic to abort fit and to prevent recurrence: its action is aided by purgatives, emetics and aromatics.
Quinine Hydrobromate: like quinine, and less liable to produce cinchonism.
Quinoidine.

Quinoline.
Quinoline Tartrate.
Resorcin.
Saccharated Lime.
Salicin.
Salipyrine.
Sodium Chloride: tablespoonful in glass of hot water at a draught on empty stomach.
Spider Web: as pill.
Stramonium.
Strychnine.
Zinc Sulphate.

Intertrigo.—See also, *Excoriations.*

Acetanilid: locally.
Acid, Boric.
Acid, Carbolic.
Aluminium Oleate.
Bismuth Subgallate.
Bismuth Subnitrate.
Calomel.
Camphor: added to dusting-powders to allay heat and itching.
Carbonate of Calcium.
Fullers' Earth.
Glycerite of Tannin.
Ichthyol.
Lead Lotion.
Lime Water.
Lycopodium.
Soap.
Tannin.
Tannoform.
Zinc Carbonate.
Zinc Ointment.
Zinc Oxide.

Intestinal Catarrh. —See Catarrh, Enteritis, etc.

Intestinal Inflammation.— See Enteritis.

Intestinal Irritation.— See Enteritis, etc.

Intestinal Obstruction.—See also, Constipation. Intussusception, Hernia.

Belladonna.
Caffeine.
Mercury.
Morphine.
Opium.
Strychnine.

Iritis. — See also, Syphilis.

Acid, Salicylic.
Acidum Hydrocyanicum.
Aconite.
Atropine.

Belladonna: internally and locally.
Bleeding.
Cantharides.
Copaiba.
Counter-irritation.
Daturine.
Dry Heat.
Duboisine: substitute for atropine.
Eserine.
Gold.
Grindelia.
Homatropine.
Hot fomentations.
Iced compresses in early stages of traumatic iritis.
Iodide of Potassium.
Iron.
Leeches.
Mercury: most serviceable.
Morphine.
Nicotine.
Opium: to lessen pain.
Paracentesis.
Pilocarpine.
Pyoktanin.
Quinine.
Saline Laxatives.
Santonin.
Scopolamine.
Sodium Salicylate.
Tro acocaine.
Turpentine Oil: in rheumatic iritis.

Irritability.—See also, Insomnia, Nervousness.

Acid, Hydrocyanic: in irritability of the stomach.
Alkaline Waters.
Almonds: as a drink in irritability of intestines and air passages.
Bromalin.
Bromide of Potassium.
Bromipin.
Bromo-hemol.
Cantharides: in irritable bladder of women and children.
Chloral Hydrate.
Cimicifuga: in uterine irritability.
Colchicine.
Colchicum: with potash in large quantity of water when gouty.
Cupro-hemol.
Hops: in vesical irritability.
Hyoscyamus: for vesical irritability with incontinence.
Ignatia: in small doses.
Laxatives: in constipation.
Opium.
Petrolatum: as a soothing agent in gastrointestinal types.

Piperazine: in bladder irritation due to excess of uric acid.

Potassium Bromide: in irritability of pharynx.

Sitz-Bath.

Strychnine: in small doses.

Itch.—*See Scabies.*

Jaundice. — *See also, Hepatic Cirrhosis, Hepatic Diseases, Calculi.*

Acid, Benzoic.
Acid, Carbolic.
Acid, Citric.
Acid, Nitrohydrochloric: internally, and as local application over liver, or as bath in catarrhal cases.
Acids, Mineral.
Alkaline mineral waters in catarrh of duodenum or bile-ducts.
Aloes.
Ammonium Chloride: in scruple doses in jaundice from mental emotio
Ammonium Iodide: when catarrh of bile-ducts.
Arsenic: in malaria.
Berberine Carbonate: in chronic intestinal catarrh.
Calcium Phosphate.
Calomel Purgative: followed by saline, often very useful.
Carlsbad Salts.
Carlsbad Waters.
Celandine.
Chelidonium.
Chloroform.
Colchicum.
Diet.
Dulcamara.
Emetics.
Enemata: cold water, one or two liters once a day.
Ether: when due to gall-stones.
Euonymin.
Hydrastine.
Hydrastis: in cases of catarrh of ducts.
Iodoform.
Ipecacuanha.
Iridin.
Iris.
Iron Succinate.
Lemon Juice.
Levico Water.
Magnesia.
Magnesium Sulphate.
Manganese: in malarial or catarrhal cases.
Mercurials.
Ox-gall.

Pichi.
Pilocarpine Hydrochlorate.
Podophyllum: in catarrhal conditions very useful.
Potassium Bicarbonate.
Potassium Carbonate.
Potassium Chloride.
Potassium Sulphate: as laxative.
Quinine: in malarial cases.
Rhubarb: in children.
Saline Purgatives.
Salol.
Sanguinaria.
Sodium Phosphate: very useful in catarrh of bile-ducts.
Stillingia: after ague.
Taraxacum.
Turpentine Oil.

Joint Affections.— *See also, Arthritis, Bursitis, Coxalgia, Gout, Rheumatism, Synovitis.*

Acetanilid.
Acid, Salicylic.
Aconite.
Ammoniac Plaster.
Aristol.
Arsenic.
Cadmium Iodide.
Digitalis.
Europhen.
Gaduol.
Ichthalbin: internally.
Ichthyol: topically.
Iodine.
Iodoform.
Iodoformogen: more diffusible and persistent than Iodoform.
Iodole.
Iron Iodide.
Levico Water.
Mercury Oleate.
Methylene Blue.
Rhus Toxicodendron.
Silver Nitrate.
Sozoiodole-Mercury.
Tartar Emetic Ointment.
Triphenin.
Turpentine Oil.
Veratrine.

Joints, Tuberculosis of.

Formaldehyde.
Iodoform.
Iodoformogen.
Iodole.

Keratitis. — *See also, Corneal Opacities.*

Aniline.
Antisyphilitic treatment.

Aristol.
Arsenic.
Atropine.
Calcium Sulphide.
Curetting.
Eserine.
Europhen.
Gallisin.
Hot Compresses.
Iron.
Leeches.
Levico Water: as alterative.
Massage of Cornea: and introduction of yellow-oxide ointment.
Mercurial Ointment.
Physostigma.
Potassium Bromide.
Potassium Iodide.
Pressure: bandages if perforation threatens.
Pyoktanin.
Quinine.
Sozoiodole-Sodium.

Kidney Disease.—*See also, Albuminuria, Bright's Disease, Calculi; Colic. Renal; Diabetes, Dropsy, Gout, Hematuria.*

Ammonium Benzoate: for atony of kidney.
Digitoxin.
Fuchsine.
Ichthalbin.
Levico Water.
Methylene Blue.
Pilocarpine.
Saliformin.
Strontium Bromide or Lactate.
Tannalbin.

Labor.—*See also, Abortion, After-Pains, False Pains, Postpartum Hemorrhage, Lactation, Puerperal Convulsions, Fever.*

Acetanilid.
Amyl Nitrite.
Anesthetics.
Antipyrine.
Belladonna.
Borax.
Cannabis Indica.
Chloral Hydrate.
Chloroform.
Cimicifuga.
Creolin.
Ethyl Bromide.
Eucalyptus Oil.
Geiseminine.
Mercury Bichloride.
Morphine.
Opium.
Pilocarpine Hydrochlorate.
Quinine.

La Grippe.—*See Influenza.*

Lactation, Defective.—*See also, Abscess, Agalactia, Mastitis, Nipples; also the list of Galactagogues.*

Ammonium Chloride.
Calabar Bean.
Castor-Oil : topically.
Gaduol.
Glycerinophosphates.
Hypophosphites.
Jaborandi.
Malt Extract, Dry.
Mustard Poultice.
Pilocarpine Hydrochlorate.
Vanilla.

Lactation, Excessive.

Agaricin.
Alcohol.
Belladonna : internally and locally.
Camphor and Glycerin.
Chloral Hydrate.
Coffee.
Conium : internally.
Electricity.
Ergot.
Galega.
Hempseed Oil.
Iodides.
Iodine.
Mercury.
Parsley.
Quinine.
Tobacco : as poultice.

Laryngeal Tuberculosis.—*See also, Phthisis.*

Formaldehyde.
Hydrogen Peroxide.
Iodole.
Sozoiodole salts.

Laryngismus Stridulus.—*See also, Croup, Laryngitis.*

Acetanilide.
Aconite.
Amyl Nitrite.
Antipyrine.
Antispasmin.
Atropine.
Belladonna.
Bromides : very useful in large doses.
Bromoform.
Chloral Hydrate.
Chloroform : as inhalation to stop spasm.
Codeine.
Cod-Liver Oil.
Cold Sponging.
Cold Water : dashed in the face.

Coniine ; pushed until physiological action observed.
Creosote.
Emetics.
Ether.
Gaduol.
Gelsemium.
Glycerinophosphates
Gold and Sodium Chloride.
Guaiacol.
Ipecacuanha : as emetic.
Lancing Gums.
Lobelia.
Mercury Sub-sulphate.
Morphine : hypodermically.
Musk.
Nitroglycerin.
Peronin.
Potassium Bromide.
Quinine.
Spinal Ice-Bag.
Tartar Emetic.
Worms, Removal of.

Laryngitis.

Acid, Camphoric.
Acid, Sulpho-anilic.
Aristol.
Aseptol.
Ammonium Chloride.
Chlorophenol.
Cocaine.
Ethyl Iodide.
Ichthyol.
Iodole.
Napthol, Camphorated.
Pilocarpine Hydrochlorate.
Potassium Iodide.
Silver Nitrate.
Sozoiodole-Sodium.
Sozoiodole-Zinc.
Thymol.

Laryngitis, Acute.—*See also, Croup, Catarrhal; Laryngismus Stridulus, Pharyngitis.*

Abstinence from talking, with bland and unirritating, but nutritious diet during attack.
Acid, Acetic : as inhalation.
Acid, Sulphurous : as inhalation or spray.
Aconite.
Antimon. Pot. Tart.
Antipyrine : as a spray.
Benzoin : as inhalation.
Bromides : in full doses.
Calomel : in small and repeated doses followed by saline purges, also hot mustard foot-bath and demulcent drinks.
Cocaine.
Copper Sulphate.

Creosote Spray : in subacute laryngitis.
Cubeb Cigarettes for hoarseness.
Dover's Powder.
Gelsemium.
Glycerin.
Inhalations.
Iodine : as inhalation and counter-irritant over neck.
Leeches : to larynx or nape of neck.
Mercury.
Morphine.
Oil of Amber.
Purgatives.
Quinine.
Scarification of Larynx.
Steam Inhalations.
Silver Nitrate : as spray.
Tracheotomy.
Veratrum Viride.
Zinc Chloride.
Zinc Sulphate : as emetic.

Laryngitis, Chronic.—*See also, Cough, Dysphagia, Laryngitis Tuberculosa, Syphilis.*

Acid, Carbolic : as spray.
Acid, Sulphurous : as fumigation, inhalation or spray.
Alum : as gargle.
Ammonium Chloride : as spray.
Bismuth : locally by insufflation.
Ferric Chloride : as spray, or brushed on interior of larynx.
Gelsemium.
Glycerin.
Guaiacum : as lozenges or mixture.
Inhalation.
Iodine : as counter-irritant.
Mercury.
Morphine : mixed with bismuth or starch as insufflation ; most useful when much irritation, as in laryngeal phthisis.
Silver Nitrate : as solution to interior of larynx.
Sozoiodole-Zinc.
Tannin : as gargle or spray.
Uranium Nitrate : as spray.

Laryngitis Tuberculosa.

Acid, Lactic.
Bismuth Subgallate.
Bismuth Subnitrate.
Cocaine Hydrochlorate

Europhen.
Ichthalbin: internally.
Iodoform.
Iodoformogen.
Iodole.
Maragliano's Serum.
Menthol.
Resorcin.
Silver Nitrate.
Sozoiodole-Sodium.
Sozoiodole-Zinc.
Xeroform.
Zinc Sulphate.

Lepra.—*See Leprosy.*

Leprosy.

Acid, Arsenous.
Acid, Gynocardic.
Ammonium Iodide.
Arsenic Iodide.
Gaduol.
Gold.
Glycerin.
Ichthalbin: internally.
Ichthyol: topically.
Iron Arsenate.
Mercury Bichloride.
Oil Chaulmoogra.
Oil Gurjun.
Potassium Iodide.
Silver Nitrate.
Solution Arsenic and Mercury Iodide.
Solution Potassa.
Sulphur Iodide.

Leucemia.—*See Leucocythemia.*

Leucocythemia.

Arsenic.
Arsen-hemol.
Hypophosphites.
Iron.
Levico Water.
Phosphorus.

Leucoplakia Buccalis.

Balsam Peru.
Pyoktanin.
Sozoiodole-Sodium.
Tannoform.

Leucorrhea.—*See also, Endometritis, Uterine Ulceration, Vaginitis.*

Acid, Boric.
Acid, Carbolic: as injection.
Acid, Chromic.
Acid, Nitric, and Cinchona.
Acid, Phosphoric.
Alkalies.
Aloes.
Alum: as injection.
Aluminium Sulphate.
Ammonio-Ferric Alum
Ammonium Chloride.
Arsenic.
Bael Fruit.

Balsam of Peru: internally.
Balsam of Tolu: internally.
Belladonna; as pessary, for over-secretion and pain.
Bismuth: as injection or pessary.
Bismuth Subnitrate.
Blister.
Borax: as injection.
Calcium Phosphate.
Cimicifuga.
Cocculus Indicus.
Cold Sponging.
Copaiba.
Copper Sulphate: as injection.
Creosote.
Dry Red Wine.
Ergot.
Glycerin.
Hamamelis.
Helenin.
Hematoxylon.
Hot Sitz Bath or Vaginal Injections of hot water: if due to uterine congestion.
Hydrastine Hydrochlorate.
Hydrastis: locally.
Ichthyol.
Iodine.
Iodoform or Iodoformogen: as local application, alone or mixed with tannic acid.
Iron Chloride.
Iron Iodide.
Iron Sulphate.
Lead salts.
Lime Water.
Monsel's Solution.
Myrrh: internally.
Oil Turpentine.
Pulsatilla.
Pyoktanin.
Quercus.
Phosphate of Calcium: internally.
Potassium Bicarbonate: dilute solution as injection.
Potassium Bromide.
Potassium Chloride.
Potassium Permanganate.
Resorcin.
Saffron.
Silver Oxide.
Sozoiodole-Sodium.
Spinal Ice-Bag.
Sumbul.
Tannin: as injection or suppository.
Tannoform.
Thymol.
Zinc Sulphate.

Lichen.

Aconite.
Alkalies.

Arsenic.
Calomel.
Cantharides.
Chloroform.
Cod-Liver Oil.
Glycerin.
Glycerite of Aloes.
Ichthalbin: internally.
Ichthyol: topically.
Levico Water.
Mercury: locally.
Naftalan.
Potassium Cyanide.
Silver Nitrate: solution locally.
Strontium Iodide.
Sulphides.
Sulphur.
Tar Ointment.
Thymol.
Warm Baths.

Lipoma.—*See Tumors.*

Lips, Cracked.—*See also, Fissures.*

Adeps Lanæ.
Ichthyol.
Lead Nitrate.

Lithemia.—*See also, Lithiasis, Calculus, Dyspepsia, Gout.*

Acid, Benzoic.
Acid, Nitric.
Acid, Salicylic.
Alkalies.
Arsenic.
Calcium Benzoate.-
Colchicum.
Formin.
Hippurates.
Ichthalbin.
Lithium Carbonate.
Lycetol.
Lysidine.
Magnesium Carbonate.
Methyl Salicylate.
Oil Wintergreen.
Piperazine.
Potassium Acetate.
Potassium Carbonate.
Potassium Citrate.
Potassium Permanganate.
Saliformin.
Sodium Benzoate.
Sodium Borate.
Sodium Carbonate.
Sodium Phosphate.
Solution Potassa.
Strontium Lactate.
Strontium Salicylate.

Liver: Cirrhosis, Congestion, Diseases of.—*See Hepatic Cirrhosis, Congestion, Diseases.*

Locomotor Ataxia.

Acetanilid.
Acid, Nitro-hydrochloric.

Amyl Nitrite.
Antipyrine.
Belladonna.
Calabar Bean.
Cannabis Indica.
Chloride of Gold.
Damiana.
Electricity.
Ergot.
Exalgine.
Gaduol.
Glycerinophosphates.
Hyoscyamus.
Mercuro-iodo-hemol.
Mercury Bichloride.
Methylene Blue.
Morphine.
Neurodin.
Phenacetin.
Phosphorus.
Physostigma.
Pilocarpine.
Potassium Bichromate.
Potassium Bromide.
Potassium Iodide: for
 syphilitic taint.
Silver Nitrate.
Silver Oxide.
Silver Phosphate.
Sodium Hypophosphite.
Sodium Salicylate.
Solanin.
Spermine.
Strychnine.
Suspension.

Lumbago. — See also,
 Myalgia, Rheuma-
 tism, Neuralgia.

Acetanilid.
Acid, Carbolic: hypo-
 dermically.
Acid, Salicylic.
Aconite: small doses
 internally, and lini-
 ment locally.
Acupuncture.
Ammonium Chloride.
Antipyrine.
Aquapuncture: some-
 times very useful.
Atropine.
Belladonna.
Camphor, Monobroma-
 ted.
Capsicum: locally.
Cautery.
Chloroform: liniment.
Cimicifuga: sometimes
 very useful internally.
Cod-Liver Oil
Electricity.
Emplastra.
Ether Spray.
Eucalyptus Oil: as lin-
 iment.
Faradization.
Foot-bath and Dover's
 Powder.
Galvanism.
Guaco.
Gaduol.
Guarana: in large doses
Glycerinophosphates.

Hot Douche or Hot
 Poultice.
Ice: rubbed over back.
Ice-bag or Ether Spray
 to loins: if hot appli-
 cations fail.
Iodide of Potassium.
Iodides.
Ironing Back with laun-
 dry iron, skin being
 protected by cloth or
 paper.
Lead Plaster.
Massage.
Morphine: hypodermi-
 cally.
Mustard or Capsicum:
 plaster or blister over
 painful spot.
Neurodin.
Nitrate of Potassium.
Oil Turpentine.
Oleoresin Capsicum.
Phenacetin and Salol:
 of each 5 grn.
Pitch: plaster.
Potassium Salicylate.
Poultices.
Quinine.
Quinine Salicylate.
Rhus Toxicodendron
Sulphur.
Thermodin.
Triphenin.
Turkish Bath.
Turpentine Oil: intern-
 ally and locally.
Veratrum Viride.

Lupus.

Acid, Carbolic.
Acid, Chromic.
Acid, Cinnamic.
Acid, Lactic.
Acid, Pyrogallic.
Acid, Salicylic.
Alumnol.
Aristol.
Arsenic.
Arsenic Iodide.
Blisters.
Calcium Chloride.
Calcium Lithio-Carbo-
 nate.
Calomel.
Cantharidin.
Cautery.
Chaulmoogra Oil.
Chrysarobin.
Cod-Liver Oil.
Creosote.
Europhen.
Formaldehyde.
Gaduol.
Galvano-Cautery.
Glycerin.
Gold Chloride.
Guaiacol.
Hydroxylamine Hydro-
 chlorate.
Ichthalbin: internally
Ichthyol: topically.
Iodine: in glycerin.
Iodoform.
Iodoformogen.

Iodole.
Iron Arsenate.
Lead Lotion.
Levico Water.
Mercuric Nitrate.
Mercury Biniodide.
Mercury: internally and
 locally.
Naftalan.
Naphtol.
Phosphorus.
Plumbic Nitrate.
Potassium Cantharidate
Potassium Chlorate.
Potassium Iodide.
Silver Nitrate.
Sodium Acetate.
Sodium Ethylate.
Sodium Salicylate.
Solution Arsenic and
 Mercury Iodide.
Sozoiodole-Sodium.
Starch, Iodized.
Strontium.
Sulphur Iodide: exter-
 nally.
Thiosinamine.
Thyraden.
Zinc Chloride.
Zinc Sulphate.

Lymphangitis. — See
 also, Bubo.

Acid, Picric.
Acid, Tannic.
Belladonna.
Gaduol.
Ichthalbin: internally.
Ichthyol: topically.
Lead.
Lime, Sulphurated.
Quinine.
Salicin.

Malaria.—See also, Intermittent Fever,
 Remittent Fever.

Acid, Arsenous, a n d
 Arsenites.
Acid, Carbolic.
Acid, Hydrofluoric.
Acid, Picric.
Ammonium Fluoride.
Ammonium Picrate.
Antipyrine.
Apiol.
Arsen-hemol.
Bebeerine.
Benzanilide.
Berberine.
Berberine Carbonate.
Calomel.
Cinchona alkaloids
 and salts.
Eucalyptol.
Gentian.
Guaiacol.
Hydrastis.
Iodine.
Iron.
Iron and Quinine
 Citrate.
Levico Water.
Manganese.

Manganese Sulphate.
Methylene Blue.
Mercury.
Phenocoll Hydrochlorate.
Pilocarpine Hydrochlorate.
Piperine.
Potassium Citrate.
Quinine.
Quinoidine.
Salicin.
Salicylates.
Sodium Chloride.
Sodium Fluoride.
Solution Potassium Arsenite.
Warburg's Tincture.

Mania.—*See also, Delirium, Insanity, Puerperal Mania.*

Acid, Hydrocyanic.
Acid, Valerianic.
Actæa Racemosa.
Alcohol.
Amylene Hydrate.
Anesthetics.
Apomorphine: inemetic dose.
Atropine.
Belladonna: useful.
Blisters.
Bromides.
Camphor.
Cannabis Indica.
Chloral: in full dose, if kidneys are healthy.
Chloral and Camphor.
Chloroform: for insomnia.
Cimicifuga: in cases occurring after confinement, not due to permanent causes.
Cold Douche: to head while body is immersed in hot water.
Conline: alone or with morphine.
Croton Oil: as purgative.
Daturine.
Digitalis: in acute and chronic mania, especially when complicated with general paralysis and epilepsy.
Duboisine: as calmative.
Ergot: in recurrent mania.
Ether: in maniacal paroxysms.
Galvanism: to head and to cervical sympathetic.
Gamboge.
Gelsemium: when much motor excitement and wakefulness.
Hyoscine Hydrobromate.

Hyoscyamine or Hyoscyamus: in hallucinations and hypochondriasis.
Iron.
Morphine.
Opium: alone or with tartar emetic.
Paraldehyde.
Physostigma.
Potassium Bromide.
Scopolamine: as a soporific.
Stramonium.
Sulfonal: as a hypnotic.
Veratrum Viride.
Wet Pack.
Zinc Phosphide.

Marasmus.—*See Adynamia, Cachexia, Emaciation, etc.*

Mastitis. — *See also, Abscess, Lactation.*

Aconite.
Ammonium Chloride: as lotion locally.
Arnica.
Belladonna: locally as liniment or ointment.
Breast-pump.
Calcium Sulphide: internally if abscess is forming.
Camphor.
Chloral Hydrate Poultice.
Conium.
Digitalis Infusion: locally as fomentation.
Friction: with oil.
Galvanism.
Hyoscyamus: as plaster to relieve painful distention from milk.
Ice.
Ichthyol topically: one of the best remedies.
Iodine.
Jaborandi.
Mercury and Morphine Oleate: locally in mammary abscess.
Phytolacca: to arrest inflammation, local application.
Plaster: to support and compress mammæ.
Potassium Bromide.
Salines.
Stramonium: fresh leaves as poultice.
Tartar Emetic: in small doses frequently repeated at commencement.
Tobacco Leaves: as poultice.

Measles.—*For Sequelæ, see Bronchitis, Cough, Ophthalmia, Otorrhea, Pneumonia, etc.*

Acid, Carbolic: internally at commencement.
Aconite.
Adeps Lanæ.
Ammonium Acetate.
Ammonium Carbonate.
Antimony.
Calcium Sulphide.
Camphor.
Cold Affusion.
Digitalis.
Fat.
Iodine.
Ipecacuanha.
Jaborandi.
Mustard Bath: when retrocession of rash.
Packing.
Potassium Bromide: when sleeplessness.
Potassium Chlorate: in adynamic cases.
Pulsatilla.
Purgatives.
Quinine.
Triphenin.
Veratrum Viride.
Zinc Sulphate.

Melancholia. — *See also, Hypochondriasis, Hysteria, Insanity.*

Acid, Hydrocyanic.
Acid, Nitrohydrochloric after meals: if associated with oxaluria.
Alcohol.
Arsenic: in aged persons along with opium
Belladonna.
Bromides.
Caffeine.
Camphor.
Cannabis Indica.
Chloral Hydrate: as hypnotic.
Chloroform: for insomnia.
Cimicifuga: in puerperal or uterine despondency.
Cocaine.
Colchicum.
Colocynth.
Galvanism.
Gold.
Ignatia.
Iron.
Morphine.
Musk.
Nitrous Oxide.
Opium: in small doses especially useful.
Paraldehyde.
Phosphorus.

Thyraden.
Turkish Bath.
Valerian: in hysterical and suicidal cases.
Zinc Phosphide.

Menière's Disease.

Bromalin.
Bromides.
Bromo-hemol.
Gelsemium.
Quinine.
Sodium Salicylate.

Meningitis, Cerebral, Spinal and Cerebro-Spinal.

—See also, Meningitis, Tubercular.

Aconite.
Alcohol.
Ammonium Carbonate.
Antimony: in cerebrospinal meningitis.
Belladonna.
Blister to nape of neck in early stage, to prevent effusion; also in comatose state.
Bromides and Chloral: to allay nervous symptoms.
Bryonia: when effusion.
Calomel with Opium: in early stages.
Cold Baths.
Digitalis.
Ergot.
Gelsemium.
Hyoscyamus.
Ice-bag to head.
Iodide of Potassium.
Jalap.
Leeches: to nape of neck.
Mercury: as ointment or internally.
Milk Diet: in second stage.
Opium: in small doses, alone or with tartar emetic.
Phosphorus: in chronic meningitis.
Pilocarpine.
Pulsatilla: in acute cases
Purgatives: at commencement; calomel and jalap most useful.
Spermine.
Turpentine Oil.
Quinine: contraindicated in acute stage.
Veratrum Viride.
Venesection: in early stage of sthenic cases, if aconite or veratrum viride is not at hand; also when much excitement.

Meningitis, Tubercular.

Croton Oil.
Iodine.
Magnesium Carbonate.
Mercury.
Potassium Bromide.
Potassium Iodide.
Purgatives.
Tartar Emetic.
Turpentine Oil.

Menorrhagia and Metrorrhagia. —

See also, Amenorrhea, Hemorrhage, Uterine Tumors.

Acid, Gallic: very useful.
Acid, Pyrogallic.
Acid, Tannic.
Acid, Sulphuric: when due to fibroid or polypus.
Actæa Racemosa.
Aloes: as adjuvant to iron.
Ammonium Acetate.
Ammonium Chloride: for headache.
Arsenic: with iron.
Atropine.
Berberine.
Bromides.
Calcium Phosphate: in anemia.
Cannabis Indica: sometimes very useful.
Cimicifuga.
Cinnamon Oil: when erigeron is not at hand, in oozing flow.
Confine.
Creosote.
Digitalis: sometimes useful.
Dry Cups over Sacrum: if due to congestion.
Ergot: most useful.
Ferri Perchloridum.
Guaiacum.
Hamamelis: useful.
Hot Water Bag: to dorsal and lumbar vertebræ.
Hydrargyri Perchloridum.
Hydrastine Hydrochlorate.
Hydrastinine Hydrochlorate.
Hydrastis.
Ice: to spine.
Iodine.
Iodoform.
Ipecacuanha: in emetic doses in evening, followed by acidulated draught in morning.
Lemons.
Levico Water.
Magnesium Sulphate: sometimes useful.

Mercury Bichloride.
Monsel's Solution.
Oil Erigeron.
Opium.
Phosphates.
Potassium Chlorate.
Quinine.
Rhus Aromatica.
Rue.
Savin.
Senega.
Stypticin.
Silver Oxide.
Turpentine Oil.
Tannin.
Urtica Urens.
Vinca Major.

Menstrual Disorders.—See also, Amenorrhea, Dysmenorrhea, Climacteric Disorders.

Aconite.
Aloes.
Cimicifuga.
Cocculus Indicus.
Opium.
Pulsatilla.

Mentagra.

Acid, Carbolic.
Acid, Sulphurous: with glycerin.
Arsenic.
Canada Balsam.
Cod-Liver Oil.
Copper: locally, as lotion.
Epilation.
Goa Powder.
Iodide of Sulphur.
Iodine.
Mercury.
Oil of Turpentine.
Oleate, Bichloride, or Nitrate of Mercury: as ointment or lotion.
Petroleum.
Silver Nitrate.
Tr. Iodine, Compound, Zinc and Copper Sulphate.
Zinc Chloride.

Mercurial Cachexia.

Gaduol.
Glycerinophosphates.
Hemogallol
Iodine and Iodides.
Iodipin.
Iodohemol.

Meteorism. — See Tympanites.

Metritis (Para- and Peri-). — See also, Puerperal Fever, Puerperal Metritis.

Acid, Carbolic.
Acid, Nitric.

Aconite.
Aloes: enema.
Creosote.
Ergotin.
Gold and Sodium Chloride.
Hydrargyri Bichloridum.
Ichthyol.
Iodine.
Iodipin.
Iodoform.
Iodoformogen.
Levico Water.
Mercury Bichloride.
Nitrate of Silver.
Opium: as suppository or enema.
Potassa Fusa.
Potassium Iodide.
Poultices.
Saline Laxatives.
Saline Mineral Waters.
Silver Nitrate.
Sozoiodole salts.
Turpentine Oil.
Turpentine Stupes.

Migraine.—*See also, Hemicrania.*

Acetanilid.
Acid, Salicylic.
Aconitine.
Antipyrine.
Amyl Nitrite.
Caffeine.
Cannabis Indica
Camphor, Monobromated.
Croton Chloral.
Eucalyptol.
Exalgin.
Ferropyrine.
Gelseminine.
Gold Bromide.
Guarana.
Ichthyol.
Methylene Blue.
Neurodin.
Phenacetin.
Picrotoxin.
Potassium Bromide with Caffeine.
Sodium Salicylate.
Triphenin.

Miliary Fever.

Aconite.
Zinc Oxide.

Mitral Disease.—*See Heart Affections.*

Mollities Ossium.—*See Bone Diseases, Rachitis.*

Morphine Habit.—*See Opium Habit.*

Mouth, Sore.—*See also, Aphthæ, Cancrum Oris, Gums, Parotitis, Ptyalism, Stomatitis, Toothache, Tongue.*

Acetanilid.
Acid, Boric.
Acid, Citric.
Pyoktanin.
Silver Nitrate.
Sodium Bisulphate.
Sodium Borate.
Sodium Thiosulphate.
Sozoiodole-Sodium.
Zinc Acetate.

Mumps.—*See Parotitis.*

Muscæ Volitantes.

Alteratives, and Correction of anomalies of refraction.
Mercury.
Blue Pill: in biliousness.
Iodide of Potassium.
Iron Perchloride: in anemia and climacteric.
Valerian.

Myalgia.—*See also, Pleurodynia, Lumbago.*

Acupuncture.
Aquapuncture.
Ammonium Chloride.
Arnica: internally and locally.
Belladonna Liniment: locally.
Belladonna Plaster.
Camphor-Chloral.
Camphor Liniment.
Camphor, Monobromated.
Chloroform Liniment: with friction.
Cimicifuga.
Clove Oil: added to liniment, as a counter-irritant.
Diaphoretics.
Electricity.
Ether.
Exalgin.
Friction.
Gelseminine.
Gelsemium: large doses.
Ichthyol.
Iodides.
Iodine.
Massage, or good rubbing, very necessary.
Oil Cajuput.
Opium.
Packing.
Potassium Acetate or Citrate.

Poultices: hot as can be borne.
Salicylates.
Salol.
Triphenin.
Veratrine: externally.
Xanthoxylum: internally and externally.

Myelitis.—*See also Meningitis, Spinal; Paralysis.*

Barium Chloride.
Belladonna.
Electricity: in chronic cases.
Ergot.
Gaduol.
Galvanism.
Glycerinophosphates.
Hydrotherapy.
Iodides.
Iodole.
Iodipin.
Massage.
Mercury.
Phosphorus: in paraplegia from excessive venery.
Picrotoxin.
Silver Nitrate: useful.
Spermine.
Strychnine.

Myocarditis.—*See Heart Affections.*

Myopia.

Atropine.
Extraction of lens.
Glasses.

Myringitis.—*See Ear Affections.*

Myxedema.—*See also, Goiter.*

Arsenic.
Iodothyrine.
Iron Salts.
Jaborandi.
Nitroglycerin.
Pilocarpine Hydrochlorate.
Strychnine Salts.
Thyraden.

Nails, Ingrowing.

Alum.
Ferri Perchloridum.
Ferri Persulphas.
Glycerin.
Iodoform.
Iodoformogen.
Iodole.
Lead Carbonate.
Liquor Potassæ.
Plumbi Nitras.
Pyoktanin.
Silver Nitrate.
Sozoiodole-Sodium.
Tannin.

Narcotism.

Apomorphine Hydro-chlorate.
Atropine.
Caffeine.
Emetics.
Exercise.
Galvanism.
Strychnine.

Nasal Diseases.—See also, Acne, Catarrh, Epistaxis, Hay Fever, Influenza, Ozena, Polypus, Sneezing.

Acid, Tannic: with glycerin.
Acid, Chromic.
Acid, Trichloracetic.
Alum.
Aluminium Aceto-tartrate.
Aluminium Tanno-tartrate.
Arsenic.
Bismuth Subgallate.
Camphor.
Cocaine Hydrochlorate
Cocaine Carbolate.
Diaphtherin.
Eucaine Hydrochlorate
Gaduol.
Glycerinophosphates.
Holocaine Hydrochlorate.
Hydrogen Dioxide.
Hydrastine Hydrochlorate.
Ichthyol.
Iodipin.
Iodoform.
Iodoformogen.
Iodole.
Levico Water.
Naphtol.
Potassium Iodide.
Pulsatilla.
Pyoktanin.
Resorcin.
Sanguinarine.
Silver Nitrate.
Sodium Borate, Neutral.
Sozoiodole salts.
Zinc Chloride.
Zinc Oxide.

Nasal Polypus.—See also, Polypus.

Ichthyol.

Nausea.— See also, Dyspepsia, Headache, Biliousness, Sea-Sickness, Vomiting, Vomiting of Pregnancy.

Acid, Carbolic.
Acid, Hydrocyanic.
Acid, Sulphuric.
Acid, Tartaric.

Aconite.
Ammonio-Citrate of Iron.
Belladonna.
Bismuth.
Calomel.
Calumba.
Cerium Oxalate.
Chloral Hydrate.
Chloroform.
Cinnamon.
Cloves.
Cocaine.
Cocculus Indicus: in violent retching without vomiting.
Codeine.
Coffee.
Creosote.
Electricity.
Ether.
Hoffmann's Anodyne: when due to excessive use of tobacco.
Ice.
Ingluvin.
Iodine.
Ipecacuanha : in sickness of pregnancy and chronic alcoholism; very small dose, 1 minim of wine.
Kumyss.
Lead Acetate.
Leeches.
Lime Water.
Liquor Potassæ.
Magnesium Carbonate.
Mercury.
Morphine.
Nux Vomica.
Nutmeg.
Orexine: when with lack of appetite.
Papain.
Pepper.
Peppermint.
Pepsin.
Pimento.
Pulsatilla ; in gastric catarrh.
Salicin.
Spt. Nucis Juglandis.
Strychnine.

Necrosis.—See Caries, Bone Disease, Syphilis, Scrophulosis.

Neoplasms.—See Tumors.

Nephritis, Acute.— See also, Albuminuria, Bright's Disease.

Acid, Gallic.
Aconite: at commencement.
Alkalies.
Ammonium Acetate.
Ammonium Benzoate.
Aqua Calcis.
Arsen-hemol.
Arsenic.

Belladonna.
Caffeine.
Camphor.
Cannabis Indica : as diuretic, especially in hematuria.
Cantharides : one minim of tincture every three hours, to stop hematuria after acute symptoms have subsided.
Cod-Liver Oil.
Copaiba.
Croton Liniment.
Cytisus Scoparius.
Digitalis : as diuretic.
Elaterium.
Eucalyptus : given cautiously.
Fuchsine.
Hyoscyamus.
Ichthalbin : internally.
Ichthyol : externally.
Incisions.
Iron.
Jaborandi.
Juniper.
Lead.
Levico Water.
Liquor Ammonii Acetatis.
Liquor Potassæ.
Methylene Blue.
Nitroglycerin.
Pilocarpine.
Potassium Bitartrate.
Potassium Bromide.
Potassium Iodide.
Potassium Sulphate.
Poultices : over loins, very useful.
Senega.
Strontium Bromide.
Strontium Lactate.
Tannalbin.
Tannin.
Theobromine and Sodium Salicylate.
Tinctura Ferri Perchloridi.
Turkish Baths.
Turpentine Oil : one minim every two to four hours.
Uropherin.
Warm Baths.

Nervous Affections.
—See also, Diabetes, Hemicrania; Headache, Nervous; Hemiplegia, Hysteria Insomnia, Locomotor Ataxia, Mania, Melancholia, Myelitis, Neuralgia, Neurasthenia, Neuritis, Nervousness, Paralysis, Paralysis Agitans, Spinal Paralysis, etc.

Acid Hypophosphorous.

Acid, Valerianic.
Arsen-hemol.
Arsenic.
Bromipin.
Bromo-hemol.
Caffeine.
Cæsium and Rubidium and Ammonium Bromide.
Cocaine.
Cupro-hemol.
Ferropyrine.
Gold and Sodium Chloride.
Glycerinophosphates.
Hyoscine.
Hyoscyamine.
Iodipin.
Neurodin.
Nux Vomica.
Opium.
Picrotoxin.
Phosphorus.
Physostigma.
Potassium Bromide.
Santonin.
Silver Chloride.
Silver Phosphate.
Sodium Arsenate.
Sodium Phosphate.
Solanine.
Spermine.
Valerianates.
Zinc Sulphate.
Zinc Valerianate.

Nervous Exhaustion.—*See Adynamia, Neurasthenia.*

Nervousness. — *See also, Insomnia, Irritability*

Aconite: one minim. of tincture at bedtime for restlessness and fidgets.
Ammonium Chloride.
Argenti Phosphas.
Bromide of Potassium: over-work and worry.
Bromo-hemol.
Caffeine: where much debility.
Camphor.
Chamomile.
Chloral Hydrate.
Chloroform.
Cod-Liver Oil.
Cold Sponging.
Conium.
Cupro-hemol.
Electricity.
Ergot.
Ether.
Hops: internally, and as pillow.
Hydrargyri Perchlorid.
Ignatia.
Lime salts.
Levico Water.
Massage.
Morphine Valerianate.
Musk: in uterine derangements.

Opium.
Phosphorus.
Pulsatilla: tincture.
Resorcin.
Rest-Cure.
Simulo: tincture.
Sodium Bromide.
Strontium Bromide.
Strychnine.
Sumbul: in pregnancy, and after acute illness.
Suprarenal Gland.
Sweet Spirit of Nitre.
Valerian.
Zinc Phosphate.

Neuralgia.—*See also, Gastralgia, Hemicrania, Hepatalgia, Otalgia, Ovarian Neuralgia, Sciatica, Tic Douloureux, etc.*

Acetanilid.
Acid, Hydrocyanic.
Acid, Perosmic.
Acid, Salicylic.
Acid, Valerianic.
Aconite: locally.
Aconitine: as ointment.
Acupuncture.
Adeps Lanæ.
Agathin.
Alcohol.
Ammonium Chloride: one-half dram doses.
Ammonium Picrate.
Ammonium Valeriamate.
Amyl Nitrite.
Anesthetics.
Aniline.
Antipyrine.
Antiseptic Oils.
Aquapuncture.
Arsenic.
Atropine: as liniment, or hypodermically near the nerve.
Auro-Terchlor. Iod.
Belladonna.
Bebeeru Bark or Bebeerine.
Berberine.
Bismuth Valerianate.
Blisters.
Bromides.
Butyl-Chloral Hydrate: for neuralgia of fifth nerve.
Cactus Grandiflorus: tincture.
Caffeine.
Camphor, Carbolated.
Camphor, Monobromated.
Cannabis Indica.
Capsicum: locally.
Carbon Disulphide.
Cautery.
Chamomile.
Chaulmoogra Oil.
Chelidonium.
Chloralamide.
Chloral-Ammonia.

Chloral and Camphor: equal parts, locally applied.
Chloral and Morphine.
Chloral-Menthol.
Chlorate of Potassium: in facial neuralgia.
Chloroform: locally, and by inhalation, when pain is very severe.
Cimicifuga: in neuralgia of fifth nerve, and ovarian neuralgia.
Cocaine.
Codeine.
Cod-Liver Oil.
Colchicine.
Colchicum.
Coniine Hydrobromate.
Conium.
Counter-irritation.
Creosote.
Cupri-Ammonii Sulphas.
Digitalis.
Dogwood, **Jamaica.**
Electricity.
Epispastics.
Ergot: in visceral neuralgia.
Ether.
Ethyl Chloride.
Eserine.
Eucalyptol.
Euphorin.
Exalgin.
Ferric Perchloride.
Ferro-Manganates.
Ferropyrine.
Freezing Parts: with ether or rhigolene spray.
Gaduol: as nerve-tonic and alterative.
Galvanism.
Gelsemium.
Gelseminine.
Gold and Sodium Chloride.
Glycerinophosphates.
Guaiacol: locally.
Guethol.
Hyoscyamus.
Ichthyol: as alterative and hematinic.
Ignatia: in hysterical and in intercostal neuralgia.
Iodides: especially when nocturnal.
Iodoform.
Kataphoresis.
Levico Water.
Massage.
Menthol.
Methacetin.
Methyl Chloride.
Methylene Blue.
Morphine: hypodermically.
Mustard: poultice.
Narceine.
Neurodin.
Nickel.

Nitroglycerin.
Nux Vomica: in visceral neuralgia.
Oil, Croton.
Oil, Mustard.
Oil of Cloves : locally.
Oil, Peppermint.
Oleoresin Capsicum.
Opium.
Peppermint: locally.
Peronin.
Phenacetin.
Phenocoll Hydrochlorate.
Phosphorus.
Potassium Arsenite Solution.
Potassium Bichromate.
Potassium Bromide.
Potassium Cyanide.
Potassium Salicylate.
Pulsatilla.
Pyoktanin.
Pyrethrum: as masticatory.
Quinine Salicylate.
Rubefacients.
Salicin.
Salophen.
Salol.
Sodium Dithio-salicylate, Beta.
Sodium Salicylate.
Sodium Sulphosalicylate.
Specific Remedies: if due to scrofula or syphilis.
Spinal Ice-bag.
Stavesacre.
Stramonium.
Strychnine.
Sumbul: sometimes very useful.
Thermo-cautery.
Thermodin.
Triphenin.
Tonga.
Turkish Bath.
Turpentine Oil.
Valerian.
Veratrine.
Vibration.
Wet Pack.
Zinc Cyanide.
Zinc Valerianate.

Neurasthenia. — *See also, Adynamia, Exhaustion, Gout, Hysteria, Spinal Irritation.*

Arsenic.
Bromalin.
Bromipin.
Bromo-hemol.
Codeine.
Cocaine.
Gaduol.
Glycerinophosphates
Gold.
Hypophosphites.
Levico Water.

Methylene Blue.
Orexine : as appetizer, etc.
Phosphorus.
Potassium Bromide.
Spermine.
Strychnine.
Sumbul.
Zinc Oxide.

Neuritis.—*See also, Alcoholism, Neuralgia, Spinal Irritation.*

Acetanilid.
Arsenic.
Benzanilide.
Gold.
Mercury.
Potassium Iodide.
Salicylates.
Strychnine.

Nevus.—*See also, Tumors, Warts.*

Acid, Carbolic.
Acid, Chromic.
Acid, Nitric.
Acid, Trichloracetic.
Aluminum Sulphate.
Antimonium Tartaratum.
Chloral Hydrate.
Collodion.
Creosote.
Croton Oil.
Electrolysis.
Galvano-Cautery.
Hydrargyri Bichloridum.
Ichthyol : topically.
Ichthalbin : internally.
Iodine : paint.
Iron Chloride.
Liquor Plumbi.
Nitrate of Mercury, Acid.
Potassium Nitrate.
Scarification.
Sodium Ethylate.
Tannin.
Zinc Chloride.
Zinc Iodide.
Zinc Nitrate.

Nightmare.

Bromide of Potassium.
Camphor Water.

Night-Sweats. — *See also, Perspiration, Phthisis, etc.*

Acid, Acetic: as a lotion.
Acid, Agaric.
Acid, Camphoric.
Acid, Gallic.
Acid, Salicylic.
Acid, Sulphuric, diluted.
Agaricin.
Alum.
Atropine.
Chloral Hydrate.

Ergotin.
Homatropine Hydrobromate.
Iron Sulphate.
Lead Acetate.
Picrotoxin.
Pilocarpine Hydrochlorate.
Potassium Ferrocyanide.
Potassium Tellurate.
Silver Oxide.
Sodium Tellurate.
Sulfonal.
Thallium Acetate.
Zinc Oleate.
Zinc Sulphate.

Nipples, Sore.—*See also, Lactation, Mastitis.*

Acid, Boric.
Acid, Carbolic.
Acid, Picric: fissures.
Acid, Sulphurous.
Acid, Tannic.
Alcohol : locally.
Arnica.
Balsam of Peru.
Balsam of Tolu.
Benzoin.
Bismuth Subgallate.
Borax : saturated solution locally.
Brandy and Water.
Breast-pump.
Catechu.
Chloral Hydrate Poultice.
Cocaine Solution (4 grn. to the ounce): applied and washed off before nursing, if breast is very painful.
Collodion.
Ferrous Subsulphate : locally.
Ichthyol: when indurated.
India Rubber.
Lead Nitrate.
Lead Tannate.
Lime Water.
Potassium Chlorate.
Rhatany : one part extract to 15 of cacao butter.
Silver Nitrate.
Sozoiodole salts.
Tannin, Glycerite of.
Yolk of Egg.
Zinc Oxide.
Zinc Shield.

Nodes.—*See also, Exostosis, Periostitis.*

Acid, Arsenous.
Arsen-hemol.
Cadmium Iodide.
Ichthalbin : internally.
Ichthyol : topically.
Iodipin.
Levico Water.
Mercury Oleate : with morphine, locally.

150

Potassium Iodide: internally and externally.

Stramonium Leaves: as poultice.

Nose-bleed. — *See Epistaxis.*

Nutrition, Defective. — *See list of Tonics, Gastric Tonics, etc.*

Nyctalopia.

Amyl Nitrite.
Blisters: small to external canthus.
Quinine.
Strychnine.

Nymphomania.

Acid, Sulphuric.
Anaphrodisiacs.
Bromide of Potassium: in large doses.
Camphor: in large doses.
Camphor, Monobromated.
Digitalis.
Hyoscine Hydrobromate.
Lupuline.
Opium.
Sodium Bromide.
Stramonium.
Sulphur: when due to hemorrhoids.
Tobacco: so as to cause nausea; effectual but depressing.

Obesity.

Acid, Hydriodic.
Acids, Vegetable.
Adonis Æstivalis: tincture.
Alkalies.
Alkaline Waters: especially those of Marienbad.
Ammonium Bromide.
Banting's System: living on meat and green vegetables, and avoiding starch, sugars and fats.
Cold Bath.
Diet.
Fucus Vesiculosus.
Iodides.
Iodoform.
Iodole.
Laxative Fruits and Purges.
Lemon Juice.
Liq. Potassæ.
Phytolacca.
Pilocarpine Hydrochlorate.
Potassium Permanganate.

Saccharin: to replace sugar in diet.
Salines.
Sodium Chloride.
Sulphurous Waters.
Thyraden.
Turkish Baths.
Vinegar very injurious.

Odontalgia. — *See also, Neuralgia.*

Acid, Carbolic: a single drop of strong, on cotton wool placed in cavity of tooth.
Acid, Nitric: to destroy exposed nerve.
Acid, Tannic.
Aconite: liniment or ointment in facial neuralgia if due to decayed teeth.
Aconitine.
Alum: a solution in nitrous ether locally applied.
Argenti Nitras: the solid applied to the clean cavity and the mouth then gargled.
Arsenic: as caustic to destroy dental nerve.
Belladonna.
Butyl-Chloral: in neuralgic toothache.
Calcium salts.
Camphor: rubbed on gum, or dropped on cotton wool and placed in tooth.
Camphor and Chloral Hydrate: liniment to relieve facial neuralgia.
Camphor, Carbolated.
Capsicum: a strong infusion on lint.
Carbon Tetrachloride.
Chamomile.
Chloral: solution in glycerin one in four, or solid, in cotton wool to be applied to the hollow tooth.
Chloral-Camphor.
Chloroform: into ear or tooth on lint; a good liniment with creosote; or injected into the gum.
Cocaine: the hydrochlorate into a painful cavity.
Colchicum: along with opium in rheumatic odontalgia.
Collodion: mixed with melted crystallized carbolic acid, and put into cavity on cotton wool; first increases, then diminishes, pain.
Coniine: solution in alcohol on cotton

wool and put into tooth.
Creosote: like carbolic acid.
Croton Oil.
Electricity.
Ethyl Chloride.
Gelsemium: to relieve the pain of a carious tooth unconnected with any local inflammation.
Ginger.
Ichthyol.
Iodine: painted on to remove tartar on teeth; and in exposure of fang due to atrophy of gum.
Menthol.
Mercury: as alterative and purgative.
Methyl Chloride.
Morphine: subcutaneously injected.
Nitroglycerin.
Nux Vomica.
Oil of Cloves: dropped into the cavity of a hollow tooth.
Opium: dropped into cavity.
Pellitory: chewed.
Potassium Bromide.
Pulsatilla: in rheumatic odontalgia.
Quinine: in full dose.
Resorcin: like creosote.
Sodium Bicarbonate: saturated solution to rinse mouth with.
Tannin: ethereal solution dropped in carious tooth.
Zinc Chloride: to destroy exposed pulp.

Œdema. — *See Dropsy.*

Œsophageal Affections. — *See also, Choking, Dysphagia.*

Anesthetics.
Belladonna.
Conium.
Hyoscyamus.
Silver Nitrate.

Onychia and Paronychia.

Cocaine.
Ichthyol.
Iodine.
Iodole.
Iodoformogen.
Morphine.
Pyoktanin.
Sodium Chloride.
Sozoiodole salts.
Turpentine Oil.

Onychia.

Acid, Carbolic: as local anesthetic.

Alum.
Aluminium Sulphate.
Arsenic.
Chloral Hydrate : locally.
Corrosive Sublimate.
Ferri Perchloridum.
Ferri Persulphas.
Iodoform : locally.
Lead Nitrate.
Mercury : as ointment, alternately with poultices.
Silver Nitrate : at commencement.
Tannin.
Tar Ointment.
Tartar Emetic.

Oöphoritis. —*See Ovaritis.*

Ophthalmia. — *See also, Blepharitis, Conjunctivitis, Keratitis.*

Acid, Boric.
Acid, Carbolic : pure, for chronic granulation ; excess removed with water.
Acid, Citric : ointment or lemon juice.
Acid, Tannic.
Alum.
Antimony.
Aristol.
Arsenic.
Atropine.
Boroglyceride (20 to 50 per cent.) : applied to chronic granulations.
Calcium Sulphide.
Calomel.
Colchicum.
Copper Sulphate.
Eserine.
Europhen.
Formaldehyde: for purulent ophthalmia.
Hot Compresses.
Ichthyol.
Iodine.
Iodoform.
Iodoformogen.
Iodole.
Jequirity : infusion painted on inner side of eyelids.
Lead Acetate.
Leeches : to temples.
Liquor Potassæ.
Mercury.
Mercury Bichloride: as lotion.
Mercury Oxide, Red: as ointment.
Naphtol.
Oil of Cade : 1 in 10.
Pulsatilla.
Pyoktanin.
Silver Nitrate.
Sozoiodole-Sodium.
Strontium Iodide.

Sulphur : insufflation for diptheritic conjunctivitis.
Tartar Emetic: as counter-irritant.
Zinc Acetate.
Zinc Chloride.
Zinc Oxide.
Zinc Sulphate.

Ophthaimia Neonatorum.—*See Ophthalmia.*

Opium Habit.

Ammonium Valerianate.
Atropine.
Bromo-hemol.
Bromalin.
Bromipin.
Bromides.
Cannabis Indica.
Capsicum.
Chloral Hydrate.
Cocaine.
Codeine.
Conium.
Cupro-hemol.
Duboisine.
Eserine.
Gelsemium.
Gold and Sodium Chloride.
Hyoscine Hydrobromate.
Iron.
Nitroglycerin.
Paraldehyde.
Sparteine Sulphate.
Sodium Bromide.
Strychnine.
Zinc Oxide.

Orchitis.—*See also, Epididymitis.*

Ammonium Chloride.
Anemonin.
Belladonna.
Calomel.
Ichthyol.
Iodine.
Iodole.
Iodoform.
Iodoformogen.
Guaiacol.
Mercury Oleate.
Morphine.
Pulsatilla.
Silver Nitrate.
Sodium Salicylate.
Strapping.
Tartar Emetic.

Osteomalacia. — *See also, Bone Diseases.*

Glycerinophosphates.
Levico Water.
Phosphates.

Osteomyelitis.—*See also, Bone Diseases.*

Europhen.
Sozoiodole-Mercury.

Otalgia.—*See also, Otitis.*

Aconite.
Atropine.
Brucine.
Chloral Hydrate.
Chloroform.
Cocaine.
Glycerin.
Oil Almonds.
Opium.
Pulsatilla.
Tincture Opium.

Otitis.—*See also Otalgia.*

Acid, Carbolic.
Aconite.
Alumnol.
Aristol.
Atropine.
Creosote.
Cocaine.
Creolin.
Diaphtnerin.
Europhen.
Ichthyol : in otitis media.
Iodole.
Naphtol.
Potassium Permanganate.
Pulsatilla Tincture.
Pyoktanin.
Resorcin.
Retinol.
Salol.
Sozoiodole salts.
Styrone.

Otorrhea.—*See also, Otitis.*

Acid, Boric.
Acid, Carbolic.
Acid, Tannic.
Aconite.
Alcohol.
Alum : insufflation.
Arsenic.
Cadmium : locally.
Cadmium Sulphate.
Caustic.
Chloral Hydrate.
Cod-Liver Oil.
Cotton Wool.
Creosote.
Diaphtherin.
Gaduol.
Hydrastine Hydrochlorate.
Hydrogen Peroxide.
Iodide : two grn. to the ounce, locally.
Iodipin.
Iodole.
Iodoform.
Iodoformogen.
Lead Acetate.
Lead Lotions.
Levico Water.
Lime Water.

Liquor Sodæ: locally when discharge is fetid.
Mercury, Brown Citrine Ointment.
Permanganate of Potassium: as injection or spray.
Pyoktanin.
Quinine.
Resorcin.
Silver Nitrate: locally.
Sozoiodole-Sodium.
Sozoiodole-Zinc.
Sulphocarbolates.
Tannin, Glycerite of: very useful.
Zinc Sulphate.

Ovarian Diseases.

Atropine.
Bromo-hemol.
Bromipin.
Bromides.
Codeine.
Conium.
Glycerinophosphates.
Ichthyol.
Ovariin.

Ovarian Neuralgia.

—See also, Dysmenorrhea, Neuralgia, Ovaritis.

Ammonium Chloride.
Atropine.
Camphor, Monobromated.
Cannabis Indica.
Codeine.
Conium.
Gelsemium.
Gold and Sodium Chloride.
Opium.
Triphenin.
Zinc Valerianate.

Ovaritis.

Anemonine.
Belladonna.
Camphor.
Cannabis Indica.
Conium.
Gold.
Ichthalbin: internally.
Ichthyol: topically.
Mercury.
Opium.
Ovariin.
Tartar Emetic: as ointment.
Turpentine Oil: as counter-irritant.

Oxaluria.

Acid, Lactic.
Acids, Mineral.
Acid, Nitric.
Acid, Nitrohydrochloric.
Zinc Sulphate.

Ozena.

—See also, Catarrh, Chronic, Nasal.

Acetate of Ammonium.
Acid, Carbolic.
Acid, Chromic.
Acid, Salicylic.
Acid, Sulphurous.
Acid, Trichloracetic.
Alum: as powder or wash.
Aluminium Acetotartrate.
Alumnol.
Aristol.
Bichromate of Potassium.
Bismuth Subgallate.
Bismuth Subnitrate.
Boroglyceride.
Bromine: as inhalation
Calcium Chloride.
Calomel Snuff.
Carbolate of Iodine.
Chlorinated Lime or Chlorinated Soda: injections of the solution.
Chlorophenol.
Creolin.
Cubeb.
Diaphtherin.
Ethyl Iodide.
Gaduol.
Glycerin and Iodine.
Gold salts.
Hydrastis: internally and locally.
Hydrogen Peroxide.
Insufflation.
Iodides.
Iodine: as inhalation. Much benefit derived from washing out the nose with a solution of common salt, to which a few drops of the tincture of iodine have been added.
Iodipin.
Iodoform.
Iodoformogen.
Iodole.
Iron.
Medicated Cotton.
Mercuric Oxide, or Ammoniated Mercury.
Naphtol.
Papain.
Potassium Chlorate.
Potassium Iodide.
Potassium Permanganate.
Salol.
Silver Nitrate.
Sodium Arseniate.
Sodium Chloride.
Sodium Ethylate.
Sozoiodole salts.
Stearates.
Tannin, Glycerite of.
Thujæ: tincture.

Pain.

—See also, After-Pains, Anesthesia, Boils, Chest Pains, Colic, Gastralgia, Headache, Hepatalgia, Inflammation, Lumbago, Myalgia, Neuralgia, Neuritis. Odontalgia, Otalgia, Ovarian Neuralgia, Rheumatism, etc. Also lists of Andlgesics, Anesthetics and Narcotics.

Acetanilid.
Acid, Carbolic.
Aconite.
Aconitine.
Ammonium Iodide.
Atropine.
Belladonna.
Camphor, Monobromated.
Camphor-phenol.
Cannabis Indica.
Chloroform.
Chloral Hydrate.
Chloral-Camphor.
Cocaine.
Codeine.
Conium.
Duboisine.
Ethyl Chloride Spray.
Exalgine.
Gelseminine.
Guaiacol.
Hyoscyamine.
Ichthyol.
Iodine.
Iodoform.
Iron.
Manganese Dioxide.
Menthol.
Methyl Chloride Spray.
Morphine.
Neurodin.
Opium.
Peronin.
Phenacetin.
Potassium Cyanide.
Solanine: in gastric pain.
Stramonium.
Triphenin.
Tropacocaine.

Pain, Muscular.

—See Myalgia.

Palpitation.

Aconite.
Belladonna.
Cactus Grandiflorus: tincture.
Convallaria.
Spirit Ether.
Sparteine Sulphate.
Strophanthus: tincture.

Papilloma.—*See Tumors, Warts.*

Paralysis Agitans.—*See also, Chorea, Tremor.*

Arsenic.
Arsen-hemol.
Borax.
Cannabis Indica.
Chloral Hydrate.
Cocaine.
Conium.
Duboisine.
Gelsemine.
Glycerinophosphates.
Hyoscine Hydrobromate.
Hyoscyamine.
Hypophosphites.
Levico Water.
Opium.
Picrotoxin.
Phosphorus.
Potassium Iodide.
Sodium Phosphates,
Sparteine.
Spermine.

Paralysis, Lead.—*See Lead Poisoning.*

Paralysis and Paresis.—*See also, Hemiplegia, Locomotor Ataxia, Paralysis Agitans.*

Ammonium Carbonate.
Ammonium Iodide.
Arnica.
Arsen-hemol.
Belladonna.
Cannabis Indica.
Calcium Lactophosphate.
Capsicum.
Colocynth.
Eserine.
Glycerinophosphates.
Levico Water.
Nux Vomica.
Phosphorus.
Picrotoxin.
Rhus Toxicodendron.
Spermine.
Strychnine.

Parametritis and Perimetritis.—*See Metritis.*

Parasites.

Acid, Sulphurous.
Anise.
Bake Clothes: to destroy ova of parasites.
Benzin.
Chloral.
Chloroform.
Chrysarobin.
Cocculus Indicus.
Creolin.

Delphinium.
Essential Oils.
Ichthyol: pure.
Insect Powder.
Laurel Leaves: decoction.
Losophan.
Mercury Bichloride: in parasitic skin diseases.
Mercury Oleate.
Mercury Oxide, red.
Naftalan.
Naphtol.
Oil Cajuput will destroy pediculi.
Oil of Cloves.
Petroleum.
Picrotoxin: against pediculi.
Pyrogallol.
Quassia.
Sabadilla.
Sodium Hyposulphite.
Sozoiodole salts.
Stavesacre.

Sulphurated Potassa.
Veratrine.

Parotitis.

Aconite.
Ammonium Acetate.
Emetics.
Gaduol: internally, as alterative.
Guaiacol.
Ichthalbin: internally, as tonic and alterative.
Ichthyol.
Jaborandi.
Leeches.
Mercury: one-half grn. of gray powder three or four times a day.
Poultice.

Parturition.

Antipyrine.
Chloral Hydrate.
Castor Oil: to relieve constipation.
Creolin: as irrigation.
Cimicifuga.
Diaphtherin.
Mercuric Chloride.
Quinine: as a stimulant to uterus.

Pediculi.—*See Parasites.*

Pelvic Cellulitis.—*See Metritis.*

Pemphigus.

Arsen-hemol.
Arsenic.
Belladonna.
Bismuth Subgallate.
Chlorate of Potassium.
Cod-Liver Oil.
Hot Bath.
Iodide of Potassium.
Levico Water.

Mercury.
Naftalan.
Naphtol.
Phosphorus.
Silver Nitrate.
Sulphides.
Tar.
Zinc Oxide.

Pericarditis.—*See also, Endocarditis.*

Aconite.
Alcohol: sometimes very useful.
Aspiration, gradual, if exudation threatens life.
Bleeding.
Blisters: near heart.
Bryonia: useful in exudation.
Calomel and Opium: formerly much used.
Digitalis: when heart is rapid and feeble with cyanosis and dropsy.
Elaterium.
Ice: bag over the precordium.
Iodides.
Iodine.
Iron.
Jalap.
Leeches.
Mercury.
Oil Gaultheria.
Opium: in grain doses every three to six hours, very useful.
Poultice.
Quinine.
Saliformin.
Sodium Salicylate.
Squill.
Veratrum Viride.

Periones.—*See Chilblains.*

Periostitis.—*See also, Nodes, Onychia.*

Calcium Phosphate.
Formaldehyde.
Ichthalbin: internally.
Ichthyol: topically.
Iodide of Potassium, or Ammonium.
Iodine: locally.
Mercury: internally.
Mercury and Morphine Oleate: externally.
Mezereon: in rheumatic and scrofulous cases.
Morphine.
Phosphates.
Poultices.
Sozoiodole-Sodium.
Sozoiodole-Zinc.
Stavesacre: when long bones affected.
Tonics and Stimulants.

Peritonitis.—*See also, Puerperal Peritonitis.*

Acetanilid.
Aconite: at commencement.
Ammonia.
Antimony.
Blisters.
Bryonia: when exudation.
Calomel.
Chloral Hydrate.
Chlorine Solution.
Cocculus Indicus: for tympanites.
Codeine.
Cold.
Hyoscyamus.
Ice.
Ichthyol: in pelvic peritonitis.
Iodine.
Ipecacuanha.
Leeches.
Mercury: when there is a tendency to fibrous exudation.
Opium: freely, most useful.
Plumbic Acetate.
Potassium salts.
Poultices.
Quinine.
Rectal Tube: milk or asafetida or turpentine injections, in tympanites.
Rubefacients.
Salines.
Steam: applied to the abdomen under a cloth when poultices cannot be borne.
Turpentine Oil: for tympanites.
Veratrum Viride.

Peritonitis, Tubercular.

Arsenic.
Creosote.
Gaduol.
Glycerinophosphates.
Guaiacol.
Ichthyol: locally.
Ichthalbin: internally
Maragliano's Serum.
Opium.
Quinine.
Spermine.

Perspiration, Excessive.—*See also, Night-Sweats, Feet.*

Acid, Agaricic.
Acid, Aromatic Sulphuric: in phthisis.
Acid, Camphoric.
Acid, Carbolic: with glycerin locally for fetid sweat.
Acid, Chromic.

Acid, Gallic: in phthisis.
Acid, Salicylic: with borax in fetid perspiration.
Agaricin: in phthisis.
Atropine: in sweating of phthisis, internally.
Belladonna: as liniment for local sweats.
Betula.
Copper salts.
Duboisine.
Ergot.
Formaldehyde.
Glycerin.
Hydrastine Hydrochlorate.
Iodoform.
Jaborandi.
Lead.
Mercury.
Muscarine.
Naphtol.
Neatsfoot Oil: rubbed over the surface.
Oils.
Opium: as Dover's powder in phthisis.
Permanganate of Potassium: locally for fetid perspiration.
Picrotoxine.
Pilocarpine.
Quinine.
Salicin: in phthisis.
Spinal Ice Bag.
Sponging: very hot.
Strychnine: in phthisis.
Tannin.
Tannoform.
Thallium.
Turpentine Oil.
Vinegar: locally.
Zinc Oxide: in phthisis.

Pertussis (*Whooping-Cough*). — *See also, Cough.*

Acetanilid.
Acid, Carbolic: as spray
Acid, Hydrobromic.
Acid, Hydrocyanic: in habitual cough when the true whooping cough has ceased.
Acid, Nitric.
Acid, Salicylic: as spray.
Aconite.
Allyl Tribromide.
Alum.
Ammonium Bromide.
Ammonium Chloride.
Ammonium Valerianate.
Amyl Nitrite.
Amylene Hydrate.
Anemonin.
Antipyrine.
Antispasmin.
Argenti Oxidum.
Arnica.
Arsenic.

Atropine.
Belladonna.
Benzin: sprinkled about the room.
Bitter Almond Water.
Blister: to nape of neck.
Bromalin.
Bromides.
Bromoform.
Butyl-Chloral.
Cantharides.
Castanea Vesca.
Cerium Oxalate.
Cheken.
Cherry-Laurel Water.
Chloral Hydrate: in spasmodic stage.
Chloroform: as inhalation during paroxysm
Clover Tea.
Cocaine Hydrochlorate.
Cochineal.
Codeine.
Cod-Liver Oil.
Coffee.
Conine.
Copper Arsenite.
Decoction of Chestnut leaves, *ad lib.* Sometimes useful.
Drosera.
Ergot.
Ether, Hydriodic.
Ether Spray.
Formaldehyde.
Gaduol.
Gelsemium: in spasmodic stage.
Grindelia.
Gold and Sodium Chloride.
Hydrogen Peroxide.
Hyoscyamus.
Inhalation of atomized fluids.
Ipecacuanha: sometimes very useful alone, or combined with bromide of ammonium.
Lactucarium.
Leeches: to nape of neck.
Levico Water.
Lobelia: in spasmodic stage.
Milk Diet.
Monobromate of Camphor.
Morphine.
Myrtol.
Naphtalin.
Oil Amber.
Opium: in convulsive conditions.
Peronin.
Phenacetin.
Potassa Sulphurata.
Potassium Cyanide.
Quinine.
Quinoline Salicylate.
Resorcin.
Silver Chloride.

155

Silver Nitrate.
Sodium Benzoate.
Sodium Carbolatum.
Sodium Salicylate.
Sozoiodole-Sodium.
Tannin.
Tar: for inhalation.
Tartar Emetic.
Terpene Hydrate.
Thymol.
Turpentine Oil.
Urtica.
Vaccination.
Valerian.
Valerianate of Atropine
Veratrum Viride.
Wild Thyme.
Zinc Oxide.
Zinc Sulphate.

Phagedena.

Acid, Nitric.
Iodoform.
Iodoformogen.
Iodole.
Sozoiodole-Zinc.
Opium.
Potassa.

Pharyngitis.—*See also, Throat, Sore; Tonsillitis.*

Acetanilid.
Acid, Nitric.
Acid, Sulphurous.
Actæa Racemosa.
Aconite.
Alcohol: dilute as gargle.
Alum: as gargle.
Alumnol.
Ammonii Acetatis, Liq.
Ammonium Chloride.
Antipyrine: in 4 per cent. spray.
Asaprol.
Belladonna.
Boroglyceride.
Capsicum: as gargle.
Catechu.
Cimicifuga: internally when pharynx is dry.
Cocaine: gives temporary relief; after-effects bad.
Copper Sulphate: locally.
Creolin.
Cubeb Powder.
Electric Cautery.
Ergot.
Ferric Chloride: locally as astringent, internally as tonic.
Glycerin: locally, alone or as glycerin and tannin.
Guaiacum.
Hamamelis.
Hydrastine Hydrochlorate.
Hydrastis: internally and locally.
Hydrogen Peroxide.
Ice.

Ichthyol.
Iodine.
Iodoform.
Iodoformogen.
Ipecacuanha: as spray.
Myrrh.
Monsel's Solution: pure, or diluted with glycerin one half, applied on pledgets of cotton or camel's hair brush.
Naphtol.
Opium.
Pomegranate Bark: as gargle.
Potassium Chlorate: locally.
Pyoktanin.
Quinine: as tonic.
Resorcin.
Salol.
Silver Nitrate: in solution locally.
Sodium Borate.
Sozoiodole-Sodium.
Sozoiodole-Zinc.
Strychnine: as tonic.
Tannin: as powder or glycerin locally.
Tropacocaine.
Zinc Sulphate: as gargle.

Phimosis.

Belladonna: locally.
Chloroform.
Cocaine.
Elastic Ligament.
Lupulin: after operation.
Sozoiodole-Potassium.
Warm Baths.

Phlebitis. — *See also, Phlegmasia, Varicocele.*

Blisters.
Calomel.
Hamamelis.
Hot Fomentations.
Ichthalbin: internally.
Ichthyol: topically.
Lead and Opium Wash.
Mercury.
Opium: to allay pain.
Rest, absolute.

Phlegmasia Alba Dolens.

Acid, Hydrochloric: with potassium chlorate, in barley water.
Ammonium Carbonate: in full doses when much prostration.
Belladonna Extract: with mercurial ointment locally.
Blisters: in early stage.
Creosote: as enemata.
Hamamelis.
Ichthalbin: internally.
Ichthyol: topically.

Leeches: during active inflammation.
Opium: internally and locally to allay pain.
Pyoktanin.

Plegmon. — *See also, Erysipelas.*

Acid, Carbolic: injections.
Aconite.
Belladonna.
Creolin.
Ichthyol.
Iodine.
Iodole.
Iodoformogen.
Pyoktanin.
Silver Nitrate.
Sozoiodole-Sodium.

Phosphaturia.

Acid, Benzoic.
Acid, Lactic.
Benzoates.
Hippurates.
Glycerinophosphates.

Photophobia.

Ammonium Chloride.
Atropine.
Belladonna: to eye.
Bromide of Potassium.
Butyl-Chloral.
Calabar Bean.
Calomel: insufflation.
Chloroform Vapor.
Cocaine.
Cold.
Coniine: in scrofulous photophobia locally.
Galvanism.
Iodine Tincture.
Mercuric Chloride: by insufflation.
Nitrate of Silver.
Opium.
Potassium Chlorate: in large doses.
Seton.
Tonga.

Phthisis.—*See also, Cough, Hemoptysis, Hectic Fever, Perspiration, Night Sweats, Laryngitis, Tubercular; Meningitis, Tubercular; Peritonitis, Tubercular; Tuberculosis, Acute; Tuberculous affections.*

Acetanilid.
Acid, Agaric.
Acid, Benzoic.
Acid, Campheric.
Acid, Carbolic.
Acid, Cinnamic.
Acid, Gallic.
Acid, Gynocardic.
Acid, Hydrochloric.

Acid, Hydrocyanic, Dil.
Acid, Lactic.
Acid, Oxalic.
Acid, Phenylacetic.
Acid, Phosphoric.
Acid, Salicylic: when breath foul and expectoration offensive.
Acid, Sulphuric.
Acid, Sulphurous: as fumigation.
Aconite.
Actæa Racemosa.
Agaricin.
Alantol.
Alcohol: along with food or cod-liver oil.
Alum.
Amylene Hydrate.
Ammonium Borate.
Ammonium Carbonate.
Ammonium Iodide.
Ammonium Urate.
Antimony Tartrate.
Antipyrine: to reduce temperature.
Antituberculous Serum
Apomorphine Hydrochlorate.
Aristol.
Arsenic: to remove commencing consolidation, and also when tongue is red and irritable.
Asaprol.
Atropine: to check perspiration.
Balsam Peru.
Belladonna: locally for pain in muscles.
Benzoin: as inhalation to lessen cough and expectoration.
Benzosol.
Bismuth Citrate.
Bismuth Subgallate.
Bitter Almond Oil.
Blisters.
Bromides.
Butyl-Chloral: to check cough.
Cantharidin.
Calcium Chloride.
Calcium Hippurate.
Camphor.
Cannabis Indica.
Carbo Ligni.
Cerium Oxalate.
Cetrarin.
Chaulmoogra Oil.
Chloralamide.
Chloral: as hypnotic.
Chlorine.
Chlorodyne.
Chloroform: as linctus to check cough.
Chlorophenol.
Cimicifugin.
Climate Treatment.
Clove Oil.
Cocaine: a solution locally to throat and mouth tends to relieve irritable condi-

tion and aphthæ, especially in later stages
Codeine.
Cod-Liver Oil: most useful as nutrient.
Conium.
Coto Bark.
Counter-Irritation.
Copper Sulphate.
Creolin.
Creosote (Beech-Wood): as inhalation, and internally.
Croton Oil: to chest as counter-irritant.
Cupro-hemol.
Digitalis.
Enemata: of starch and opium, to control diarrhea.
Ether.
Ethyl Iodide.
Eucalyptus Oil.
Eudoxin.
Eugenol.
Euphorbia Pilulifera.
Europhen.
Gaduol.
Gelsemium.
Glycerin: as nutrient in place of cod-liver oil, locally to mouth in the last stages to relieve dryness and pain.
Glycerinophosphates.
Gold Iodide.
Guaiacol and salts.
Guaiacum.
Guethol.
Homatropine Hydrobromate.
Hydrastinine Hydrochlorate.
Hydrogen Dioxide.
Hypnal.
Hypophosphites: very useful in early stage.
Ichthalbin: internally, to regulate digestive functions, increase food-assimilation and act as reconstitutive.
Ichthyol: by inhalation.
Inulin: possibly useful.
Iodine: liniment as a counter-irritant to remove the consolidation in early stage, and to remove pain and cough later; as inhalation to lessen cough and expectoration.
Iodine Tincture.
Iodipin.
Iodoform: inhalation.
Iodole.
Iron Iodide.
Iron Sulphate.
Kumyss.
Lactophosphates.
Lead Acetate.
Lead Carbonate.

Magnesium Hypophosphite.
Manganese Iodide.
Menthol.
Mercury Bichloride: in minute doses for diarrhea.
Mercury Bichloride Solution (1:10,000): heat, and inhale steam, stopping at first sign of mercurial effect. In laryngeal phthisis: precede inhalation with cocaine spray (4 per cent. sol.).
Methacetin.
Methylene Blue.
Mineral Waters.
Morphine, with Starch or Bismuth: locally to larynx and in laryngeal phthisis most useful.
Mustard Leaves: most useful to lessen pain and prevent spread of subacute intercurrent inflammation.
Myrtol.
Naphtol.
Nuclein.
Ol. Pini Sylvestris.
Ol. Lini and Whisky.
Opium: to relieve cough, and, with ipecacuanha and Dover's powder, to check sweating.
Orexine Tannate: as appetizer and indirect reconstituent.
Oxygen.
Ozone.
Pancreatin.
Peronin.
Phellandrium.
Phenacetin.
Phenocoll Hydrochlorate.
Phosphate of Calcium: as nutrient, and to check diarrhea.
Picrotoxin: to check perspiration.
Pilocarpine: to check sweats.
Podophyllum.
Potassæ Liquor.
Potassium Cantharidate.
Potassium Chloride.
Potassium Cyanide.
Potassium Hypophosphite.
Potassium Iodide.
Potassium Phosphate.
Potassium Tellurate.
Prunus Virginiana: tincture.
Pyridine.
Quinine: as tonic to lessen temperature, to check sweat,

Raw Meat and Phosphates.
Salicin.
Salophen.
Sanguinaria.
Sea Bathing.
Sea Voyage.
Serum, Antitubercular.
Silver Nitrate.
Snuff.
Sodium Arsenate.
Sodium Benzoate.
Sodium Chloride.
Sodium Hypophosphite.
Sodium Hyposulphite.
Sodium Phosphate.
Sodium Tellurate.
Spermine.
Sponging : very hot.
Stryacol.
Strychnine.
Sulphaminol.
Sulphur.
Sunbul.
Tannalbin : as antidiarrheal and indirect reconstitutive.
Tannoform.
Tar.
Terebene.
Terpene Hydrate.
Thallium Acetate.
Thermodin.
Thiocol.
Thymol.
Transfusion.
Tuberculin.
Turpentine Oil.
Vinegar.
Xeroform.
Zinc Sulphate.

Piles.—*See Hemorrhoids.*

Pityriasis.—*See also, Seborrhea ; and for Pityriasis Versicolor, see Tinea Versicolor.*

Acid, Acetic.
Acid, Carbolic : with glycerin and water locally.
Acid, Sulphurous : locally.
Alkalies and Tonics.
Anthrarobin.
Arsen-hemol.
Arsenic.
Arsenic and Mercury : internally.
Bichloride of Mercury.
Borax : saturated solution or glycerite locally.
Cajuput Oil.
Chrysarobin.
Citrine Ointment.
Gaduol.
Glycerin.
Glycerinophosphates.
Ichthalbin : internally.
Ichthyol : topically.

Lead : locally.
Levico Water.
Mercury Ointment.
Myrtol.
Naftalan.
Oleate of Mercury.
Resorcin.
Sapo Laricis.
Solution Arsenic and Mercury Iodide.
Sulphides : locally.
Sulphides.
Sulphur.
Thyraden.

Pityriasis Capitis.— *See Seborrhea.*

Pleurisy.—*For Chronic Pleurisy, see Empyema. See also, Hydrothorax, Pleuro-Pneumonia.*

Acid, Hydriodic.
Aconite : in early stage.
Antimony.
Antipyrine.
Aspiration.
Belladonna Plaster : most useful to relieve pain in old adhesions.
Blisters.
Blood-letting.
Bryonia : after aconite.
Calomel.
Cantharides.
Chloral Hydrate.
Cod-Liver Oil.
Coniine.
Cotton Jacket.
Digitalis : when much effusion.
Diuretin.
Elaterium.
Gaduol.
Gelsemium.
Glycerinophosphates.
Guaiacol.
Ice Poultice or Jacket : in sthenic cases.
Iodide of Potassium : to aid absorption.
Iodides.
Iodine : as a liniment to assist absorption, or as a wash or injection to cavity after tapping.
Jaborandi.
Jalap.
Leeches.
Local Wet Pack.
Mercury Salicylate.
Morphine.
Neurodin.
Oil Gaultheria.
Oil Mustard.
Orexine: for anorexia.
Paraldehyde.
Pilocarpine.
Poultices.
Purgative salts.
Quinine.
Sinapisms.

Sodium Chloride.
Sodium Salicylate.
Sodium Sulphosalicylate.
Strapping Chest : if respiratory movements are very painful.
Strontium Salicylate.
Thermodin.
Triphenin.
Veratrum Viride.

Pleuritic Effusions.
Iodine.

Pleurodynia.—*See also, Neuralgia.*

Acid, Carbolic.
Acupuncture.
Belladonna . plaster or liniment very useful.
Blistering.
Chloral Hydrate: with camphor locally.
Cimicifuga.
Croton Oil : locally in obstinate cases.
Ether : as spray, locally.
Gelsemium.
Iodine : locally.
Iron: when associated with leucorrhea.
Morphine.
Mustard Leaves.
Nerve-stretching.
Opium : liniment rubbed in after warm fomentations or hypodermic injections. Internally, most useful to cut short attack and relieve pain.
Pilocarpine.
Plasters : to relieve pain and give support.
Poultices.
Quinine.
Sanguinaria.
Strapping.
Turpentine Oil.
Veratrum Viride.
Wet-cupping : when pain severe and fever high.

Pleuro-Pneumonia.

Acid, Carbolic : two per cent. solution injected locally.
Bryonia.
Sanguinaria.
Turpentine Oil : locally.

Pneumonia.—*See also, Pleuro-Pneumonia.*

Acid, Hydriodic.
Acid, Phosphoric.
Acid, Salicylic.
Aconite : very useful, especially at commencement.
Alantol.
Alcohol

Ammonia.
Ammonium Carbonate: as stimulant.
Ammonium Chloride.
Antimony.
Antipyrine.
Arnica.
Belladonna: at commencement.
Benzanilide.
Bleeding.
Blisters: at beginning to lessen pain.
Bryonia: when pleurisy present.
Caffeine.
Calomel.
Camphor.
Carbonate of Sodium.
Chloral Hydrate.
Chloroform.
Codeine.
Cold Bath.
Cold Compress to Chest
Cold Sponging.
Confine.
Copper Acetate.
Copper Sulphate.
Cups, dry and wet: in first stage.
Digitalis: to reduce temperature.
Dover's Powder: for pain at onset.
Ergot.
Ether.
Eucalyptus.
Expectorants.
Gelsemium.
Gin.
Guaiacol.
Hoffman's Anodyne.
Ice-bag: to heart, if fever be high and pulse tumultuous.
Ice Poultice or Jacket: in first stage of sthenic cases.
Iodides.
Mercury.
Morphine.
Muscarine.
Naphtol.
Neurodin.
Nitroglycerin.
Nux Vomica: tincture.
Opium.
Oxygen Inhalations.
Phosphorus.
Pilocarpine.
Plumbi Acetas.
Potassium Chlorate
Potassium Citrate.
Potassium Nitrate.
Poultices: to lessen pain.
Quinine: to lower temperature.
Salicylate of Sodium: as antipyretic.
Senega: as expectorant.
Sanguinaria.
Serpentaria: with carbonate of ammonium as stimulant.

Sinapisms.
Stimulants.
Strychnine.
Sodium Bicarbonate.
Sodium Carbonate.
Sodium Paracresotate.
Sweet Spirit of Nitre.
Tartar Emetic.
Thermodin.
Triphenin.
Turpentine Oil: as stimulant at crisis.
Veratrine.
Veratrum Viride.
Wet Pack.

Podagra, Acute and Chronic.—See Arthritis.

Ichthyol.

Polypus.

Acid, Acetic.
Acid, Carbolic, and Glycerin
Alcoholic Spray.
Alum: as insufflation.
Aluminium Sulphate.
Iodole.
Iodoformogen.
Iron.
Sanguinaria.
Sesquichloride of Iron.
Sodium, Ethylate.
Sozoiodole salts.
Tannin: as insufflation.
Tr. Opii Crocata.
Zinc Chloride.
Zinc Sulphate.

Porrigo.—See also, Impetigo, Alopecia Areata, Tinea, etc.

Acid, Carbolic.
Acid, Sulphurous.
Ammoniated Mercury.
Ammonium Acetate.
Bismuth Subgallate.
Creolin.
Levico Water.
Losophan.
Manganese Dioxide.
Mercuric Nitrate Ointment.
Naftalan.
Picrotoxin.
Red Mercuric Oxide Ointment.
Solution Arsenic and Mercuric Iodide.
Sulphites.

Pregnancy, Disorders of.—See also, Albuminuria, Nephritis, Nervousness, Ptyalism, Vomiting of Pregnancy.

Acid, Tannic.
Aloes.
Alum.
Antispasmodics.
Berberin.

Bismuth.
Bromo-hemol.
Calcium Bromide.
Calcium Phosphate.
Camphor.
Chloroform Water.
Cocculus Indicus.
Digitalis.
Iodine.
Mercury.
Opium.
Orexine: for the vomiting; most efficacious.
Potassium Acetate.
Potassium Bromide.
Sumbul.

Proctitis.—See Rectum.

Prolapsus Ani.

Acid, Nitric.
Aloes.
Alum: in solution locally.
Bismuth.
Electricity.
Ergotin.
Glycerinophosphates.
Hydrastis: as enema or lotion.
Ice: when prolapsed parts inflamed.
Ichthyol.
Injections of hot or cold water.
Iron Sulphate.
Nutgall.
Nux Vomica.
Opium.
Pepper: confection.
Podophyllum: in small doses.
Silver Nitrate.
Stearates.
Strychnine: as adjunct to laxatives.
Sulphur.
Tannin: as enema.

Prolapsus Uteri.

Alum: as hip-bath and vaginal douche.
Astringents.
Bromide of Potassium
Cimicifuga: to prevent miscarriage and prolapsus.
Electricity.
Galls: decoction of, as injection.
Glycerin Tampon.
Ice: locally when part inflamed, and to spine.
Oak Bark: as injection.
Secale.
Tannin.

Prostate, Enlarged.—See also, Cystitis.

Alkalies: when irritation of the bladder, with acid urine.

159

Ammonium Benzoate: for cystitis with alkaline urine.
Ammonium Chloride.
Colchicum.
Conium.
Ergot.
Ichtalbin: internally.
Ichthyol: topically.
Iodine: to rectum.
Iodoform or Iodoformogen: as suppository very useful.
Iodole.
Prostaden.
Sulphides.

Prostatitis.—See also, *Prostatorrhea ; and Prostate, Enlarged.*

Blisters to Perineum: in chronic cases.
Buchu.
Cantharides: small doses of tincture.
Cold Water: injections and perineal douches.
Cubebs.
Hot Injections.
Hydrastis: internally and locally.
Ichthalbin: internally, as vaso-constrictor or tonic.
Ichthyol.
Iron.
Juniper Oil.
Local treatment to prostatic urethra,and use of cold steel sounds,in chronic types.
Perineal incision to evacuate pus if abscess forms.
Rest in bed, regulation of bowels, leeches to perineum, medication to render urine alkaline, and morphine hypodermically or in suppository.
Silver Nitrate: locally.
Soft Catheter: allowed to remain in bladder if retention of urine.
Turpentine Oil.

Prostatorrhea. — See also, *Prostatitis.*

Atropine.
Cantharides.
Hydrastis.
Iron.
Lead.
Potassium Bromide.

Prurigo. — See also, *Pruritus.*

Acid, Boric.
Acid, Carbolic: internally and locally, especially in prurigo senilis.

Acid, Citric.
Acid, Hydrocyanic: locally.
Acid, Salicylic.
Aconite: externally.
Adeps Lanæ, Benzoated
Alkaline Lotions.
Alkaline Warm Baths.
Alum: a strong solution for pruritus vulvæ.
Aluminium Nitrate.
Arsen-hemol.
Arsenic: internally.
Atropine.
Balsam of Peru.
Belladonna.
Borax: saturated solution.
Bromide of Potassium.
Brucine.
Calcium Chloride.
Calomel: ointment very useful in pruritus ani.
Camphor, Carbolated.
Cantharides.
Chloral and Camphor.
Chloroform Ointment.
Cocaine
Cod-Liver Oil: as inunction.
Cold Douche.
Corrosive Sublimate: for pruritus vulvæ.
Cyanide of Potassium: as lotion or ointment, to be used with care.
Electricity.
Gaduol.
Gallanol.
Gelsemium.
Glycerin.
Glycerite of Tar.
Goulard's Extract.
Hot Water.
Ice.
Ichthalbin: internally.
Ichthyol: topically.
Iodide of Sulphur, Ointment of.
Iodoform: as ointment.
Levico Water.
Losophan.
Mercury Oleate with Morphine.
Mercury Bichloride.
Naftalan.
Naphtol.
Opium.
Oil of Cade.
Petroleum.
Phosphorus.
Pilocarpine.
Potassium Carbonate.
Quinine.
Resorcin.
Sapo Viridis.
Silver Nitrate.
Sodium Carbonate.
Sodium Iodide.
Stavesacre.
Strychnine.
Sulphate of Zinc.
Sulphides.

Sulphites.
Sulphur and compounds.
Tar Ointment.
Tobacco: useful but dangerous.
Tonics.
Turkish Baths.
Warm Baths.

Pruritus.—See also, *Eczema, Erythema, Parasites, Prurigo, Scabies, Urticaria.*

Lead Water.
Menthol.
Mercury Bichloride.
Mercury Oleate with Morphine.
Oil Amond, Bitter.
Potassium Cyanide.
Resorcin.
Sodium Salicylate.
Strychnine.
Sozoiodole-Zinc.
Zinc Sulphate.

Psoriasis.

Acid, Carbolic.
Acid, Chromic: ten grn. to the ounce in psoriasis of tongue.
Acid, Chrysophanic.
Acid, Gallic.
Acid, Hydriodic.
Acid, Hydrochloric.
Acid, Pyrogallic.
Acids, Mineral.
Acids, Nitric and Nitrohydrochloric: when irruption is symptomatic or indigestion.
Aconite.
Adeps Lanæ.
Alkaline Baths.
Alumnol.
Ammonium Carbonate.
Ammonium Chloride.
Ammonium Iodide.
Anthrarobin.
Aristol.
Arsen-hemol.
Arsenic.
Arsenic and Mercuric Iodides, Solution of.
Baths: alkaline, to remove scales.
Berberine.
Bleeding.
Cajeput Oil.
Calcium Lithio-carbonate.
Calomel: locally as ointment.
Cantharides.
Chlorinated Lime or Chlorinated Soda, Solution of.
Chrysarobin.
Cod-Liver Oil.
Copaiba.
Copper Sulphate.
Corrosive Sublimate Bath.

160

Creosote Baths.
Electricity: constant current.
Eugallol.
Europhen.
Fats and Oils.
Formaldehyde.
Galium.
Gallanol.
Glycerin.
Glycerite of Lead.
Gold.
Hepar Sulphuris.
Hydroxylamine Hydrochlorate.
Ichthalbin: internally.
Ichthyol: topically.
India-Rubber Solution.
Iodine.
Iodole.
Iris.
Iron Arsenate.
Lead.
Lead Iodide: locally.
Levico Water.
Liq. Potassæ.
Mercury: locally as ointment.
Mercury Ammoniated.
Mezereon.
Myrtol.
Naftalan.
Naphthalene.
Naphtol.
Oil Cade.
Oil Chaulmoogra.
Oleate of Mercury.
Phosphorus.
Pitch.
Potassa, Solution of.
Potassium Acetate.
Potassium Iodide.
Resorcin.
Sapo Laricis.
Silver Nitrate: in psoriasis of tongue.
Soap.
Sodium Arseniate.
Sodium Ethylate.
Sodium Iodide.
Sozoiodole-Mercury.
Stearates.
Sulphides.
Sulphur: internally.
Sulphur Baths.
Sulphur Iodide: internally and externally (ointment.)
Sulphurated Potassa.
Tar: as ointment.
Terebinthinæ Ol.
Thymol.
Thyraden.
Traumaticin.
Turkish Baths.
Ulmus.
Vaselin.
Warm Baths.

Pterygium.

Cocaine.
Eucaine, Beta-Holocaine.
Silver Nitrate.
Tropacocaine.

Ptosis.

Acid, Salicylic.
Arseniate of Sodium.
Ergot.
Tr. Iodi.
Veratrine: to the eyelids and temples.
Zinc Chloride.

Ptyalism. — *See also,
Mouth Sores; also
list of Sialogogues
and Antisialogogues.*

Acids: in small doses internally and as gargles.
Alcohol: dilute as gargle.
Alum.
Atropine: hypodermically.
Belladonna: very useful.
Borax.
Brandy.
Calabar Bean.
Chlorate of Potassium: as gargle.
Chloride of Zinc.
Ferropyrine.
Hyoscine Hydrobromate.
Iodide of Potassium.
Iodine: as gargle, one of tincture to 30 of water.
Myrrh.
Naphtol.
Opium.
Potassium Bromide.
Purgatives.
Sodium Chloride.
Sozoiodole-Sodium.
Sulphur.
Tannin.
Vegetable Astringents.

Puerperal Convulsions.—*See also,
After-Pains, Hemorrhage, Labor, Lactation, Mastitis.
Nipples, Phlegmasia
Alba Dolens, etc*

Acid, Benzoic.
Aconite: in small doses frequently.
Anesthetics.
Belladonna: useful.
Bleeding.
Bromides.
Camphor.
Chloral: in full doses.
Chloroform: by inhalation.
Cold: to abdomen
Dry Cupping: over loins
Ether.
Ice: to head.
Morphine: hypodermically, very useful.
Mustard · to feet.

Nitrite of Amyl: of doubtful utility.
Nitroglycerin.
Ol. Crotonis.
Opium.
Potassium Bitartrate.
Pilocarpine.
Saline Purgatives.
Urethane.
Veratrum Viride: pushed to nausea, very useful.

Puerperal Fever.—
*See also, Puerperal
Peritonitis.*

Acid, Boric, or Creolin (2 per cent.), or Bichloride (1:8000) Solutions: as injections into bladder, to prevent septic cystitis.
Acid, Carbolic.
Acid, Salicylic.
Aconite: useful at commencement.
Alkaline Sulphates: in early stages.
Ammoniæ Liq.
Blisters.
Borax.
Calumba: as tincture.
Camphor.
Chloroform.
Creolin see under "Acid, Boric," above.
Creosoted Oil.
Curette or Placental forceps: to remove membranes if fever continues after antiseptic injections.
Digitalis.
Emetics.
Epsom Salts: if peritonitis develops.
Ergot.
Ice.
Iodine.
Ipecacuanha.
Laparotomy.
Mercury Bichloride: see under "Acid, Boric," above.
Nutriment and Stimulants.
Opium: for wakefulness and delirium, very useful.
Permanganate of Potassium
Plumbi Acetas
Potassium Oxalate.
Purgatives.
Quinine: in large doses.
Resorcin.
Silver Nitrate or Zinc Chloride: to unhealthy wounds.
Sodium Benzoate.
Sodium Sulphite.
Stimulants.
Stramonium: when cerebral excitement.
Sulphocarbolates.

Terebene.
Tr. Ferri Perchloridi.
Turpentine Oil: when much vascular depression and tympanites.
Venesection.
Veratrum Viride.
Warburg's Tincture.

Puerperal Mania.

Aconite: when much fever.
Anesthetics: during paroxysm.
Bromides.
Camphor.
Chalybeates.
Chloral Hydrate.
Chloroform.
Cimicifuga: useful in hypochondriasis.
Duboisine.
Hyoscyamus in mild cases.
Iron: in anemia.
Morphine.
Opium.
Poultices.
Quinine: when much sickness.
Stramonium: when delirium furious but intermittent, or suicidal, or when impulse to destroy child.
Tartar Emetic: frequently repeated.

Puerperal Peritonitis.—See also, Puerperal Fever.

Aconite: at commencement.
Antimony.
Cathartics: recommended by many; condemned by many; evidence in favor of mild aperients combined with Dover's powder or hyoscyamus.
Chlorine Water.
Cimicifuga: in rheumatic cases.
Heat to Abdomen.
Ice to Abdomen.
Mercury.
Opium: very useful.
Quinine: in large doses.
Turpentine Oil: as stimulant, 10 m. frequently repeated.

Pulmonary Affections.—See Lung Diseases.

Pulpitis.—See also, Inflammation.

Formaldehyde.
Thymol.

Purpura.—See also, Hemorrhage, Scurvy

Acid, Gallic.
Acid, Sulphuric.
Acid, Tannic.
Agrimonia.
Alum: locally with brandy
Arsenic.
Digitalis.
Electricity.
Ergot: very useful.
Hamamelis.
Iron: internally.
Lead Acetate.
Lime Juice.
Malt Extract, Dry.
Milk.
Molasses.
Nitrate of Potassium.
Nux Vomica.
Oil Turpentine.
Phosphates.
Potassium Binoxalate.
Potassium Chlorate.
Potassium Citrate.
Quinine.
Strontium Iodide.
Styptics.
Suprarenal Gland.
Tr. Laricis.

Pyelitis.—See also, Bright's Disease, etc.

Acid, Camphoric.
Arbutin.
Buchu.
Cantharides.
Copaiba.
Juniper.
Methylene Blue.
Myrtol.
Oil Sandal.
Pareira.
Pichi.
Saliformin.
Salol.
Uva Ursi.

Pyelonephritis.

Acid, Gallic.
Cantharides.
Erigeron.
Eucalyptus.
Hydrastis.
Pipsissewa (Chimaphila).
Potassa Solution.
Turpentine Oil.

Pyemia.

Acid, Boric.
Acid, Salicylic.
Alcohol.
Alkalies.
Ammonium Carbonate.
Bleeding.
Ergotin.
Ferri Chloridum.
Iodine.
Jaborandi.
Malt Liquor.
Oil of Cloves: locally.

Oil Turpentine: as stimulant.
Potassium Permanganate: internally.
Quinine: in large doses.
Resorcin.
Salicin.
Tannin.

Pyemia and Septicemia.

Manganese Dioxide.
Sodium Thiosulphate.
Sulphites.

Pyrosis.—See also Pyrosis and Cardialgia (below).

Acid, Carbolic.
Acid, Gallic.
Acid, Nitric.
Acid, Sulphuric.
Bismuth.
Camphor.
Creosote.
Glycerin.
Lead.
Manganese Oxide.
Nitrate of Silver.
Nux Vomica.
Oxide of Silver.
Pulvis Kino Compositus.
Strychnine.

Pyrosis and Cardialgia.—See also, Acidity, Dyspepsia.

Bismuth Subnitrate.
Bismuth Valerianate.
Calcium Carbonate, Precipitated.
Capsicum.
Cerium Oxalate.
Kino.
Melissa Spirit.
Opium.
Podophyllin.
Pulsatilla.
Silver Oxide.
Sodium Bicarbonate.

Quinsy.—See Tonsillitis.

Rachitis.

Acid, Gallic.
Acids, Mineral.
Calcium Bromo-iodide.
Calcium Lactophosphate.
Calcium Phosphate.
Cinchona.
Cod-Liver Oil.
Cool Sponging or Rubbing with salt and whisky.
Copper Arsenite.
Digestive Tonics.
Gaduol.
Glycerinophosphates.
Hypophosphites
Iodoform.

162

Iodole.
Iron Iodide.
Lactophosphates.
Levico Water.
Lime Salts.
Massage and Passive Movements.
Nux Vomica.
Phosphates.
Phosphorus.
Physostigma.
Quinine.
Simple Bitters.
Sodium salts.
Strychnine.
Thyraden.

Rectum, Diseases of.
—See also, Anus, Diarrhea, Dysentery; Rectum, Ulceration of; Hemorrhage, Intestinal; Hemorrhoids, Prolapsus.

Acetanilid.
Acid, Tannic.
Belladonna.
Bismuth Subnitrate.
Cocaine Hydrochlorate.
Conium.
Ichthyol.
Iodoform.
Iodoformogen.
Naphtol.
Phosphorus.
Podophyllin.
Potassium Bromide.
Purgatives.
Stramonium.
Sulphur.

Rectum, Ulceration of.

Belladonna.
Chloroform.
Copper Sulphate.
Iodoform.
Iodoformogen.
Iodole.
Mercury Oxide, Red.
Opium.
Phosphorus.
Quinine.
Silver Nitrate.

Relapsing Fever.—
See also, Typhus Fever.

Acid, Salicylic.
Calomel.
Carthartics.
Potassium Citrate.
Laxatives.
Leeches: as cupping for headache.
Quinine.

Remittent Fever.

Acid, Gallic.
Acid, Nitric.
Acid, Salicylic.
Acid, Tannic.
Aconite.

Antipyrine: or cold pack if fever is excessive.
Arsen-hemol.
Arsenic.
Benzoates.
Chloroform.
Cinchonidine.
Cinchonine.
Cold Affusion.
Diaphoretics.
Emetics.
Eupatorium.
Gelsemium: in bilious remittents.
Hyposulphites.
Ipecacuanha.
Levico Water.
Methylene Blue.
Monsel's Salt.
Morphine: hypodermically.
Myrrh.
Oil Eucalyptus.
Packing: useful.
Phenocoll.
Potassium Salts.
Purgatives.
Quinidine.
Quinine: twenty to thirty grn. for a dose, once or twice daily.
Quinoidine.
Resorcin.
Resin Jalap.
Silver Nitrate.
Sodium Chloride.
Tonics.
Turpentine Oil.
Warburg's Tincture.

Renal Calculi. — See Calculi.

Retina, Affections of.—*See also Amaurosis.*

Atropine: dark glasses, and later suitable lenses, in retinitis due to eye strain.
Eserine.
Ichthalbin: internally, as alterant and hematinic.
Ichthyol.
Iron.
Mercury.
Pilocarpine.
Potassium Bromide.
Potassium Iodide.
Pyoktanin.
Sozoiodole-Sodium.

Rheumatic Arthritis.—*See also, Rheumatism.*

Aconite: locally.
Actæa Racemosa.
Arnica: internally and externally.
Arsenic.
Buckeye Bark.
Chaulmoogra Oil.

Cimicifuga: when pains are nocturnal.
Cod-Liver Oil.
Colchicine.
Colchicum.
Cold Douche.
Electricity.
Formin.
Guaiacum.
Ichthyol.
Iodides.
Iodine: internally as tonic.
Iodoform.
Levico Water.
Lithium Salts.
Methylene Blue.
Morphine.
Potassium Bromide: sometimes relieves pain.
Quinine Salicylate.
Sodium Phosphate.
Sodium Salicylate.
Stimulants.
Strychnine.
Sulphides.
Sulphur.
Turkish Bath.

Rheumatism, Acute and Chronic.—*See also, Arthritis, Lumbago, Myalgia, Pleurodynia, Sciatica.*

Absinthin.
Acetanilid.
Acid, Benzoic.
Acid, Carbolic.
Acid, Citric.
Acid, Diiodo-Salicylic.
Acid, Gynocardic.
Acid, Hydriodic.
Acid, Perosmic.
Acid, Salicylic.
Aconite.
Actæa Racemosa.
Acupuncture.
Agathin.
Alcohol.
Alkaline Baths.
Alkaline Mineral Waters.
Alkalies.
Amber, Oil of.
Ammonium Benzoate.
Ammonium Bromide.
Ammonium Chloride.
Ammonium Iodide.
Ammonium Phosphate.
Ammonium Salicylate.
Antimony Sulphide.
Antipyrine.
Aquapuncture.
Arnica.
Arsen-hemol.
Arsenic.
Arsenic and Mercury Iodides, Solution.
Asaprol.
Atropine.
Belladonna.
Benzanilide.

Benzoates.
Betol.
Blisters : very efficient.
Bryonia.
Burgundy Pitch.
Cactus Grandiflorus : tincture.
Caffeine and Sodium Salicylate.
Cajeput Oil.
Capsicum.
Chaulmoogra Oil.
Chimaphila.
Chloral.
Chloroform.
Cimicifuga.
Cimicifugin.
Cocaine Carbolate.
Cod-Liver Oil.
Colchicine.
Colchicum.
Cold Baths.
Cold Douche.
Conium.
Creosote.
Digitalis.
Dover's Powder.
Dulcamara : in persons liable to catarrh.
Eserine.
Ethyl Iodide.
Eucalyptus.
Euphorin.
Europhen.
Faradization.
Fraxinus Polygamia.
Gaduol.
Galvanism.
Gelseminine.
Glycerinophosphates.
Gold and Sodium Chloride.
Guaiacol.
Guaiacum.
Guarana.
Horse-Chestnut Oil.
Hot Pack.
Ice : cold compresses may relieve inflamed joints.
Ice and Salt.
Iodide of Potassium : especially when pain worst at night.
Ichthalbin : internally.
Ichthyol : topically.
Iodides.
Iodine : locally.
Iodoform.
Iron.
Jaborandi.
Lactophenin.
Leeches.
Lemon Juice.
Levico Water.
Lime Juice.
Lithium Bromide : especially when insomnia and delirium present.
Lithium Carbonate.
Lithium Iodide.
Lithium Salicylate.
Lupulin.
Magnesia.

Magnesium Salicylate.
Manaca.
Manganese Sulphate.
Massage.
Mercury Bichloride.
Mercury and Morphine Oleate : locally.
Mezereon.
Mineral Baths.
Morphine.
Mustard Plasters.
Neurodin.
Oil Croton.
Oil Gaultheria.
Oil Mustard.
Oil Turpentine.
Oleoresin Capsicum.
Opium : one grn. every two or three hours, especially when cardiac inflammation.
Orexine : for anorexia.
Packing.
Pellitory.
Permanganate of Potassium.
Phenacetin : alone or with salol.
Phytolacca.
Pilocarpine Hydrochlorate.
Pine-Leaf Baths.
Potassa, Sulphurated.
Potassio - Tartrate of Iron
Potassium Acetate.
Potassium Arsenite : solution.
Potassium Bicarbonate
Potassium Iodide and Opium.
Potassium Nitrate.
Potassium Oxalate.
Potassium Phosphate.
Potassium Salicylate.
Potassium and Sodium Tartrate.
Poultices.
Propylamine (see Trimethylamine).
Pyoktanin.
Quinine Salicylate.
Quinoline Salicylate.
Rhus Toxicodendron : exceedingly useful in after-stage and subacute forms.
Saccharin : to replace sugar in diet.
Salicin.
Salicylamide.
Salicylates.
Salipyrine.
Salol.
Salophen.
Sodium Dithio-salicylate.
Sodium Paracresotate.
Spiræa Ulmaria.
Splints for fixation of limb may relieve.
Steam Bath.
Stimulants.
Stramonium.
Strontium Iodide.

Strontium Lactate.
Strontium Salicylate.
Sulphur.
Tetra-ethyl-ammonium Hydroxide : solution.
Thuja Occidentalis.
Thymol.
Trimethylamine Solution.
Triphenin.
Turkish Bath.
Turpentine Oil.
Veratrine.
Veratrum Viride.
Xanthoxylum.
Zinc Cyanide.
Zinc Oxide.

Rheumatism, Gonorrheal.

Ammonium Chloride.
Ichthalbin.
Opium.
Phenacetin.
Potassium Chlorate.
Potassium Iodide.
Rubidium Iodide.

Rheumatism, Muscular. — See also, Lumbago, Myalgia, Neuritis, Pleurodynia ; Rheumatism, Acute and Chronic ; Torticollis.

Ammonium Chloride.
Atropine.
Capsicum.
Chloral Hydrate.
Cimicifuga.
Colchicine.
Croton-Oil Liniment.
Diaphoretics.
Dover's Powder : with hot drinks and hot foot bath.
Euphorin.
Gold.
Jaborandi.
Lithium Bromide.
Methylene Blue.
Morphine.
Mustard.
Phenacetin.
Potassium Iodide.
Potassium Nitrate.
Salol.
Salipyrine.
Triphenin.
Veratrine Ointment.

Rhinitis. — See also, Catarrh, Acute Nasal ; Influenza, Nasal Affections.

Alumnol.
Aristol.
Bismuth Subgallate.
Camphor.
Creolin : (1:1000) **as a** nasal douche.
Diaphtherin.

Europhen.
Fluid Cosmoline in Spray.
Menthol.
Potassium Permanganate.
Retinol.
Sozoiodole-Sodium and Sozoiodole - Zinc in atrophic rhinitis.
Stearates.

Rickets.—*See Rachitis.*

Ring-Worm. — *See also, Tinea, etc.*

Acid, Boric.
Chrysarobin.
Formaldehyde.
Ichthyol.
Iron Tannate.
Mercury, Ammoniated.
Mercury Bichloride.
Mercury Oxide, Red.
Naftalan.
Picrotoxin.
Sulphites.
Tincture Iodine: topically.

Rosacea. — *See Acne Rosacea.*

Roseola. — *See also, Measles.*

Aconite.
Ammonium Acetate.
Ammonium Carbonate.
Belladonna.

Rubeola.—*See Measles*

Salivation.—*See Ptyalism.*

Sarcinae. — *See also, Dyspepsia, Cancer, Gastric Dilatation.*

Acid, Carbolic.
Acid, Sulphuric.
Calcium Chloride.
Creosote.
Formaldehyde.
Gastric Siphon: to wash out stomach.
Hyposulphites.
Naftalan.
Sodium Thiosulphate.
Sulphites.
Wood Spirit.

Satyriasis. — *See also, Nymphomania, and list of Anaphrodisiacs.*

Bromipin.
Bromo-hemol.
Ichthalbin.
Levico Water.
Potassium Bromide.
Sodium Bromide.

Scabies.

Acid, Benzoic: as ointment or lotion.

Acid, Carbolic: dangerous.
Acid, Sulphuric: internally as adjuvant.
Acid, Sulphurous.
Alkalies.
Ammoniated Mercury.
Anise: as ointment.
Arsenic.
Baking of clothes to destroy ova.
Balsam of Peru: locally; agreeable and effective.
Calcium Sulphide.
Chloroform.
Coal-Tar Naphta.
Cocculus Indicus: as ointment.
Copaiba.
Copper Sulphate.
Corrosive Sublimate.
Creolin.
Glycerin.
Hydroxylamine Hydrochlorate.
Ichthyol.
Iodine.
Kamala: as ointment.
Levico Water.
Liq. Potassæ.
Losophan.
Manganese Dioxide.
Mercury Bichloride.
Mercury: white precipitate ointment.
Naftalan.
Naphtol.
Oil Cade.
Oil Cajuput.
Oily Inunction.
Petroleum.
Phosphorated Oil.
Potassium Iodide.
Soft Soap.
Sozoiodole-Potassium.
Stavesacre: as ointment.
Storax: with almond oil, when skin cannot bear sulphur.
Sulphides.
Sulphides.
Sulphur: as ointment.
Sulphur and Lime.
Sulphurated Potassa.
Sulphur Baths.
Tar: ointment.
Vaselin.

Scalds.—*See Burns and Scalds.*

Scarlet Fever. — *See also, Albuminuria, Bright's Disease, Uremia.*

Acetanilid.
Acid, Acetic.
Acid, Carbolic: as gargle.
Acid, Gallic.
Acid, Salicylic.
Acids, Mineral: internally and as gargle.

Acid, Sulphurous: inhalation when throat much affected.
Aconite: harmful if constantly employed.
Adeps Lanæ.
Alcohol: indicated in collapse.
Ammonium Acetate: solution.
Ammonium Benzoate.
Amyl Hydride.
Antipyrine.
Arsenic: if tongue remains red and irritable during convalescence.
Baptisin.
Belladonna.
Benzoate of Sodium.
Bromine.
Calcium Sulphide.
Carbonate of Ammonium: greatly recommended in frequent doses given in milk or cinnamon water.
Chloral.
Chlorine Water: as gargle.
Chloroform.
Cold Compress: to throat.
Cold Affusion.
Copaiba.
Digitalis.
Fat: as inunction to hands and feet during the rash, and over the whole body during desquamation.
Ferric Perchloride: in advanced stage with albuminuria and hematuria; very useful.
Hot Bath.
Hydrogen Peroxide.
Ice: applied externally to throat, and held in mouth, to prevent swelling of throat.
Ice Bag, or rubber head-coil: to head, if very hot.
Ice: to suck, especially at commencement.
Iodine.
Jalap: compound powder, with potassium bitartrate, or hot dry applications, to produce sweat in nephritis.
Juniper Oil: as diuretic when dropsy occurs.
Lactophenin.
Mercury: one-third of a grn. of gray powder every hour to lessen inflammation of tonsils.
Mustard Bath: when rash recedes.
Naphtol.

165

Neurodin.
Oil Gaultheria.
Packing: useful and comforting.
Philocarpine Hydrochlorate.
Potassium Chlorate.
Potassium Iodide.
Potassium Permanganate: as gargle to throat.
Purgatives: most useful to prevent albuminuria.
Quinine.
Resorcin.
Rhus Toxicodendron.
Salicylate of Sodium as antipyretic.
Salol.
Sodium Bromide: with chloral, when convulsions usher in attack.
Sodium Sulphocarbolate.
Strychnine: hypodermically in paralysis.
Sulphate of Magnesium.
Sulphur.
Thermodin.
Tr. Ferri Chloridi.
Triphenin.
Veratrum Viride.
Warm Wet Pack.
Water.
Zinc Sulphate.

Scars, to Remove.

Thiosinamine.

Sciatica. — See also, Neuralgia, Rheumatism.

Acetanilid: absolute rest of limb in splints very needful.
Acid, Perosmic.
Acid, Salicylic.
Acid, Sulphuric.
Aconite: as ointment or liniment.
Actæa Racemosa.
Acupuncture.
Antipyrine.
Aquapuncture.
Apomorphine.
Asaprol.
Atropine.
Belladonna.
Benzanilide.
Blisters.
Cautery: exceedingly useful; slight application of Paquelin's thermo-cautery.
Chloride of Ammonium
Chloral.
Chloroform: locally as liniment; inhalation when pain excessive.
Cimicifuga.
Cod-Liver Oil.
Colchicine.

Coniine Hydrobromate.
Conium.
Copaiba Resin.
Counter-Irritation.
Croton Oil; internally as purgative.
Duboisine.
Electricity.
Ether: as spray.
Euphorin.
Galvanism.
Gelsemium.
Gold.
Guaiacol.
Guaiacum.
Glycerinophosphates.
Iodides.
Iodipin.
Massage of Nerve with Glass Rod.
Menthol.
Methylene Blue.
Morphine: hypodermically, most useful.
Nerve Stretching.
Neurodin.
Nitroglycerin.
Nux Vomica.
Opium.
Phosphorus.
Plasters.
Potassium Bitartrate or Citrate: 40 grn. thrice daily, in plenty of water, to regulate kidneys.
Poultices.
Rhus Toxicodendron.
Salicylate of Sodium.
Salol.
Salophen.
Sand Bath.
Secale.
Silver Nitrate.
Sodium Dithiosalicylate
Stramonium: internally, pushed until physiological action appears
Sulphur: tied on with flannel over painful spot.
Triphenin.
Tropacocaine.
Turkish Bath.
Turpentine Oil: in ½ oz. doses internally for three or four nights successively.
Veratrine: as ointment.
Wet or Dry Cups over course of nerve.

Sclerosis. — See also, Locomotor Ataxia, Atheroma, Paralysis Agitans

Acetanilid.
Antipyrine.
Arsenic.
Gaduol.
Glycerinophosphates.
Gold and Sodium Chloride.
Hyoscyamine.
Ichthalbin.

Mercuro-iodo-hemol.
Phenacetin.
Physostigma.
Silver Oxide.
Sozoiodole-Mercury.
Spermine.

Sclerosis, Arterial.

Barium Chloride.
Digitoxin.
Glycerinophosphates.
Iodo-hemol.

Scorbutus.—See Scurvy

Scrofula.—See also, Cachexiæ, Coxalgia, Glands, Ophthalmia.

Acacia Charcoal.
Acid, Hydriodic.
Acid, Phosphoric.
Alcohol.
Antimony Sulphide.
Arsenic.
Barium Chloride.
Barium Sulphide.
Blisters: to enlarged glands.
Bromine.
Cadmium Iodide.
Calcium Benzoate.
Calcium Chloride.
Calcium Sulphide.
Calomel.
Chalybeate Waters.
Cod-Liver Oil: exceedingly serviceable.
Copper Acetate.
Cupro-hemol.
Ethyl Iodide.
Excision, or scraping gland, and packing with iodoform gauze.
Extract Malt, Dry.
Fats: inunction.
Gaduol.
Galium Aparinum.
Glycerinophosphates.
Gold salts.
Hyoscyamus: tincture.
Hypophosphites.
Ichthalbin: internally.
Ichthyol: ointment.
Iodides.
Iodine: locally to glands, and internally.
Iodipin.
Iodoform.
Iodoformogen.
Iodo-hemol.
Iodole.
Iron.
Lactophosphates.
Manganese Iodide.
Mercury Bichloride.
Milk and Lime Water.
Peroxide of Hydrogen.
Pipsissewa.
Phosphates.
Phosphorus.
Potassium Chlorate.
Sanguinaria.
Sanguinarine.

Sarsaparilla.
Sodium Bromide.
Sodium Hyposulphite.
Soft Soap.
Solution Potassa.
Stillingia.
Sulphides.
Thyraden.
Walnut Leaves.
Zinc Chloride.

Scurvy.—*See also, Cancrum Oris, Purpura*

Acid, Citric or Tartaric: as preventive in the absence of lime-juice.
Aconite: in acute stomatitis with salivation in scorbutic conditions.
Agrimony: useful in the absence of other remedies.
Alcohol: diluted, as gargle.
Alum: locally with myrrh for ulcerated gums.
Ammonium Carbonate: in scorbutic diathesis.
Arsen-hemol.
Arsenic: in some scorbutic symptoms.
Atropine: hypodermically when salivation.
Cinchona: as decoction, alone or diluted with myrrh, as gargle.
Ergot.
Ergotin Hypodermic, or Ergot by Mouth: to restrain the hemorrhage.
Eucalyptus.
Ferri Arsenias: as a tonic where other remedies have failed.
Ferri Perchloridi, Tinctura: to restrain hemorrhage.
Laricis, Tinctura: like Ferri Perchl., Tinct.
Lemon Juice: exceedingly useful as preventive and curative.
Liberal Diet often sufficient.
Liquor Sodæ Chlorinatæ: locally to gums.
Manganese Dioxide.
Malt: an antiscorbutic.
Oil Turpentine.
Oranges: useful.
Phosphates: when non-assimilation a cause.
Potassium Binoxalate: in doses of four grn. three times a day; if not obtainable sorrel is useful instead.
Potassium Chlorate.
Potassium Citrate: substitute for lime-juice.
Pyrethrum.

Quinine: with mineral acids internally.
Silver Nitrate.
Tartar Emetic.
Vegetable Charcoal: as tooth-powder to remove fetid odor.
Vinegar: very inferior substitute for lime-juice.

Sea-Sickness. — *See also, Nausea, Vomiting.*

Acetanilid
Acid, Hydrocyanic.
Acid, Nitro-Hydrochloric: formula: Acidi nitro-hydrochlorici, dil. 3 fl. drams; Acidi hydrocyanici dil. half fl. dram; Magnesii sulphatis, 2 drams; Aq. 8 fl. oz.: \ fl. oz. 3 times a day.
Amyl Nitrite: a few drops on handkerchief inhaled; the handkerchief must be held close to the mouth.
Atropine: one-hundredth grn. hypodermically.
Bitters: calumba, etc.
Bromalin.
Bromides.
Caffeine Citrate: for the headache..
Cannabis Indica: one-third to one-half grn. of the extract to relieve headache.
Capsicum.
Champagne, Iced: small doses frequently repeated.
Chloralamide and Potassium Bromide.
Chloral Hydrate: fifteen to thirty grn. every four hours most useful; should be given before nausea sets in; the combination with potassium bromide, taken with effervescing citrate of magnesia, is very good.
Chloroform: pure, two to five minims on sugar.
Coca: infusion quickly relieves.
Cocaine.
Counter-irritation: mustard plaster or leaf to epigastrium.
Creosote.
Hyoscyamine: one-sixtieth grain with the same quantity of strychnine.
Hyoscyamus.
Ice: to spine.

Kola.
Magnetic Belt.
Morphine: hypodermically.
Neurodin.
Nitroglycerin.
Nux Vomica: when indigestion with constipation.
Orexine Tannate.
Potassium Bromide: should be given several days before voyage is begun.
Resorcin.
Levico Water.
Salt and Warm Water.
Sodium Bromide: like potassium salt.
Strychnine.
Triphenin.

Seborrhea.—*See also, Acne, Pityriasis.*

Acid, Boric.
Acid, Salicylic.
Alumnol.
Borax: with glycerin and lead acetate, as a local application.
Euresol.
Glycerin.
Hydrastine Hydrochlorate.
Ichthalbin: internally, Ichthyol: topically.
Iodine.
Lead Acetate: with borax and glycerin as above.
Liquor Potassæ: locally to hardened secretion.
Mercury.
Naphtol.
Resorcin.
Sodium Chloride.
Zinc Oxide: in inflammation the following formula is useful: Take Zinci oxidi, 1 dram; Plumbi carbonat. 1 dram : Cetacei, 1 oz.: Ol. olivæ q. s.; ft. ung.

Septicemia.—*See Pyemia etc.*

Sexual Excitement. —*See Nymphomania, Satyriasis.*

Shock.

Alcohol.
Ammonia.
Amyl Nitrite.
Atropine.
Blisters.
Codeine.
Digitalin.
Digitalis.
Ergotin.
Erythrol Tetranitrate.
Heat.
Hypodermoclysis.

Nitroglycerin.
Oxygen.
Strychnine.

Skin Diseases. — *See
the titles of the va-
rious diseases in
their alphabetic
order.*

Small - Pox. —*See Va-
riola.*

Sleeplessness. — *See
Insomnia, Nervous-
ness.*

Sneezing. —*See also,
Catarrh, Hay Fever,
Influenza.*

Arsen-hemol.
Arsenic : in paroxysmal
sneezing as usually
ushers-in hay fever.
Belladonna.
Camphor : as powder,
or strong tincture in-
haled in commencing
catarrh.
Chamomile Flowers : in
nares.
Cotton Plug : in nares.
Gelsemium : in exces-
sive morning sneez-
ings with discharge.
Iodine : inhalation.
Iodipin.
Levico Water.
Menthol.
Mercury : when heavi-
ness of head and pain
in limbs.
Potassium Iodide : ten
grn. doses frequently
repeated.
Pressure beneath Nose,
over the termination
of the nasal branch of
the ophthalmic di-
vision of the fifth.

Somnambulism. —
See also, Nightmare

Bromides.
Bromipin.
Bromalin.
Bromo-hemol.
Glycerinophosphates.
Opium.

Somnolence.

Arsen-hemol.
Caffeine.
Coca.
Glycerinophosphates.
Kola.
Levico Water.
Spermine.

**Spasmodic Affec-
tions.**—*See list of
Antispasmodics;*

*also Angina Pec-
toris, Asthma,
Chorea, Colic,
Cough, Convulsions,
Croup, Dysuria,
Epilepsy, Gastro-
dynia, Hydropho-
bia, Hysteria,
Laryngismus, Per-
tussis, Stammering,
Tetanus, Torticollis,
Trismus, etc.*

Spermatorrhea. —
*See also, Emissions,
Hypochondriasis,
Impotence; also list
of Anaphrodisiacs.*

Acetanilid.
Acid, Camphoric.
Antispasmin.
Arsenic : in functional
impotence ; best com-
bined with iron as the
arsenate, and with
ergot.
Atropine.
Belladonna : in relaxa-
tion of the genital
organs where there is
no dream nor orgasm;
one-fourth grain of
extract, and a grain
and a half of zinc sul-
phate.
Bladder to be emptied
as soon as patient
awakes.
Bromain.
Bromides : when it is
physiological in a ple-
thoric patient ; not
when genitalia are re-
laxed.
Bromipin.
Bromo-hemol.
Calomel : ointment ap-
plied to urethra.
Camphor Bromide : or
camphor alone ;
diminishes venereal
excitement.
Cantharides : in cases of
deficient tone, either
from old age, excess,
or abuse ; should be
combined with iron.
Chloral Hydrate : to ar-
rest nocturnal emis-
sions.
Cimicifuga : where
emission takes place
on the least excite-
ment.
Cold Douching and
Sponging.
Cornutine.
Digitalis : in frequent
emissions with lan-
guid circulation; with
bromide in plethoric
subjects.

Electricity.
Ergot : deficient tone in
the genital organs.
Gold Chloride.
Hydrastis : local appli-
cation to urethra.
Hygienic Measures.
Hyoscine Hydrobro-
mate.
Hypophosphites : ner-
vine tonic.
Iron : where there is
anemia only.
Levico Water.
Lupulin : oleoresin, to
diminish nocturnal
emissions.
Nitrate of Silver : vesi-
cation by it of the
perineum ; and local
application to the
prostatic portion of
the urethra.
Nux Vomica : nervine
tonic and stimulant.
Phosphorus: in physical
and mental debility.
Potassium Citrate.
Quinine : as a general
tonic.
Solanine.
Spermine.
Spinal Ice-Bag.
Strychnine.
Sulfonal.
Sulphur : as a laxative,
especially if sequent
to rectal or anal
trouble.
Tetronal.
Turpentine Oil: in sper-
matorrhea with im-
potence.
Warm bath before re-
tiring.
Zinc Oxide.

Spina Bifida.

Calcium Phosphate.
Collodion : as means of
compression.
Cotton Wool over
tumor.
Glycerin : injection
after tapping.
Iodine : injection. For-
mula : Iodine, 10 grn.;
Potassium Iodide, 30
grn.; Glycerin, 1 fl. oz.
Potassium Iodide.
Tapping : followed by
compression.

Spinal Concussion. —
See also, Myelitis.

Arnica.
Bleeding : to relieve
heart.
Lead Water and Opium:
as lotion.
Leeches.
Vinegar : to restore con-
sciousness.

Spinal Congestion.—
See also, Meningitis, Myelitis.

Aconite.
Antiphlogistic Treatment.
Cold Affusions: to spine.
Ergot: in large doses.
Gelsemium.
Nux Vomica.
Turpentine Oil.
Wet Cupping.

Spinal Irritation.—
See also, Meningitis, Myelitis, Neuritis, Neurasthenia.

Aconite Ointment: locally.
Acid, Phosphoric.
Arsen-hemol.
Arsenic.
Atropine.
Belladonna: gives way to this more readily than to aconite.
Blisters: to spine.
Bromalin.
Bromides: to lessen activity.
Bromo-hemol.
Cimicifuga.
Cocculus Indicus: like strychnine.
Codeine.
Conium.
Counter-irritation.
Digitalis.
Electricity: combined with massage and rest.
Ergot: when spinal congestion.
Glycerinophosphates.
Ignatia.
Leeches.
Nux Vomica.
Opium: in small doses.
Phosphorus.
Picrotoxin.
Sinapis Liniment: counter-irritant.
Sodium Hypophosphite
Spermine.
Strychnine: to stimulate the depressed nerve centres.
Veratrum Viride.

Spinal Paralysis and Softening.— *See also, Locomotor Ataxia, Myelitis.*

Argentic Nitrate: in chronic inflammation of the cord or meninges.
Belladonna: in chronic inflammatory conditions.
Cod-Liver Oil: as a general nutrient.

Electricity: combined with massage and rest.
Ergot: in hyperemia of the cord.
Hyoscyamus: in paralysis agitans to control tremors.
Iodide of Potassium: in syphilitic history.
Mercury: temporarily cures in chronic inflammation of the cord and meninges.
Phosphorus: as a nervine tonic.
Physostigma: in a few cases of progressive paralysis of the insane, in old-standing hemiplegia, in paraplegia due to myelitis, and in progressive muscular atrophy it has done good service.
Picrotoxin: spinal stimulant after febrile symptoms have passed off.
Spermine.
Strychnine: like picrotoxin.

Spleen, Hypertrophied.—*See also, Malaria, Leucocythemia.*

Ammonium Fluoride.
Arsenic.
Arsen-hemol.
Bromides.
Ergot.
Levico Water.
Methylene Blue.
Quinine.
Salicin.

Sprains.

Aconite Liniment: well rubbed in.
Ammonia.
Ammonium Chloride: prolonged application of cold saturated lotion.
Arnica: much vaunted, little use.
Bandaging: to give rest to the injured ligaments.
Calendula: as a lotion.
Camphor: a stimulating liniment.
Cold Applications.
Cold Douche.
Collodion: a thick coating to exert a firm even pressure as it dries.
Croton-Oil Liniment.
Hamamelis.
Hot Foot-bath: prolonged for hours, for sprained ankle.

Hot Fomentations: early applied.
Ichthyol: ointment.
Inunction of Olive Oil: with free rubbing.
Iodine: to a chronic inflammation after a sprain.
Lead Lotion: applied at once to a sprained joint.
Lead Water and Laudanum.
Oil of Bay.
Rest.
Rhus Toxicodendron: as lotion.
Shampooing: after the inflammation has ceased, to break down adhesions.
Soap Liniment.
Soap Plaster: used as a support to sprained joints.
Sodium Chloride.
Soluble Glass.
Strapping: to give rest.
Turpentine Liniment: a stimulant application to be well rubbed in.
Vinegar: cooling lotion.
Warming Plaster.

Stammering.

Hyoscyamus.
Stramonium.
Vocal Training: the rythmical method most useful.

Sterility.

Alkaline Injections: in excessively acid secretions from the vagina.
Aurum: where due to chronic metritis, ovarian torpor or coldness; also in decline in the sexual power of the male.
Borax: vaginal injection in acid secretion.
Cantharides: as a stimulant where there is impotence in either sex.
Cimicifuga: in congestive dysmenorrhea.
Dilatation of Cervix: in dysmenorrhea; in pinhole os uteri; and in plugging of the cervix with mucus.
Electrical Stimulation of Uterus: in torpor.
Gossypii Radix: in dysmenorrhea with sterility.
Guaiacum: in dysmenorrhea with sterility.

Intra-uterine Stems : to stimulate the lining membrane of the uterus.

Key-tsi-ching : a Japanese remedy for female sterility.

Phosphorus : functional debility in the male.

Potassium Iodide : as emmenagogue.

Spermine.

Stings and Bites.— *See also, Wounds.*

Acid, Carbolic : mosquito-bites and scorpion-stings.

Acid, Salicylic.

Aconite.

Alum : for scorpion-sting.

Ammonia or other Alkalies : in stings of insects to neutralize the formic acid ; and in snake-bite.

Ammonium Carbonate.

Aqua Calcis : in stings of bees and wasps.

Arsenic : as a caustic.

Calcium Chloride.

Camphor.

Chloroform : on lint.

Creolin.

Essence of Pennyroyal : to ward off mosquitoes.

Eucalyptus : plant in room to keep away mosquitoes.

Hydrogen Dioxide.

Ichthyol.

Ipecacuanha : leaves as poultice for mosquito and scorpion-bites.

Ligature, or cleansing of wound, at once, to prevent absorption, in snake-bites.

Menthol.

Mercury Bichloride.

Mint Leaves.

Oil of Cinnamon : 1 dram with 1 oz. of spermaceti ointment, spread over hands and face, to ward off mosquitoes.

Oil of Cloves : the same.

Potassa Fusa : in dog-bites a most efficient caustic.

Potassium Permanganate : applied and injected around snake-bite, followed by alcohol in full doses.

Removal of Sting.

Rosemary.

Sage.

Silver Nitrate : a caustic, but not sufficiently strong in dog-bites.

Soap : to relieve itching of mosquito-bites.

Stimulants.

Sugar : pounded, in wasp-stings.

Vinegar.

Stomach, Catarrh of.— *See Catarrh, Gastric.*

Stomach, Debility of.— *See List of Gastric Tonics and Stomachics.*

Stomach, Dilatation of.— *See Gastric Dilatation.*

Stomach, Sour.— *See Acidity.*

Stomach, Ulcer of.— *See Gastric Ulcer.*

Stomatitis.— *See also, Aphthæ, Cancrum Oris ; Mouth, Sore.*

Acid, Boric : lotion of 1 in 50.

Acid, Carbolic : strong solution locally to aphthæ.

Acid, Hydrochloric : concentrated in gangrenous stomatitis ; dilute in mercurial, aphthous, etc.

Acid, Nitric.

Acid, Nitrohydrochloric : as gargle or internally in ulcerative stomatitis.

Acid, Salicylic : one part in sufficient alcohol to dissolve, to 50 of water, in catarrhal inflammation to ease the pain.

Acid, Sulphurous.

Acid, Tannic.

Alcohol : brandy and water, a gargle in mercurial and ulcerative stomatitis.

Alum, or Burnt Alum : locally in ulcerative stomatitis.

Argentic Nitrate : in thrush laver.

Bismuth : in aphthæ of nursing children, sore mouth, dyspeptic ulcers, mercurial salivation ; locally applied.

Borax : in thrush and chronic stomatitis.

Cleansing Nipples : in breast-fed babies.

Cocaine : before cauterization.

Copper Sulphate : locally in ulcerative

stomatitis, and to indolent ulcers and sores.

Cornus : astringent.

Eucalyptus : tincture, internally.

Glycerite of Tannin : in ulcerative stomatitis.

Hydrastis : fluid extract locally.

Hydrogen Dioxide.

Iris : in dyspeptic ulcer.

Krameria : local astringent.

Lime Water : in ulcerative stomatitis.

Mercury : in dyspeptic ulcers, gray powder.

Myrrh : tincture, with borax, topically.

Papain.

Potassium Bromide : for nervous irritability.

Potassium Chlorate : the chief remedy, locally and internally.

Potassium Iodide ; in syphilitic ulceration.

Rubus : astringent.

Sodium Bromide.

Sodium Chlorate.

Sozoiodole-Sodium.

Sozoiodole-Zinc.

Sweet Spirit of Niter.

Thymol.

Tonics.

Strabismus.

Atropine : to lessen converging squint when periodic in hypermetropia.

Cocaine.

Eserine : to stimulate the ciliary muscles in deficient contraction.

Electricity.

Eucaine.

Holocaine.

Hyoscyamus.

Mercury : like Iodide of Potassium.

Operation.

Potassium Iodide : in syphilitic history if one nerve only is paralyzed.

Shade over one Eye : in children to maintain acuity of vision.

Suitable Glasses : to remedy defective vision.

Tropacocaine.

Stricture, Urethral. *—See Urethral Stricture.*

Strophulus.— *See also, Lichen.*

Antimonium Crudum.

Adeps Lanæ.

Borax and Bran Bath : if skin is irritable.

Carbonate of Calcium.
Chamomile.
Glycerin.
Ichthyol.
Lancing the Gums.
Lead Lotion : to act as astringent.
Magnesia.
Mercury : gray powder if stools are pale.
Milk Diet.
Pulsatilla.
Spiritus Ætheris Nitrosi: where there is deficient secretion of urine.
Zinc Oxide.

Struma.—*See Scrofula.*

Stye.—*See Hordeolum.*

Summer Complaint. —*See Cholera Infantum, Diarrhea, etc.*

Sunstroke.

Aconite : not to be used with a weak heart.
Alcohol : is afterwards always a poison.
Ammonia, for its diaphoretic action.
Amyl Nitrite.
Apomorphine : one-sixteenth grn. at once counteracts symptoms.
Artificial Respiration.
Belladonna.
Bleeding : in extreme venous congestion.
Brandy : in small doses in collapse.
Camphor.
Chloroform : in convulsions.
Digitalis : to stimulate heart.
Ergot : by the mouth or subcutaneously.
Gelsemium.
Hot baths (105°–110° F.), or hot bottles or bricks, in heat exhaustion, and in collapse.
Ice : application to chest, back, and abdomen, as quickly as possible, in thermic fever, and to reduce temperature ; ice drinks as well.
Leeches.
Nitroglycerin.
Potassium Bromide to relieve the delirium.
Quinine : in thermic fever.
Scutellaria.
Tea : cold, as beverage instead of alcoholic drinks.
Tonics : during convalescence.

Venesection : best treatment if face be cyanosed and heart laboring, and if meningitis threaten after thermic fever (Hare).
Veratrum Viride.
Water : cold affusion.
Wet Sheet : where the breathing is steady; otherwise cold douche.

Suppuration. — *See also, Abscess, Boils, Carbuncle, Pyemia.*

Acid, Carbolic : lotion and dressing.
Acid, Gallic.
Alcohol : to be watched.
Ammonium Carbonate: in combination with Cinchona.
Bismuth Oxyiodide.
Calcium salts : to repair waste.
Calcium Sulphide.
Cinchona : as tonic, fresh infusion is best.
Creolin.
Gaduol.
Glycerinophosphates.
Hypophosphites : tonic.
Ichthalbin : internally.
Ichthyol.
Iodole.
Iodoformogen.
Iodipin.
Iron Iodide : tonic.
Manganese Iodide: tonic.
Mercury.
Phosphates : like the hypophosphites:
Pyoktanin.
Quinine : tonic.
Sarsaparilla : tonic.
Sulphides : when a thin watery pus is secreted, to abort, or hasten suppuration.

Surgical Fever.

Acid, Salicylic.
Aconite.
Chloral.
Quinine.
Tinctura Ferri Perchloridi : as a prophylactic.
Veratrum Viride : to reduce the circulation and fever.

Surgical Operations.—*See also, List of Antiseptics.*

Acid, Carbolic.
Acid, Salicylic.
Acid, Oxalic.
Aristol.
Chloroform.
Creolin.

Diaphtherin.
Europhen.
Iodoform.
Iodoformogen.
Iodole.
Mercuric Chloride.
Mercury and Zinc Cyanide.
Tribromphenol.

Sweating.—*See Perspiration, Night-Sweats, Bromidrosis, etc.*

Sweating, Colliquative.—*See Night-Sweats.*

Sycosis.—*See also, Condylomata, Mentagra.*

Acid, Boric.
Acid, Sulphurous : in parasitic sycosis.
Arsen-hemol.
Arsenici et Hydrargyri Iodidi Liquor : when much thickening.
Arsenic.
Alumnol.
Canada Balsam and Carbolic Acid : in equal parts, to be applied after epilation in tinea sycosis.
Chloride of Zinc : solution in tinea sycosis.
Chrysarobini Ung. : in parasitic sycosis.
Cod-Liver Oil : in chronic non-parasitic.
Copper Sulphate.
Europhen.
Euresol.
Hydrargyri Acidi Nitratis : as ointment.
Hydrargyri Ammoniatum Ung. : in parasitic.
Hydrargyri Oxidi Rubri Ung.
Hydroxylamine Hydrochlorate.
Ichthalbin : internally.
Ichthyol.
Iodide of Sulphur Ointment : in non-parasitic.
Levico Water.
Losophan.
Naftalan.
Naphtol.
Oleate of Mercury : in parasitic.
Oleum Terebinthinæ : in parasitic.
Phytolacca.
Salol.
Shaving.
Sodium Sulphite.
Sozoiodole-Sodium.
Thuja.
Zinc Sulphate.

Syncope.—*See also, Heart Affections.*

Acid, Acetic.
Aconite.
Alcohol: sudden, from fright or weak heart.
Ammonia: inhaled cautiously.
Ammonium Carbonate.
Arsenic: nervine tonic; prophylactic.
Atropine.
Belladonna: in cardiac syncope.
Camphor: cardiac stimulant.
Chloroform: transient cardiac stimulant; mostly in hysteria.
Cold Douche.
Counter-irritation to Epigastrium: in collapse.
Digitalis: in sudden collapse after hemorrhage; the tincture by the mouth, digitalin hypodermically.
Duboisine.
Ether: in collapse from intestinal colic.
Galvanism.
Heat to Epigastrium.
Lavandula.
Musk.
Nitrite of Amyl: in sudden emergency, in fatty heart, in syncope during anesthesia, and in hemorrhage.
Nux Vomica.
Position: head lowest and feet raised.
Stimulants: undiluted.
Veratrum Album: an errhine.
Veratrum Viride.

Synovitis.—*See also, Coxalgia, Joint Affections.*

Acid, Carbolic: injections of one dram of a two per cent. solution into the joint.
Aconite.
Alcohol and Water: equal parts.
Antimony: combined with saline purgatives.
Arnica.
Bandage or Strapping: Martin's elastic bandage in chronic.
Blisters: fly blisters at night in chronic synovitis; if not useful, strong counter-irritation.
Calcium Sulphide: as an antisuppurative
Carbonate of Calcium.

Cod-Liver Oil: tonic.
Conium: in scrofulous joints.
Counter-irritation.
Gaduol: as alterative and reconstitutive.
Glycerinophosphates.
Heat.
Ichthalbin: as tonic and alterative.
Ichthyol.
Iodine: injection in hydrarthrosis after tapping; or painted over.
Iodoform: solution in ether, 1 in 5, injected into tuberculous joints; also as a dressing after opening.
Iodoformogen.
Iodole.
Mercury: Scott's dressing in chronic strumous disease; internally in syphilitic origin.
Morphine.
Oleate of Mercury: to remove induration left behind,
Potassium Iodide.
Pressure: combined with rest.
Quinine.
Shampooing and Aspiration.
Silver Nitrate: ethereal solution painted over.
Splints.
Sulphur.

Syphilis.—*See also, Chancre, Condylomata, Ptyalism, Ulcers.*

Acid, Acetic: caustic to sore.
Acid, Arsenous.
Acid, Boric: like benzoin.
Acid, Carbolic: to destroy sore, mucous patches, condylomata, etc.; as bath in second stage.
Acid, Chromic.
Acid, Dichlor-acetic.
Acid, Gynocardic.
Acid, Hydriodic.
Acid, Nitric: in primary syphilis, to destroy the chancre, especially when phagedenic.
Acid, Salicylic: antiseptic application.
Antimony Sulphide, Golden.
Arsen-hemol.
Arsenic and Mercury Iodides: solution of.
Aristol.
Aurum: in recurring syphilitic affections

where mercury and iodide of potassium fail.
Barium Chloride.
Barium Sulphide.
Benzoin: antiseptic dressing for ulcers.
Bicyanide of Mercury: to destroy mucous tubercles, condylomata, and to apply to syphilitic ulceration of the tonsils and tongue.
Bismuth and Calomel: as a dusting powder.
Bromine.
Cadmium Sulphate.
Calcium Sulphide.
Calomel: for vapor bath in secondary; dusted in a mixture with starch or oxide of zinc over condylomata will quickly remove them.
Camphor: dressing in phagedenic chancres.
Cauterization.
Cod-Liver Oil: tonic in all stages.
Copper Sulphate.
Creosote: internally in strumous subjects, and where mercury is not borne.
Denutrition: hunger-cure of Arabia.
Ethyl Iodide.
Europhen.
Expectant plan of treatment.
Formaldehyde Solution: useful for cauterizing sores.
Gaduol.
Glycerin.
Guaiacum: alterative in constitutional syphilis.
Hot Applications.
Hydriodic Ether.
Ichthalbin: internally.
Ichthyol.
Iodides: followed by mercury.
Iodipin.
Iodoform or Iodoformogen: dressing for chancre and ulcers.
Iodole.
Iron: in anemia, the stearate, perchloride, and iodide are useful.
Lotio Flava: dressing for syphilitic ulcers, and gargle in sore throat and stomatitis.
Manganese: in cachexia.
Manganese Dioxide.
Mercuro - iodo - hemol: anti-syphilitic and hematinic at the same time.

Mercury: the specific remedy in one or other of its forms in congenital and acquired syphilis in primary or secondary stage.

Mixed Treatment.

Oil of Mezereon: in constitutional syphilis.

Oil of Sassafras: in constitutional syphilis.

Ointments and Washes of Mercury.

Phosphates: in syphilitic periostitis, etc.

Pilocarpine Hydrochlorate.

Podophyllum: has been tried in secondary, with success after a mercurial course.

Potassium Bichromate.

Potassium Bromide.

Potassium Chlorate: local application of powder to all kinds of syphilitic ulcers; gargle in mercurial and specific stomatitis.

Pressure bandage and mercurial inunctions for periostitis.

Pulsatilla: tincture.

Pyoktanin.

Retinol.

Rubidium Iodide.

Shampooing and local applications of croton oil or cantharides as a lotion, to combat alopecia.

Sarsaparilla: alterative in tertiary.

Silver Chloride.

Silver Nitrate.

Silver Oxide.

Soft Soap: to syphilitic glandular swellings.

Stillingia: most successful in cases broken down by a long mercurial and iodide course which has failed to cure; improves sloughing phagedenic ulcers.

Stramonium: tincture.

Suppositories of Mercury.

Thyraden.

Tonic and general treatment.

Turkish and Vapor Baths: to maintain a free action of the skin.

Wet Pack.

Zinc Chloride: locally to ulcers as caustic.

Tabes Dorsalis.—*See Locomotor Ataxia.*

Tabes Mesenterica.
—*See also, Scrophulosis.*

Acid, Gallic: astringent in the diarrhea.

Acid, Phosphoric.

Alcohol.

Arsenic: in commencing consolidation of the lung.

Barium Chloride: in scrofula.

Calcium Chloride: in enlarged scrofulous glands.

Calcium Phosphate

Cod-Liver Oil.

Diet, plain and nourishing.

Fatty Inunction.

Ferri Pernitratis Liquor: hematinic and astringent.

Gaduol.

Gelsemium: in the reflex cough.

Glycerinophosphates.

Gelseminine.

Iodine.

Iodo-hemol.

Iodipin.

Iron.

Mercury.

Oil Chaulmoogra.

Olive Oil: inunction.

Phosphates: as tonic.

Sarsaparilla.

Tape-Worm.—*See also, Worms.*

Acid, Carbolic.

Acid, Filicic.

Acid, Salicylic: followed by purgative.

Acid, Sulphuric: the aromatic acid.

Alum: as injection.

Ammonium Embelate.

Areca Nut.

Balsam of Copaiba: in half-ounce doses.

Chenopodium Oil: ten drops on sugar.

Cocoa Nut: a native remedy.

Cod-Liver Oil: tonic.

Creosote.

Ether: an ounce and a half a dose, followed by a dose of castor oil in two hours.

Extract Male Fern: followed by purgative.

Iron: tonic.

Kamala.

Kousso.

Koussein.

Mucuna: night and morning for three days, then brisk purgative.

Naphtalin.

Pelletierine: the tannate preferably.

Pumpkin Seeds: pounded into an electuary, 2 oz. at dose.

Punica Granatum: acts like its chief alkaloid, pelletierine.

Quinine: as tonic.

Resorcin: followed by purgative.

Thymol.

Turpentine Oil.

Valerian: in convulsions due to the worms.

Tenesmus.—*See Dysentery.*

Testicle, Diseases of.
—*See also, Epididymitis, Hydrocele, Orchitis, Varicocele.*

Acid, Phosphoric, and Phosphates: in debility.

Aconite: in small doses frequently repeated in acute epididymitis.

Ammonium Chloride: solution in alcohol and water; topical remedy.

Antimony: in gonorrheal epididymitis.

Belladonna: in neuralgia of the testis; as an ointment with glycerin in epididymitis or orchitis.

Collodion: by its contraction to exert pressure, in gonorrheal epididymitis.

Compression: at the end of an acute and beginning of a subacute attack, as well as in chronic inflammation.

Conium: poultice of leaves in cancer.

Copaiba: in orchitis.

Digitalis: in epididymitis.

Gold salts: in acute and chronic orchitis.

Hamamelis: in some patients gives rise to seminal emissions.

Hot Lotions: in acute inflammation.

Ice Bag: in acute orchitis.

Ichthyol

Iodine: injection into an encysted hydrocele; local application in orchitis after the acute symptoms have passed off.

Iodoform or Iodoformogen: dressing in ulceration.

Magnesium Sulphate with Antimony: in epididymitis.

Mercury Bichloride.

Mercury and Morphine Oleate: in syphilitic enlargement and chronic inflammation.

Nitrate of Silver: ethereal solution painted around an enlarged testis better than over.

Nux Vomica: in debility.

Potassium Bromide.

Potassium Iodide: in syphilitic testicle.

Pulsatilla: in very small doses along with aconite.

Suspension: in orchitis and epididymitis.

Traumaticin.

Tetanus.—*See also, Spasmodic Affections.*

Acetanilid.

Aconite: in large doses to control muscular spasm.

Acupuncture: on each side of the spines of the vertebræ.

Alcohol: will relax muscular action, also support strength.

Anesthetics: to relax muscular spasm.

Antimonium Tartaratum: in large doses, along with chlorate of potassium.

Antipyrine.

Apomorphine: as a motor paralyzer.

Arsenic.

Atropine: local injection into the stiffened muscles to produce mild poisoning. Useful in both traumatic and hysterical tetanus.

Belladonna.

Bromides: in very large doses frequently repeated.

Cannabis Indica: serviceable in many cases; best combined with chloral.

Chloral Hydrate: in large doses; best combined with bromide or cannabis indica.

Chloroform.

Cocaine Hydrochlorate

Coniine Hydrobromate.

Conium

Curare: an uncertain drug.

Curarine.

Duboisine: like atropine.

Eserine.

Freezing the Nerve: in traumatic tetanus has been proposed.

Gelsemium: in a few cases it has done good

Heat to Spine: will arrest convulsions.

Hyoscyamus: in traumatic.

Ice-bag to Spine.

Lobelia: a dangerous remedy.

Morphine: injected into the muscles gives relief.

Nerve-stretching: where a nerve is implicated in the cicatrix, has done good.

Neurotomy: in the same cases.

Nicotine: cautiously administered relieves the spasm; best given by rectum or hypodermically; by the mouth it causes spasm which may suffocate.

Nitrite of Amyl: in some cases it cures.

Nitroglycerin: like the preceding.

Opium: alone or with chloral hydrate.

Paraldehyde.

Physostigma: the liquid extract pushed to the full. Given by the mouth, or rectum, or hypodermically.

Physostigmine.

Quinine: in both idiopathic and traumatic tetanus.

Strychnine: the evidence, which is doubtful, seems to show that it is beneficial in chronic and idiopathic tetanus: should be given only in a full medicinal dose.

Tetanus Antitoxin.

Urethane.

Vapor Baths.

Warm Baths.

Tetter.—*See Herpes.*

Throat, Sore.—*See also, Diphtheria, Pharyngitis, Tonsillitis.*

Acid, Camphoric.

Acid, Carbolic: as a spray in relaxed sore throat and in coryza.

Acid, Gallic.

Acid, Nitric: as alterative with infusion of cinchona.

Acid, Sulphurous: spray

Acid, Tannic.

Acid, Trichloracetic.

Aconite: in acute tonsillitis with high temperature; in the sore-throat of children before running on to capillary bronchitis; best given frequently in small doses.

Alcohol: gargle in relaxed throat.

Alum: gargle in chronic relaxed throat, simple scarlatinal and diphtheritic sore-throat.

Aluminium Aceto-tartrate.

Ammonium Acetate.

Arsenic: in coryza and sore throat simulating hay fever; in sloughing of the throat.

Balsam of Peru.

Balsam of Tolu.

Belladonna: relieves spasm of the pharyngeal muscles; also when the tonsils are much inflamed and swollen.

Calcium Bisulphite Solution.

Capsicum: as gargle in relaxed sore throat.

Catechu: astringent gargle.

Chloral Hydrate.

Chlorine Water: gargle in malignant sore throat.

Cimicifuga: in combination with opium and syrup of tolu in acute catarrh.

Cocaine Carbolate.

Cold Compresses: in tendency to catarrh.

Creosote.

Electric Cautery: in chronic sore throat to get rid of thickened patches.

Ferri Perchloridum: gargle in relaxed sore throat.

Ferropyrine: as a styptic in throat operations.

Gaduol.

Glycerite of Tannin: to swab the throat in relaxed sore throat.

Glycerinophosphates.

Guaiacol.

Guaiacum: sucking the resin will abort or cut

short the commencing quinsy.

Hydrastis: gargle in follicular pharyngitis and chronic sore throat.

Ice: sucked, gives relief.

Ichthyol.

Iodine: locally to sores and enlarged tonsil.

Iodole.

Levico Water: as alterative tonic.

Liq. Ammonii Acetatis: in full doses.

Magnesium Sulphate: to be given freely in acute tonsillitis.

Mercury: in very acute tonsillitis, gray powder or calomel in small doses.

Mercury and Morphine Oleate: in obstinate and painful sore throat.

Myrrh: gargle in ulcerated sore throat.

Methylene Blue.

Phytolacca: internally, and as gargle.

Podophyllum: cholagogue purgative.

Potassium Chlorate: chief gargle.

Potassium Nitrate: a ball of nitre slowly sucked.

Pulsatilla: in acute coryza without gastric irritation.

Pyoktanin.

Resorcin.

Sanguinaria: the tincture sprayed in extended chronic nasal catarrh.

Silver Nitrate: solution in sloughing of the throat or chronic relaxation; saturated solution an anesthetic and cuts short inflammation.

Sodium Borate: in clergyman's sore throat.

Sodium Chlorate.

Sodium Salicylate: in quinsy.

Sozoiodole salts.

Steam: of boiling water; and vapor of hot vinegar.

Sumach: the berries infused, with addition of potassium chlorate, a most efficient gargle.

Terpin Hydrate.

Tracheotomy.

Veratrum Viride: to control any febrile change.

Zinc Acetat.

Zinc Chloride.

Zinc Sulphate: a gargle.

Thrush.—*See Aphthæ.*

Tic Douloureux. —
See also, Hemicrania, Neuralgia, Neuritis, Odontalgia.

Acetanilide.

Aconite.

Aconitine: formula: Aconitinæ (Duquesnel's) 1-10 grn.; Glycerini, Alcoholis, aa, 1 fl. oz.; Aq. menth. pip., ad 2 fl. oz.; 1 dram per dose, cautiously increased to 2 drams.

Ammonium Chloride: in large dose.

Amyl Nitrite: in pale anemic patients.

Anesthetics quickly relieve.

Antipyrine.

Arsen-hemol.

Arsenic: occasionally useful.

Atropine: hypodermically, and ointment.

Bromo-hemol.

Butyl-Chloral Hydrate.

Caffeine.

Cannabis Indica.

Cautery in Dental Canal: where pain radiated from mental foramen.

Chamomile.

Chloroform: inhalation; also hypodermically.

Counter-irritation.

Cupric Ammonio - Sulphate: relieves the insomnia.

Delphinine: externally.

Electricity.

Exalgin.

Gelseminine.

Gelsemium: valuable.

Heat.

Hyoscyamus.

Ichthyol.

Iron: in combination with strychnia ; the following formula is good: Ferri potassio-tartaratis, 4 scruples ; Vin. opii, 1½ drams ; Aq. cinnam, ad 8 fl. oz. 1 fl. oz. ter in die.

Laurocerasi Aqua.

Ligature of the Carotids: in obstinate cases a last resort; has done good.

Methylene Blue.

Morphine: hypodermically.

Nitroglycerin: in obstinate cases.

Neurodin.

Ol. Crotonis: sometimes cures; will relieve.

Phosphorus: in obstinate cases.

Physostigma.

Physostigmine.

Potassium Iodide: the following formula relieves: take Chloralis hydrati 5 grn.; Potassii iodidi, 3 grn.; Sp. ammoniæ comp, 1 fl. dr.; Infusum gentianæ, ad 1 fl. oz. The salt alone in syphilitic history.

Pulsatilla: relieves.

Quinine.

Salicin: instead of quinine, where pain is periodic.

Salicylates.

Stramonium.

Triphenin.

Turpentine Oil.

Veratrine: ointment

Zinc Valerianate: with extract hyoscyamus.

Tinea Circinata (*Ringworm of the Body*). — *See also, Ringworm.*

Acid, Acetic.

Acid, Boric: in simple or æthereal solution.

Acid, Carbolic: solution, or glycerite.

Acid, Chromic.

Adeps Lanæ.

Anthrarobin.

Aristol.

Arsenic.

Borax.

Chrysarobin.

Cocculus Indicus.

Cod-Liver Oil.

Copper Acetate.

Copper Carbonate.

Creolin.

Gaduol.

Gallanol.

Goa Powder: as ointment, or moistened with vinegar.

Glycerinophosphates.

Iodine.

Iodole.

Kamala.

Levico Water.

Losophan.

Mercury Bichloride.

Naftalan.

Naphtol.

Oil Cade.

Resorcin.

Sodium Chloride.

Sulphites: or sulphurous acid.

Sulphur.

Sulphur Baths: faithfully carried out.

Thymol.

Turpentine Oil.

Tinea Decalvans

(*Alopecia Areata*) — See also, *Tinea Circinata*.

Parasiticides.
Tonics.

Tinea Favosa.

Acid, Carbolic: lotion.
Acid, Nitric: caustic after the crust has been removed.
Acid, Sulphurous: 1 part to 2 parts glycerin, assisted by epilation.
Calcium Sulphide.
Cleanliness.
Epilation: followed up by using a parasiticide.
Hyposulphites.
Iron.
Mercury: a lotion of the bichloride, 2 grn. to the oz; or the oleate-of-mercury ointment.
Oil: to soften and remove scabs.
Oleander.
Petroleum: one part to two of lard after crusts are gone.
Sulphides.
Turkish Bath: followed by the use of carbolic soap, instead of ordinary.
Viola Tricolor.
Zinc Chloride: dilute watery solution.

Tinea Sycosis. — See Mentagra.

Tinea Tarsi.

Blisters to Temple.
Copper Sulphate.
Epilation, removal of of scabs, and application of stick of lunar caustic.
Lead Acetate.
Mercury: after removal of scabs, Ung. hydrargyri nitratis diluted to half its strength. Also take Plumbi acetatis, 1 dram; Ung. hydrargyri oxidi rubri, 1 dram; Zinci oxidi, 1 dram; Calomelanos, half dram; Adipis, 2 drams; Olei palmat., 5 drams; ft ung. Also Oleate.
Mercury Oxide, Red.
Tinct. Iodi: after removal of scabs, followed by application of glycerin.

Ung. Picis: touched along edge of tarsi.
Silver Nitrate, Molded.

Tinea Tonsurans.

(*Ringworm of the Scalp*). — See also, *Porrigo, Tinea Circinata*.

Acetum Cantharidis.
Acid, Acetic: strong, locally.
Acid, Boric: ethereal solution after head is thoroughly cleansed.
Acid, Carbolic: in early stages.
Acid, Chrysophanic: 30 grn. to the oz., as ointment.
Acid, Salicylic: strong solution in alcohol, 40 grn. to the oz.; or vaselin ointment of same strength.
Acid, Sulphurous
Anthrarobin.
Arsenic: tonic.
Borax.
Cocculus Indicus.
Cod-Liver Oil.
Coster's Paste: Iodine 2 drams, Oil cade, 3 drams.
Creosote.
Croton Oil: liniment followed by a poultice.
Epilation.
Iodine: the tincture in children.
Lime Water.
Menthol: parasiticide and analgesic.
Mercury: white precipitate lightly smeared over; the oleate, pernitrate, and oxide, as ointments. The bichloride as a lotion 2 grn. to the dram.
Naftalan.
Oil Cajeput.
Potassium Sulphocyanide.
Quinine.
Resorcin.
Sodium Chloride.
Sodium Ethylate.
Thymol: like menthol.

Tongue, Diseases of.

Acid, Nitric: in dyspeptic ulcers the strong acid as caustic.
Bi-Cyanide of Mercury: in mucous tubercles.
Borax: in chronic superficial glossitis; and in fissured tongue.
Cloves: as gargle.
Cochlearia Armoracia (Nasturtium Armoracia): as gargle.
Conium.

Frenulum: should be divided in tongue-tie.
Ginger: as masticatory.
Hydrastis: in stomatitis.
Iodine.
Iodoform or Iodoformogen: to ulcers.
Mercury: in syphilitic disease.
Mezereon, Oil of: sialagogue.
Nux Vomica.
Pepper: condiment.
Phytolacca.
Potassium Bromide.
Potassium Chlorate: in aphthous ulceration, chronic superficial glossitis, stomatitis.
Potassium Iodide: in tertiary specific ulceration, and in macroglossia.
Pyrethrum: masticatory.
Rhus Toxicodendron.
Silver Nitrate. caustic to ulcers.
Xanthoxylum: in lingual paralysis.
Zinc Chloride: caustic.

Tonsillitis. — See also, Throat, Sore.

Acetanilid: internally.
Acid, Salicylic: internally.
Acid, Tannic.
Aconite: internally.
Alum.
Alumnol.
Aluminium Acetotartrate.
Belladonna: internally.
Capsicum and Glycerin.
Cocaine Hydrochlorate.
Creolin.
Emetics.
Ferric Chloride.
Guaiacum.
Hydrogen Peroxide.
Ice-bag.
Ichthyol.
Iodole.
Iron Chloride, Tincture: locally.
Mercury.
Monsel's Solution: locally.
Myrtol.
Opium.
Potassium Chlorate.
Potassium Iodide: internally.
Pyoktanin.
Quinine: internally.
Salicylates: internally.
Salol: internally.
Saline purgatives.
Silver Nitrate.
Sodium Bicarbonate.

Tonsils, Enlarged.

Acid, Citric
Acid, Tannic

Alumnol.
Aluminium Acetotartrate.
Aluminium Sulphate: locally applied.
Ammonium Iodide.
Barium Iodide.
Catechu: astringent gargle.
Excision.
Fel Bovinum, Inspissated: rubbed up with conium and olive oil as an ointment to be painted over.
Ferric Chloride: astringent in chronically enlarged tonsils.
Gaduol.
Ichthalbin: internally.
Ichthyol: topically.
Iodine Tincture: to cause absorption.
Iodo-hemol.
Iodipin.
Massage: of the tonsils.
Silver Nitrate: caustic.
Tannin: saturated solution.
Zinc Chloride.

Tonsils, Ulcerated.

Acid, Carbolic.
Acid, Sulphurous, mixed with equal quantity of glycerin, and painted over.
Cantharides: as vesicant.
Cimicifuga.
Coptis: gargle.
Iodoformogen.
Iodole.
Iron: gargle.
Lycopodium: to dust over.
Magnesium Sulphate: free purgation with.
Mercuric Iodide: in scrofulous and syphilitic ulceration.
Potassium Chlorate: gargle.
Potassium Iodide: in tertiary syphilis.
Pyoktanin.
Sozoiodole-Potassium.
Silver Nitrate.

Toothache.—*See Odontalgia.*

Torticollis.

Aconite: liniment externally; and tincture internally.
Arsenic: controls and finally abolishes spasm.
Atropine.
Belladonna.
Capsicum: strong infusion applied on lint and covered with oiled silk.

Cimicifuga.
Conium: when due to spasmodic action of the muscles.
Electricity: galvanic to the muscles in spasm; faradic to their paretic antagonists.
Gelseminine.
Gelsemium.
Local Pressure.
Massage.
Nerve-stretching.
Nux Vomica.
Opium.
Potassium Bromide.
Strychnine.
Water: hot douche.

Tremor.—*See also, Chorea, Delirium Tremens, Paralysis Agitans.*

Arsenic.
Arsen-hemol.
Bromalin.
Bromo-hemol.
Calcium salts.
Cocaine Hydrochlorate.
Conline.
Gelseminine.
Glycerinophosphates.
Hyoscine Hydrobromate.
Hyoscyamus.
Phosphorus.
Silver Nitrate.
Sparteine Sulphate.
Zinc Phosphide.

Trichinosis.

Acid, Arsenous.
Acid, Picric.
Benzene.
Glycerin.

Trismus.

Aconite.
Anesthetics: to allay spasm.
Atropine.
Belladonna: extract in large doses.
Cannabis Indica.
Chloral Hydrate: in T. neonatorum, one grn. dose by mouth, or two by rectum when spasms prevent swallowing.
Conium: the succus is the most reliable preparation.
Ether.
Gelseminine.
Gelsemium.
Opium.
Physostigma.
Physostigmine.

Tuberculous Affections.—*See Laryngitis, Tubercular; Lupus; Meningitis, Tubercular; Peritonitis, Tubercular; Phthisis; Scrophulosis; Tabes Mesenterica.*

Tumors. — *See also, Cancer, Cysts, Glandular Enlargement, Goiter, Polypus, Uterine Tumors, Wen.*

Acid, Perosmic.
Ammoniacum and Mercury Plaster.
Ammonium Chloride.
Anesthetics: to detect the presence of phantom tumors; also to relax abdominal walls to permit deep palpation of abdomen.
Codeine: for pain.
Electricity.
Gaduol.
Iodine.
Iodipin.
Iodo-hemol.
Eserine: in phantom.
Hyoscyamus.
Iodoform.
Iodoformogen.
Lead Iodide.
Methylene Blue.
Papain.
Pyoktanin.
Silver Oxide.
Sodium Ethylate.
Stypticin.
Thiosinamine.
Zinc Chloride.
Zinc Iodide.

Tympanites. — *See also, Flatulence, Typhoid Fever, Peritonitis.*

Acid, Carbolic, or Creosote: in tympanites due to fermentation.
Acids: after meals.
Alkalies: before meals with a simple bitter.
Arsenic.
Asafetida: as an enema.
Aspiration: to relieve an over-distended gut
Bismuth.
Capsicum.
Chamomile: enema.
Chloral Hydrate: as an antiseptic to fermentation in the intestinal canal.
Cocculus Indicus.
Colchicine.
Colchicum.
Cubeb: powdered, after strangulated hernia.

Gaduol.
Galvanism: in old cases, especially of lax fibre.
Ginger.
Glycerin: when associated with acidity.
Glycerinophosphates.
Hyoscyamus.
Ice Poultice: prepared by mixing linseed meal and small pieces of ice, in tympanites of typhoid fever.
Ichthalbin.
Iris.
Nux Vomica.
Ol. Terebinthinæ: very efficient as enema, not for external application.
Plumbi Acetas: when due to want of tone of intestinal muscular walls.
Rue: very effective.
Sumbul.
Vegetable Charcoal in gruel: in flatulent distention of the colon associated with catarrh; dry, in flatulent distention of the stomach.

Typhlitis.

Aristol.
Arsen-hemol.
Arsenic.
Belladonna.
Ice Bag: or poultice over the cecum.
Leeches: at once as soon as tenderness is complained of, unless subject is too feeble.
Levico Water.
Magnesium Sulphate: only when disease is due to impaction of cecum.
Metallic Mercury.
Opium: better as morphine subcutaneously
Purgatives.
Veratrum Viride.

Typhoid Fever.—*See also, Hemorrhage, Intestinal; Rectal Ulceration; Tympanites.*

Acetanilid.
Acid, Carbolic.
Acid, Hydrochloric: to diminish fever and diarrhea.
Acid, Phosphoric: cooling drink.
Acid, Salicylic: some hold that it is good in the typhoid of children, many that it does great harm.
Acid, Sulphuric, Diluted.

Aconite: to reduce the pyrexia.
Alcohol: valuable, especially in the later stages.
Alum: to check the diarrhea.
Antipyrine: to lower the temperature.
Argenti Nitras: to check diarrhea; in obstinate cases along with opium; should not be given until the abdominal pain and diarrhea have begun.
Aristol.
Arnica: antipyretic.
Arsenic: liquor arsenicalis with opium to restrain the diarrhea.
Asafetida.
Asaprol.
Bath: agreeable to patient, and reduces hyperpyrexia.
Belladonna: during the pyrexial stage it lowers the temperature, cleans the tongue, and steadies the pulse; afterwards brings on irritability of heart.
Benzanilide: antipyretic.
Bismuth Subnitrate: to check diarrhea.
Bismuth Subgallate.
Brand's method of cold bathing.
Calomel: 10 grn. first day, and eight each day after, the German specific treatment. Or: in small continuous doses without producing stomatitis.
Calx Saccharata: in milk, when the tongue is black and parched.
Camphor
Carbolate of Iodine: one drop of tincture of iodine and of liquefied carbolic acid, in infusion of digitalis, every two or three hours.
Carbonate of Ammonium.
Cascara Sagrada.
Charcoal: to prevent fetor of stools, accumulation of fetid gas, and to disinfect stools after passage.
Chloral Hydrate.
Chlorine Water.
Chloroform Water.
Copper Arsenite.
Copper Sulphate.
Creosote.
Creolin.

Digitalis: to lower temperature and pulse-rate; death during its use has been known to occur suddenly.
Enemas: to be tried first, if constipation lasts over two days.
Ergot: for intestinal hemorrhage.
Eucalyptol.
Eucalyptus: thought to shorten disease.
Ferri Perchloridi Tinctura.
Glycerin and Water, with lemon juice, as mouth wash.
Guaiacol.
Guaiacol Carbonate.
Hydrastine.
Hyoscyamus.
Iodine: specific German treatment; use either liquor or tincture.
Iron.
Lactophenin.
Lead Acetate: to check diarrhea.
Lime Water.
Licorice Powder.
Magnesium Salicylate.
Mercury Bichloride: 10 min. of solution ⅙ grn. in 1 oz. water, every two or three hours.
Milk Diet.
Morphine: in large doses, if perforation occur.
Naphtalene.
Naphtol.
Naphtol Benzoate.
Neurodin.
Opium: to check delirium and wakefulness at night, and to relieve the diarrhea.
Phosphorus: if nervous system is affected.
Potassium Iodide: alone or with iodine.
Quinidine: equal to quinine.
Quinine: in large doses to reduce the temperature.
Resorcin: antipyretic.
Rest and Diet.
Salol.
Sodium Benzoate: antipyretic.
Sodium Paracresotate.
Sodium Thiosulphate.
Starch, Iodized.
Tannalbin: with calomel.
Tannopin.
Tartar Emetic: in pulmonary congestion.
Thalline Sulphate.
Thermodin.

Thymol.
Tribromphenol.
Triphenin.
Turpentine Oil: at end of the second week, 10 minims every two hours, and every three hours in the night; specific if the diarrhea continues during convalescence.
Veratrum Viride.
Xeroform.
Zinc Sulphocarbolate.

Typhus Fever.—See also, *Delirium*, *Typhoid Fever*.

Acid, Phosphoric: agreeable drink.
Acid, Salicylic: antipyretic.
Aconite.
Alcohol: where failure of the vital powers threatens.
Antimony with Opium: in pulmonary congestion, wakefulness, and delirium.
Antipyrine.
Arnica: antipyretic.
Baptisia.
Baths: to reduce temperature. Instead of baths, cold compresses may be used.
Belladonna: cleans the tongue, steadies and improves the pulse; too long usage makes the heart irritable.
Calx Saccharata: in milk, when the tongue is black and coated.
Camphor.
Chloral Hydrate: in wild delirium in the early stages of the fever, but not in the later.
Chlorine Water: not much used now.
Coca: tentative.
Cod-Liver Oil.
Counter-irritation.
Diet: nutritious.
Digitalis: to increase the tension of the pulse and prevent delirium; if a sudden fall of pulse and temperature should occur during its administration it must be withheld.
Expectant Treatment.
Hyoscyamus.
Musk.
Oil Valerian.
Opium.
Podophyllum.
Potassium Chlorate: in moderate doses.

Potassium Nitrate: mild diuretic and diaphoretic.
Quinine: in full doses to pull down temperature.
Strychnine: where the circulatory system is deeply involved.
Tartar Emetic.
Turpentine Oil: in the stupor.
Yeast: accelerates the course of the disease.

Ulcers and Sores.— See also, *Chancre, Chancroid, Bedsores, Throat; Gastric, Intestinal and Uterine Ulceration; Syphilis.*

Acetanilid.
Acid, Arsenous.
Acid, Boric.
Acid, Carbolic.
Acid, Chromic.
Acid, Gallic.
Acid, Nitric.
Acid, Phenyloboric.
Acid, Pyrogallic.
Acid, Salicylic.
Acid, Sulphuric.
Acid, Tannic.
Acid, Trichloracetic.
Alcohol: a useful application.
Alum: crystals, burnt, or dried.
Aluminium Sulphate.
Alumnol.
Ammonium Chloride.
Aniline.
Aristol.
Arsenic.
Balsam Peru.
Belladonna.
Benzoin Tincture.
Bismuth Benzoate.
Bismuth Oxyiodide.
Bismuth Subgallate.
Bismuth Subnitrate.
Borax.
Bromine.
Calcium Bisulphite: solution.
Calcium Carbonate, Precipitated.
Camphor.
Chimaphila.
Chloral Hydrate.
Chlorinated Lime.
Cocaine.
Conium.
Copper Sulphate.
Creolin.
Creosote.
Diaphtherin.
Ethyl Iodide.
Europhen.
Formaldehyde.
Gold Chloride.
Hamamelis.
Hot Pack.

Hydrastine Hydrochlorate.
Hydrogen Peroxide.
Ichthyol.
Iodine.
Iodoform.
Iodoformogen.
Iodole.
Iron Arsenate.
Iron Ferrocyanide.
Lead Carbonate.
Lead Iodide.
Lead Nitrate.
Lead Tannate.
Lime.
Magnesia.
Mercury Bichloride.
Mercury Iodide, Red.
Mercury Oxide, Red.
Methylene Blue: in corneal ulcers.
Morphine.
Naphtol.
Opium.
Papain.
Potassium Chlorate.
Potassium Permanganate.
Potassa Solution.
Pyoktanin.
Quinine.
Resorcin.
Silver Nitrate.
Sozoïodole salts.
Starch, Iodized.
Stearates.
Tannoform.
Turpentine Oil.
Zinc salts.

Ulcus Durum.—See *Chancre.*

Ulcus Molle.—See *Chancroid.*

Uremia.—See also, *Coma, Convulsions, Bright's Disease, Scarlet Fever; and the lists of Diaphoretics and Diuretics.*

Amyl Nitrite.
Bromides.
Caffeine.
Chloroform.
Chloral Hydrate.
Colchicine.
Digitalis.
Elaterin.
Hot Pack.
Hypodermoclysis.
Morphine.
Naphtalene.
Nitroglycerin.
Oil Croton.
Pilocarpine Hydrochlorate.
Saline or Hydragogue Cathartics.
Sodium Benzoate.
Strychnine.
Transfusion.
Urethane.
Venesection.

Urethra, Stricture of.

Electrolysis.
Silver Nitrate.
Thiosinamine.

Urethritis.—*See also, Gonorrhea; and list of Astringents.*

Acetanilid.
Acid, Tannic.
Aconite.
Alkalies: internally.
Alumnol.
Arbutin.
Argentamine.
Argonin.
Borax.
Calomel.
Europhen.
Ichthyol.
Methylene Blue.
Myrtol.
Potassium Chlorate.
Potassium Permanganate.
Protargol.
Pyoktanin.
Resorcin.
Silver Citrate.
Silver Nitrate.
Sodium Chlorate.
Sodium Salicylate.
Strophanthus.
Sozoiodole-Sodium.
Sozoiodole-Zinc.
Zinc Acetate.
Zinc Permanganate.
Zinc Sulphate.

Uric-Acid Diathesis. —*See Lithemia.*

Urinary Calculi.— *See Calculi.*

Urinary Disorders.— *See lists of Diuretics and of other agents acting on the Urine. Also, see Bladder; Albuminuria; Bright's Disease; Chyluria; Cystitis; Diabetes; Dysuria; Dropsy; Enuresis; Hematuria; Lithiasis; Nephritis; Oxaluria; Uremia; Urethral Stricture; Urine, Incontinence of; Urine, Phosphatic.*

Urine, Incontinence of.

Acid, Benzoic.
Antipyrine.
Belladonna.
Bromalin.

Bromo-hemol.
Buchu.
Cantharides.
Chloral Hydrate.
Collinsonia.
Gaduol.
Glycerinophosphates.
Hyoscyamus.
Ichthalbin.
Rhus Toxicodendron.
Strychnine.

Urine, Phosphatic.

Acid, Benzoic.
Acid, Hydrochloric, Dil.
Acid, Lactic.
Acid, Phosphoric, Dil.
Acid, Sulphuric, Dil.
Ammonium Benzoate.

Urticaria.—*See also, Prurigo*

Acetanilid.
Alkalies.
Alumnol.
Arsenic.
Arsen-hemol.
Benzoin.
Calcium Chloride: to prevent.
Chloroform.
Colchicum.
Gaduol.
Glycerinophosphates.
Ichthalbin: internally.
Ichthyol: externally.
Iodides.
Iodipin.
Iodo-hemol.
Lead.
Levico Water.
Menthol.
Sodium Salicylate.
Strychnine.

Uterine Affections.— *See Abortion, Amenorrhea, Climacteric, Dysmenorrhea, Endometritis, Hemorrhage Post-Partum, Leucorrhea, Menorrhagia, Menstrual Disorders, Metritis, Metrorrhagia, Prolapsus Uteri, Uterine Cancer, etc.*

Uterine Cancer.

Acid, Carbolic.
Acid, Tannic.
Arsenic.
Cannabis Indica.
Chloral Hydrate.
Conium.
Glycerin.
Glycerite of Tannin: mixed with iodine, to check discharge and remove smell.
Gossypium.
Hydrastinine Hydrochlorate.

Iodine.
Iodoform.
Iodoformogen.
Morphine.
Opium.
Pyoktanin.
Sozoiodole-Zinc.
Stypticin.
Thyroid preparations.

Uterine Congestion and Hypertrophy.

Acid, Carbolic.
Acid, Chromic.
Digitalis.
Ergotin.
Glycerin.
Gold salts.
Ichthalbin: internally.
Ichthyol: topically.
Iodine.
Iodoform.
Iodoformogen.
Iron.
Potassium Bromide.
Quinine.
Zinc Valerianate.

Uterine Dilatation.

Acid, Carbolic, Iodized.

Uterine Tumors.— *See also, Cysts, Tumors.*

Ammonium Chloride.
Calcium Chloride.
Iodine.
Iron Sulphate.
Mercury
Opium.
Pyoktanin.
Silver Oxide.
Thiosinamine.

Uterine Ulceration. —*See also, Ulcers.*

Acid, Carbolic.
Acid, Nitric.
Acid, Tannic.
Alum.
Aluminium Sulphate.
Bismuth Subnitrate.
Creosote.
Glycerin.
Hydrastis.
Iodoform.
Iodoformogen.
Iodole.
Mercury Nitrate Solut.
Pyoktanin.
Silver Nitrate.

Uterine Hemorrhage.—*See also, Hemorrhage.*

Hydrastinine Hydrochlorate.
Ice.
Stypticin.

Uvula, Relaxed.

Acid, Tannic.
Ammonium Bromide.
Capsicum.
Kino.
Pyrethrum.
Zinc salts.

Vaginismus.

Antispasmin.
Belladonna.
Cocaine.
Collinsonia.
Conium.
Iodoform.
Iodoformogen.
Hyoscyamine.
Morphine.
Piperin.
Sozoiodole-Zinc.
Tropacocaine.

Vaginitis.—See also, Gonorrhea, Leucorrhea.

Acetanilid.
Acid, Tannic.
Calcium Bisulphite.
Chlorine Water.
Copper Sulphate.
Eucalyptus.
Formaldehyde.
Grindelia.
Hydrastis.
Ichthyol.
Potassium Chlorate.
Potassium Silicate.
Resorcin.
Retinol.
Silver Nitrate.
Sodium Salicylate.
Sozoiodole-Potassium.
Sozoiodole-Sodium.

Varicella.—See Chicken Pox.

Varicosis.—See also, Hemorrhoids, Ulcers.

Arsen-hemol.
Bandaging.
Barium Chloride.
Digitalis.
Ergotin.
Glycerinophosphates.
Hamamelis.
Hemo-gallol.
Ichthalbin: internally.
Ichthyol: topically.
Phytolacca.

Variola (Small-Pox).

Acid, Carbolic, and Sweet Oil.
Acid, Salicylic.
Acid, Sulphurous.
Aconite.
Adeps Lanæ.
Ammonium Carbonate.
Antipyrine.
Belladonna.
Brandy and Whiskey.
Bromides.

Camphor.
Chloral Hydrate.
Cocaine.
Collodion.
Cimicifuga.
Ether.
Flexible Collodion, Glycerite of Starch, or Simple Cerate: locally applied.
Ichthyol: to prevent pitting.
Iodine.
Iodole.
Iron.
Mercury: to prevent pitting.
Opium.
Oil Eucalyptus.
Potassium Permanganate.
Quinine.
Silver Nitrate.
Sodium Benzoate.
Sulphocarbolates.
Traumaticin.
Triphenin.
Turpentine Oil.
Zinc Carbonate.
Zinc Oxide.

Vegetations.—See also, Tumors.

Acid, Chromic.
Acid, Carbolic.
Caustics: in general.
Potassium Bichromate.
Sozoiodole-Zinc.

Venereal Diseases.—See Gonorrhea, Syphilis, etc.

Vertigo.

Alkalies.
Amyl Nitrite.
Bromalin.
Bromipin.
Bromo-hemol.
Digitalis.
Erythrol Tetranitrate.
Glycerinophosphates.
Gold.
Hemo-gallol.
Iron Citrate.
Mercury Bichloride.
Nitroglycerin.
Potassium Bromide.
Quinine.
Strychnine.

Vomiting.—See list of Anti-emetics; also Cholera, Hematemesis, Nausea, Sea-Sickness, Vomiting of Pregnancy.

Acetanilid.
Acid, Carbolic: in irritable stomach along with bismuth; alone if due to sarcinæ or other ferments; in Asiatic cholera and cholera infantum.

Acid, Hydrochloric.
Acid, Hydrocyanic: in cerebral vomiting, vomiting of phthisis and of acute disease of the stomach.
Acid, Sulphurous: if due to sarcinæ.
Acids: in acid eructations; given immediately after food.
Aconite with Bismuth.
Alcohol: iced champagne, in sea-sickness, etc. Hot brandy is also useful.
Alkalies: especially effervescing drinks.
Alum: in doses of five to ten grn. in phthisis when vomiting is brought on by cough.
Ammonium Carbonate.
Ammonio-Citrate of Iron: in the vomiting of anemia, especially of young women.
Amyl Nitrite.
Apomorphine: to empty the stomach of its contents.
Arsenic: in the vomiting of cholera; in chronic gastric catarrh, especially of drunkards; chronic, not acute gastric ulcer; and chronic painless vomiting.
Atropine.
Bicarbonate of Sodium: in children half to one dram to the pint of milk. If this fails, stop milk. Also, in acute indigestion with acid vomiting.
Bismuth Subnitrate: in acute and chronic catarrh of the stomach or intestine.
Bismuth Subgallate.
Blisters: in vomiting due to renal and hepatic colic.
Brandy.
Bromides: in cerebral vomiting and cholera infantum.
Calcium Phosphate.
Calomel: in minute doses in cholera infantum and similar intestinal troubles.
Columba: a simple bitter and gastric sedative.
Carbonic Acid Waters: with milk.
Cerium Oxalate: in doses of 1 grn. in sympathetic vomiting.
Chloral Hydrate: in sea-sickness and reflex vomiting.

181

Chloroform: drop doses in sea-sickness, and in reflex vomiting such as on passage of calculi.

Cocaine.

Cocculus Indicus.

Codeine.

Creosote (Beech-wood).

Electricity: in nervous vomiting; the constant current positive pole on last cervical vertebra, and negative over stomach.

Emetics: if due to irritating substances.

Enema of Laudanum and Bromide of Sodium.

Erythrol Tetranitrate.

Ether: like chloroform.

Eucalyptus: in vomiting due to sarcinæ.

Faradism.

Gelatin: to the food of babies who suffer from chronic vomiting of lumps of curded milk.

Horseradish.

Ice: sucked.

Ice Bag: to spine or epigastrium.

Iodine: compound solut. in 3-to-5-minim doses.

Iodine and Carbolic Acid.

Ipecacuanha; in sympathetic nervous vomiting in very small doses; in the vomiting of children from catarrh and the vomiting of drunkards.

Iris.

Kumyss: in obstinate cases.

Leeches: to epigastrium if tender, especially in malarial vomiting.

Lime Water: with milk in chronic vomiting, especially in the case of children. Saccharated lime is laxative.

Magnesia: in sympathetic vomiting.

Magnesium Carbonate.

Menthol.

Mercury: in vomiting with clayey stools; see Calomel.

Methyl Chloride: spray to spine.

Morphine: hypodermically injected in the epigastrium in persistent seasickness.

Mustard Plaster: over stomach.

Nitrite of Amyl: in concentrated form in sea-sickness.

Nitroglycerin: like nitrite of amyl.

Nutrient Enemata: in persistent vomiting.

Nux Vomica: in atonic dyspepsia.

Oil Cloves.

Opium: as a suppository in severe acute vomiting, especially associated with obstinate constipation, which is relieved at the same time.

Orexine Tannate: a specific when simple, asthenic, or anemic anorexia the cause. Also, in incipient or chronic phthisis.

Oxygen Water.

Pepsin: in the vomiting of dyspepsia.

Peptonized Milk.

Podophyllin.

Potassium Iodide: in very small doses.

Potassium Nitrate.

Pulsatilla: in catarrh.

Quinine: in sympathetic vomiting.

Rectal Medication: if vomiting is uncontrollable.

Resorcin.

Seidlitz Powder.

Silver Nitrate: in nervous derangement.

Sodium Bicarbonate.

Sodium Bisulphite.

Sodium Sulphite.

Strychnine.

Veratrum: in vomiting of summer diarrhea.

Zinc Sulphate: emetic.

Vomiting of Pregnancy.

Acid, Carbolic: an uncertain remedy.

Acid, Hydrocyanic: sometimes useful; often fails.

Aconite: in full doses, so long as physiological effect is maintained.

Arsenic: where the vomit is blood, or streaked with blood, drop doses of Fowler's solution.

Atropine.

Belladonna: either internally, or plaster over the hypogastrium.

Berberine.

Berberine Carbonate.

Bismuth: along with pepsin.

Bromalin.

Bromide of Potassium: controls in some cases in large doses.

Bromo-hemol.

Calcium Phosphate.

Calomel: in small doses to salivate, or one large dose of 10 grn.

Calumba: occasionally successful.

Caustics: to the cervix if abraded.

Cerium Oxalate: the chief remedy besides orexine tannate.

Champagne.

Chloral.

Chloroform Water.

Cocaine: ten minims of a 3 per cent. solution will relieve in a few doses.

Coffee: before rising.

Copper Sulphate.

Creosote.

Dilatation of the Os Uteri.

Electricity: same as in nervous vomiting.

Hydrastine Hydrochlorate.

Ingluvin.

Iodine: a drop of the tincture or liquor sometimes a last resort.

Ipecacuanha: in minim doses often relieves.

Kumyss: as diet.

Menthol.

Methyl Chloride: spray to spine.

Morphine: suppository introduced into the vagina: no abrasion should be present, or there may be symptoms of poisoning.

Naphta: one or two drops.

Nux Vomica: one and one-half drop doses of tincture.

Orexine Tannate: extremely efficacious and prompt, after few doses, except where actual gastric lesion.

Pepsin: like ingluvin but not so successful.

Plumbic Acetate: in extreme cases.

Potassium Iodide: like iodine.

Quinine: sometimes useful.

Salicin.

Spinal Ice-bag.

Vulvitis. — See also, Pruritus, Prurigo, Vaginitis.

Acid, Carbolic.

Alum.

Arsenic.

Ichthyol.

Lead Acetate.

Naphtol.

Sodium Thiosulphate.

Sozoiodole-Sodium.

Warts.—*See also, Condylomata.*

Acid, Acetic: touch with the glacial acid.
Acid, Arsenous.
Acid, Carbolic.
Acid, Chromic.
Acid, Nitric.
Acid, Phosphoric.
Acid, Salicylic: saturated solution in collodion, with extract of Indian hemp.
Acid, Tannic.
Acid, Trichloracetic.
Alkalies.
Alum: saturated solution in ether.
Alum, Burnt.
Antimonic Chloride.
Chloral Hydrate.
Copper Oleate.
Corrosive Sublimate.
Creosote.
Fowler's Solution: locally applied.
Ferric ChlorideTincture
Ichthyol.
Mercuric Nitrate.
Papain.
Permanganate of Potassium.
Potassæ Liquor.
Potassium Bichromate.
Poultice.
Rue.
Savine.
Silver Nitrate: in venereal warts, along with savine.
Sodium Ethylate.
Stavesacre.
Sulphur.
Zinc Sulphate.

Wasting Diseases.
See Emaciation.

Weakness, Senile.—*See also, Adynamia, etc.*

Glycerinophosphates.
Muira Puama.
Spermine.

Wen.

Extirpation.

Whites. — *See Leucorrhea, Cervical Catarrh, Endometritis, etc*

Whooping-Cough.—*See Pertussis.*

Worms.—*See also, Chyluria, Tape Worm; and list of Anthelmintics.*

Acid, Filicic
Acid, Picric
Acid, Santoninic.
Acid, Tannic

Aloes.
Alum.
Ammonium Chloride.
Ammonium Embelate.
Apocodeine.
Chloroform.
Creolin.
Eucalyptus.
Gaduol.
Ichthalbin: as tonic.
Iron.
Koussein.
Male Fern.
Myrtol.
Naphtalin.
Oil Turpentine.
Papain.
Pelletierine.
Petroleum.
Potassium Iodide.
Quinine.
Quassin: infusion enemas in thread worms.
Santonin.
Strontium Lactate.
Thymol.
Valerian.

Worms, Thread, (*Ascaris Vermicularis*).

Acid, Carbolic: solution, 2 grn. to the oz ; in doses of 1 dram ; or as enema.
Aconite: in the fever produced.
Aloes: enema.
Alum: injections.
Asafetida with Aloes.
Castor Oil.
Chloride of Ammonium: to prevent accumulation of intestinal mucus, which serves as nidus.
Common Salt: along with antimony, to remove catarrhal state of intestine; or alone as enema.
Ether: injection of solution of 15 minims in water.
Eucalyptol: injection.
Ferri Perchloridi, Tinct.: enema.
Lime Water: enema.
Mercurial Ointment: introduced into rectum relieves itching and is anthelmintic.
Oleum Cajuputi.
Ol. Terebinthinæ.
Quassia: enema ; or infusion by mouth.
Santonica.
Santonin.
Scammony: for threadworms in rectum.
Tannin: enema.
Tonics.
Vinegar: enema, diluted with twice its bulk of water.

Wounds.— *See also, Bed Sores, Gangrene, Hemorrhage, Inflammation, Pyemia, Surgical Fever, Ulcers ; also, list of Antiseptics.*

Acetanilid.
Acid, Boric.
Acid, Carbolic.
Acid, Chromic.
Acid, Nitric.
Acid, Salicylic.
Acid, Sulphurous.
Aconite.
Airol.
Alcohol: in pyrexia, as an antiseptic and astringent dressing; and very useful in contused wounds.
Aluminium Acetate.
Aluminium Chloride.
Ammonium Carbonate.
Anhydrous Dressings.
Aristol.
Balsam of Peru.
Benzoin.
Bismuth Oxyiodide.
Bismuth Subgallate.
Bismuth Subnitrate.
Blotting Paper: as lint, saturated with an antiseptic.
Borax.
Calamin.
Calcium Bisulphite: solution.
Calendula.
Carbolated Camphor.
Charcoal.
Chaulmoogra Oil.
Chloral Hydrate: antiseptic and analgesic.
Cinnamon Oil.
Collodion: to exclude air.
Conium.
Copper Sulphate.
Creolin.
Creosote.
Diaphtherin.
Eucalyptus.
Euphorin.
Europhen.
Formalbumin.
Formaldehyde.
Glycerin.
Hamamelis: on lint to restrain oozing.
Heat.
Hydrogen Peroxide.
Iodine.
Iodoform.
Iodoformogen.
Iodole.
Loretin.
Naftalan.
Nitrate of Silver: to destroy unhealthy granulations.
Nosophen.
Oakum.
Opium.

Orthoform : as local anodyne.
Petroleum.
Permanganate of Potassium.
Potassium Bichromate.
Potassium Chlorate.
Poultices.
Pyoktanin.
Quinine.
Salol.
Sodium Chloride : one-half per cent. solution.
Sodium Fluoride.
Sozoiodole-Potassium, -Sodium, and -Zinc.
Stearates.
Styptic Collodion : to prevent bedsores, etc.
Sugar.
Tannin.
Tannoform.
Thymol.
Tribromphenol.
Turkish Baths.
Turpentine Oil.
Xeroform.
Yeast : in h o s p i t a l phagedena.
Zinc Carbonate.

Zinc Oxide.
Zinc Sulphate.

Yellow Fever.—*See also, Remittent Fever.*

Acid, Carbolic : subcutaneously and by the stomach.
Acid, Nitrohydrochloric.
Acid, Salicylic.
Acid, Tannic.
Aconite.
Antipyrine.
Arsenic.
Belladonna.
Calomel.
Camphor.
Cantharides.
Capsicum.
Champagne : iced.
Chlorate of Potassium.
Chloroform.
Chlorodyne.
Cimicifuga.
Cocaine.
Diaphoretics (see list of).
Diuretics (see list of).

Duboisine.
Ergot : to restrain the hemorrhage.
Gelsemium.
Iodide of Potassium.
Ipecacuanha.
Lead Acetate.
Liquor Calcis.
Mercury.
Nitrate of Silver.
Nux Vomica.
Pilocarpine.
Potassium Acetate.
Quinine : in some cases good, in others harmful.
Salines.
Sodium Benzoate : by subcutaneous injection.
Sodium Salicylate.
Stimulants.
Sulphur Baths.
Sulphurous-Acid Baths.
Tartar Emetic.
Triphenin.
Turpentine Oil : f o r vomiting.
Vegetable Charcoal.
Veratrum Viride.

PART III—CLASSIFICATION OF MEDICAMENTS

ACCORDING TO THEIR PHYSIOLOGIC ACTIONS.

Alteratives.
Acid, Arsenous.
Acid, Hydriodic.
Acid, Perosmic.
Ammonium Benzoate.
Ammonium Chloride.
Antimony salts.
Antimonauro.
Arsenauro.
Arsenic and Mercury
Iodide Solution.
Arsen-hemol.
Arsenites; and Arsenates.
Calcium Chloride.
Calcium Hippurate.
Chrysarobin.
Colchicum or Colchicine.
Copper salts.
Cupro-hemol.
Ethyl Iodide.
Firwein.
Gaduol.
Glycerin Tonic Comp.
Gold salts.
Guaiac.
Ichthalbin.
Iodia.
Iodides.
Iodipin.
Iodo-bromide Calcium
Comp.
Iodoform.
Iodoformogen.
Iodo-hemol.
Iodole.
Levico Water.
Manganese Dioxide.
Mercauro.
Mercurials.
Potassium Bichromate.
Potassium Chlorate.
Potassa, Sulphurated.
Pulsatilla.
Sanguinaria.
Silver salts.
Sozoiodole-Mercury.
Stillingia.
Sulphur.
Thiocol.
Thyraden.
Xanthoxylum.
Zinc salts.

Analgesics.—See Anodynes, General.

Anaphrodisiacs.
Belladonna.
Bromalin.
Bromides.
Bromipin.
Camphor.
Cocaine.
Conium.
Coniine Hydrobrom.
Digitalis.
Gelseminine.
Gelsemium.
Hyoscine Hydrobrom
Hyoscyamus.
Iodides.
Opium.
Purgatives.
Stramonium.

Anesthetics, General.—See also,
Anodynes, General.
Chloroform.
Ether.
Ethyl Bromide.
Nitrous Oxide.

Anesthetics, Local.
—See also, Anodynes, Local.
Camphor, Carbolated.
Camphor, Naphtolated.
Cocaine.
Creosote.
Ether Spray.
Ethyl Chloride Spray.
Eucaine.
Eugenol.
Erythrophleine Hydrochlorate.
Ethyl Chloride.
Guaiacol.
Guethol.
Holocaine.
Menthol.
Methyl Chloride.
Orthoform.
Tropacocaine.

Anodynes, General.
Acetanilid.
Acid, Di-iodo-salicylic.
Acid, Salicylic; and
Salicylates.
Aconitine.
Ammonol.
Antikamnia.
Antipyrine.
Asaprol.
Atropine.
Bromides.
Butyl-chloral Hydrate.
Caffeine.
Camphor, Monobrom.
Chloroform.
Codeine.
Colchi-sal.
Dioviburnia.
Euphorin.
Gelseminine.
Kryofine.
Lactophenin.
Methylene Blue
Morphine salts
Narceine.
Neurodin.
Neurosine.
Oil Gaultheria.
Papine.
Peronin.
Phenacetin.
Solanin.
Svapnia.
Thermodin.
Tongaline.
Triphenin.

Anodynes, Local.—
See also, Anesthetics.
Acid, Carbolic.
Aconite: tincture.

Aconitine.
Ammonia Water.
Atropine.
Belladonna.
Chloroform.
Chloral Hydrate.
Ichthyol.
Naftalan.
Oil Hyoscyamus.
Pyoktanin.

Antacids or Alkalines.
Calcium Carbonate.
Calcium Saccharate.
Lime Water.
Lithium Carbonate.
Magnesia.
Magnesium Carbonate.
Potassium Bicarbonate.
Potassium Hydrate.
Potassium Carbonate.
Sodium Bicarbonate.
Sodium Carbonate.
Sodium Hydrate.

Anthelmintics.
Acid, Filicic.
Acid, Tannic.
Alum.
Ammonium Embelate
Arecoline Hydrobromate.
Aspidium.
Chenopodium.
Chloroform.
Creolin.
Creosote.
Eucalyptol.
Koussein.
Naphtalin.
Oil Turpentine.
Oleoresin Male Fern.
Pelletierine Tannate.
Pumpkin Seed.
Quassia Infusion.
Resorcin.
Santonin (with calomel)
Sodium Santoninate.
Spigelia.
Thymol.

Anti-emetics.
Acid, Hydrocyanic.
Bismuth Subcarbonate.
Bismuth Subgallate.
Bismuth Subnitrate.
Bromalin.
Bromides.
Carbonated Water.
Cerium Oxalate.
Chloral Hydrate.
Chloroform.
Codeine.
Creosote.
Ether.
Ichthalbin.
Menthol.
Orexine Tannate.
Strontium Bromide.

Antigalactagogues.

Agaricin.
Belladonna.
Camphor: topically.
Conium.
Ergot.
Iodides.
Saline Purgatives.

Antigonorrhoics (or Antiblennorrhagics).

Acid, Tannic.
Airol.
Alum.
Alumnol.
Argentamine.
Argonin.
Aristol.
Bismuth Subgallate.
Bismuth Oxyiodide.
Copaiba.
Creolin.
Cubebs.
Europhen.
Hydrastine Hydrochlor.
Ichthyol.
Largin.
Potassium Permangan.
Protargol.
Pyoktanin.
Salol.
Silver Citrate.
Silver Nitrate.
Sozoiodole-Sodium.
Thalline Sulphate.
Zinc salts.

Antihidrotics.

Acid, Agaricic.
Acid, Camphoric.
Acid, Carbolic.
Acid, Gallic.
Acid, Tannic.
Agaricin.
Atropine.
Cocaine Hydrochlorate.
Duboisine Sulphate.
Lead Acetate.
Muscarine Nitrate.
Picrotoxin.
Pilocarpine Hydrochlor.
Potassium Tellurate.
Quinine.
Salicin.
Sodium Tellurate.
Thallium Acetate.

Antilithics.

Acid, Benzoic; and Benzoates.
Ammonium Benzoate.
Calcium Hippurate.
Colchi-sal.
Formin.
Lithium salts.
Lysidine.
Lycetol.
Magnesium Citrate.
Magnesium Oxide.
Piperazine.
Potassium Bicarbonate.
Potassium Carbonate.
Potassium Citrate.
Saliformin.

Sodium Bicarbonate.
Sodium Phosphate.
Sodium Pyrophosphate
Sodium Salicylate.
Uricedin.

Antiparasitics.—See Parasiticides.

Antiperiodics.

Acid, Arsenous; and Arsenites.
Acid, Picric.
Acid, Salicylic; and Salicylates.
Ammonium Fluoride.
Ammonium Picrate.
Arsen-hemol.
Berberine Carbonate
Cinchona; and alkaloids of.
Eucalyptol.
Euquinine.
Guaiaquin.
Levico Water.
Methylene Blue.
Piperine.
Quinidine.
Quinine.
Quinoidine.
Salicin.

Antiphlogistics.—See also, Antipyretics.

Acid, Tannic.
Aconite: tincture.
Antimony and Potassium Tartrate.
Digitoxin.
Gelsemium.
Ichthalbin: internally.
Ichthyol.
Lead salts.
Mercury.
Naftalan.
Opium.
Resinol.
Unguentine.

Antipyretics.

Acetanilid.
Acetylphenylhydrazine.
Acid, Benzoic.
Acid, Carbolic.
Acid, Di-iodo-salicylic.
Acid, Salicylic.
Aconite: tincture.
Ammonium Acetate: solution.
Ammonium Benzoate.
Ammonium Picrate.
Ammonol.
Antikamnia.
Asaprol.
Benzanilide.
Cinchonidine.
Cinchonine; and salts.
Colchicine.
Creosote.
Euphorin.
Euquinine.
Guaiacol.
Kryofine.

Lactophenin.
Methyl Salicylate.
Neurodin.
Phenacetin.
Phenocoll Hydrochlor.
Quinidine.
Quinine and salts.
Quinoline Tartrate.
Resorcin.
Salicin.
Salicylates.
Salol.
Sodium Paracresotate.
Thalline.
Thalline Sulphate.
Thermodin.
Thymol.
Triphenin.
Veratrum Viride: tr.

Antiseptics.—See also, Disinfectants.

Acetanilid.
Acid, Benzoic; and Benzoates.
Acid, Boric; and Borates.
Acid, Carbolic.
Acid, Oxy-Naphtoic, Alpha.
Acid, Paracresotic.
Acid, Picric.
Airol.
Ammonium Benzoate.
Antinosin.
Anthrarobin.
Aristol.
Asaprol.
Aseptol.
Betol.
Bismal.
Bismuth Benzoate.
Bismuth Naphtolate.
Bismuth Oxyiodide.
Bismuth Salicylate.
Bismuth Subgallate.
Boro-fluorine.
Borolyptol.
Cadmium Iodide.
Calcium Bisulphite.
Chlorine Water.
Creolin.
Creosote.
Eucalyptol.
Eudoxine.
Eugenol.
Euphorin.
Europhen.
Formaldehyde.
Galianol.
Gallobromol.
Glycozone.
Hydrogen Peroxide.
Hydrozone.
Ichthyol.
Iodoform.
Iodoformogen.
Iodole.
Largin.
Listerine.
Loretin.
Losophan.
Magnesium Salicylate.
Magnesium Sulphite.

186

Menthol.
Mercury Benzoate.
Mercury Bichloride.
Mercury Chloride.
Mercury Cyanide.
Mercury Oxycyanide.
Naftalan.
Naphtalin.
Naphtol.
Naphtol Benzoate.
Nosophen.
Oil Cade.
Oil Eucalyptus.
Oil Gaultheria.
Oil Pinus Pumilio.
Oil Pinus Sylvestris.
Oil Turpentine.
Paraformaldehyde.
Potassium Chlorate.
Potassium Permangan.
Potassium Sulphide.
Protonuclein.
Pyoktanin.
Pyridine.
Quinine.
Resorcin.
Retinol.
Salol.
Silver Citrate.
Silver Nitrate.
Sodium Biborate.
Sodium Bisulphite.
Sodium Borate, Neutral.
Sodium Carbolate.
Sodium Fluoride.
Sodium Formate.
Sodium Paracresotate.
Sodium Salicylate.
Sodium Sulphocarbol.
Sodium Thiosulphate.
Sozoiodole salts.
Styrone.
Tannoform.
Terebene.
Terpinol.
Thalline Sulphate.
Thiosinamine.
Thymol.
Tribromphenol.
Vitogen.
Xeroform.
Zinc Carbolate.
Zinc Permanganate.
Zinc Sulphocarbolate.

Antisialagogues.

Atropine..
Belladonna.
Cocaine Hydrochlorate.
Myrrh.
Opium.
Potassium Chlorate.
Sodium Borate.

Antispasmodics.

Acid, Camphoric.
Aconite: tincture.
Ammoniac.
Ammonium Valerian.
Amylene Hydrate.
Amyl Nitrite.
Anemonin.
Antispasmin.

Asafetida.
Atropine.
Benzene.
Bromoform.
Bismuth Valerianate.
Bitter-Almond Water.
Bromalin.
Bromides.
Bromoform.
Camphor.
Camphor, Monobrom.
Cherry-Laurel Water.
Chloral Hydrate.
Chloroform.
Conine Hydrobromate.
Curare.
Dioviburnia.
Eserine.
Ether.
Ethyl Bromide.
Ethyl Iodide.
Hyoscine Hydrobrom.
Hyoscyamus.
Lactucarium.
Lobelia.
Lupulin.
Morphine.
Musk.
Nitrites.
Nitroglycerin.
Opium.
Paraldehyde.
Potassium Iodide.
Pulsatilla: tincture.
Stramonium.
Sulfonal.
Urethane.
Zinc Valerianate..

Antituberculars.

Acid, Cinnamic.
Acid, Gynocardic.
Antituberculous Serum
Cantharidin.
Creosote and salts.
Eugenol.
Gaduol.
Guaiacol and salts.
Ichthalbin.
Iodoform or Iodoform-
  ogen: topically.
Iodole.
Methylene Blue.
Oil Chaulmoogra.
Oil Cod-Liver.
Potassium Canthari-
  date: subcutaneously.
Sodium Cinnamate.
Sodium Formate: sub-
  cutaneously.
Spermine.
Thiocol.

Antizymotics. — *See*
  Antiseptics and Dis-
  infectants.

Aperients. — *See Ca-*
  thartics.

Aphrodisiacs.—

Cantharides.
Damiana.
Gaduol.
Glycerinophosphates.
Gold.
Muira Puama: fl. ext.
Nux Vomica.
Phosphorus.
Spermine.
Strychnine.

Astringents.

Acid, Chromic.
Acid, Gallic.
Acid, Lactic.
Acid, Tannic.
Acid, Trichloracetic.
Alum, Burnt.
Aluminium Acetate:
  solution.
Aluminium Acetotart.
Aluminium Chloride.
Aluminium Sulphate.
Alumnol.
Baptisin.
Bismuth Subgallate, and
  other bismuth salts.
Cadmium Acetate.
Cadmium Sulphate.
Copper Acetate,
Copper Sulphate.
Eudoxine.
Ferropyrine.
Gallobromol.
Hydrastine Hydrochlor.
Hydrastis (Lloyd's).
Ichthyol.
Iron Sulphate, and
  other iron salts.
Lead Acetate, and other
  lead salts.
Potassium Bichromate.
Resinol.
Silver Citrate.
Silver Nitrate.
Sozoiodole-Sodium.
Sozoiodole-Zinc.
Tannoform.
Unguentine.
Xeroform.
Zinc Acetate.
Zinc Sulphate.

Astringents, Intestinal.

Acid, Agaricic.
Acid, Lactic.
Bismal.
Bismuth Naphtolate.
Bismuth Subgallate, and
  other bismuth salts.
Blackberry.
Bursa Pastoris.
Catechu.
Eudoxine.
Geranium.
Hematoxylon.
Kino.
Krameria.
Lead Acetate.
Monesia.
Silver Nitrate.

Tannalbin.
Tannigen.
Tannopine.
Xeroform.

Cardiac Sedatives.

Acid, Hydrocyanic.
Aconite.
Antimony preparations.
Chloroform.
Digitalis.
Gelsemium.
Muscarine.
Pilocarpine.
Potassium salts.
Veratrine.
Veratrum Viride.

Cardiac Stimulants.

Adonidin.
Adonis Vernalis.
Ammonia.
Ammonium Carbonate.
Anhalonine Hydrochlorate.
Atropine.
Cactus Grandiflorus.
Caffeine.
Convallaria.
Convallarin.
Digitalin.
Digitalis.
Digitoxin.
Erythrol Tetranitrate.
Ether.
Nerium Oleander: tr.
Nitroglycerin.
Oxygen.
Sparteine Sulphate.
Strophanthin.
Strophanthus.
Strychnine.

Carminatives.

Anise.
Calumba.
Capsicum.
Cardamom.
Caraway.
Cascarilla.
Chamomile.
Cinchona.
Chirata.
Cinnamon.
Cloves.
Gentian.
Ginger.
Nutmeg.
Nux Vomica.
Oil Cajuput.
Oil Mustard.
Orange Peel.
Orexine Tannate.
Pepper.
Pimenta.
Quassia.
Sassafras.
Serpentaria.
Valido l.

Cathartics.

LAXATIVES:
Cascara Sagrada.
Figs.

Glycerin.
Magnesium Oxide.
Manna.
Mannit.
Melachol.
Oil Olive.
Sulphur.

SIMPLE PURGATIVES:

Aloes.
Calomel.
Oil Castor.
Rhubarb.
Senna.

SALINE PURGATIVES:

Magnesium Citrate.
Magnesium Sulphate.
Potassium Bitartrate,
Potassium Tartrate.
Potassium and Sodium Tartrate.
Sodium Phosphate.
Sodium Pyrophosphate.
Sodium Sulphate.
Sodium Tartrate.

DRASTIC CATHARTICS:

Acid, Cathartinic.
Baptisin.
Colocynth.
Colocynthin.
Elaterin.
Elaterium.
Euonymin.
Gamboge.
Jalap.
Jalapin.
Oil, Croton.
Podophyllin.
Podophyllotoxin.
Podophyllum.
Scammony.

HYDRAGOGUES:

Drastic Cathartics in large doses.
Saline Purgatives.

CHOLAGOGUES:

Aloin.
Euonymin.
Iridin.
Leptandrin.
Mercurials.
Ox-Gall.
Podophyllum.

Caustics. — *See Escharotics.*

Cerebral Depressants. — *See also, Narcotics.*

Anesthetics, general.
Antispasmodics: several.
Hypnotics.
Narcotics.

Cerebral Stimulants.

Alcohol.
Amyl Nitrite.
Atropine.

Belladonna.
Caffeine.
Cannabis.
Coca.
Cocaine.
Coffee.
Erythrol Tetranitrate.
Ether.
Kola.
Nicotine.
Nitroglycerin.
Strychnine.

Cholagogues. — *See Cathartics; also, Stimulants, Hepatic.*

Cicatrizants. — *See Antiseptics.*

Constructives. — *See Tonics.*

Counter-Irritants.— *See Irritants.*

Demulcents.

Acacia.
Albumen.
Althea.
Cetraria.
Chondrus.
Elm.
Flaxseed.
Gelatin.
Glycerin.
Oil Olives.
Salep.
Starch.

Deodorants. — *See also, Disinfectants.*

Acid, Carbolic.
Ammonium Persulph.
Calcium Permanganate.
Chlorine Water.
Creolin.
Formaldehyde.
Hydrogen Peroxide.
Hydrozone.
Iron Sulphate.
Listerine.
Potassium Permangan.
Tannoform.
Vitogen.
Zinc Chloride.

Deoxidizers (*Reducing Agents or Reactives*).

Acid, Pyrogallic.
Anthrarobin.
Chrysarobin.
Eugallol.
Euro bin.
Eure sol.
Ichthyol.
Lenigallol.
Lenirobin.
Resorcin.
Saligallol.

Depilatories.

Barium Sulphide.
Calcium Oxide.
Calcium Sulphydrate.
Cautery.
Iodine.
Sodium Ethylate.
Sodium Sulphide.

Depressants, various. — *See Cerebral, Hepatic, Motor, Respiratory.* — *Also, Cardiac Sedatives.*

Diaphoretics and Sudorifics.

Acid, Salicylic; and Salicylates.
Aconite.
Alcohol.
Ammonium Acetate.
Camphor.
Cocaine.
Dover's Powder.
Ether.
Guaiac.
Oil of Turpentine.
Opium.
PilocarpineHydrochlor.
Potassium Citrate.
Potassium Nitrate.
Sodium Nitrate.
Spirit Nitrous Ether.
Tongaline.
Veratrum Viride.

Digestives.

Acid, Hydrochloric.
Acid, Lactic.
Diastase of Malt.
Extract Malt.
Ingluvin.
Lactopeptine.
Maltzyme.
Orexine Tannate: indirectly by increasing peptic secretion and gastric peristalsis.
Pancreatin.
Papain.
Pepsin.
Peptenzyme.
Ptyalin.

Discutients.—*See Resolvents.*

Disinfectants. — *See also, Deodorants.*

Acid, Boric.
Acid, Carbolic.
Acid, Sulphurous.
Aluminium Chloride.
Ammon. Persulphate.
Aseptol.
Bensolyptus.
Borates.
Boro-fluorine.
Borolyptol.

Calcium Bisulphite.
Calcium Permangan.
Chlorine Water.
Creolin.
Eucalyptol.
Formaldehyde.
Glyco-thymoline.
Glycozone.
Hydrogen Peroxide.
Hydrozone.
Iron Sulphate.
Lime, Chlorinated.
Mercury Bichloride.
Naphtol.
Oil Eucalyptus.
Potassium Permangan.
Pyoktanin.
Sodium Naphtolate.
Solution Chlorinated Soda.
Sozoiodole salts.
Thymol.
Zinc Chloride.

Diuretics.

Adonidin.
Adonis Vernalis.
Ammonium Acetate.
Apocynum.
Arbutin.
Atropine.
Belladonna.
Cactus Grandiflorus.
Caffeine.
Cantharides.
Chian Turpentine.
Colchicine.
Convallamarin.
Copaiba.
Cubebs.
Digitalis preparations.
Digitoxin.
Formin.
Juniper.
Kava Kava.
Lithium salts.
Lycetol.
Lysidine.
Matico.
Nitrites.
Oil Juniper.
Oil Santal.
Oil Turpentine.
PilocarpineHydrochlor.
Piperazine.
Potassium Acetate.
Potassium Bitartrate.
Potassium Citrate.
Potassium Nitrate.
Saliformin.
Scoparin.
Sodium Acetate.
Sodium Nitrate.
Sparteine Sulphate.
Spirit Nitrous Ether.
Squill.
Strophanthus.
Theobromine.
Theobromine and Sodium Salicylate.
Tritipalm.
Uropherin.

Ecbolics.—*See Oxytocics.*

Emetics.

Alum.
Antimony Sulphide, Golden.
Antimony and Potassium Tartrate.
Apomorphine Hydrochlorate.
Copper Sulphate.
Emetine.
Ipecac.
Mercury Subsulphate.
Mustard, with tepid water.
Sanguinarine.
Saponin.
Zinc Sulphate.

Emmenagogues.

Acid, Oxalic.
Aloes.
Apiol.
Apioline.
Cantharides.
Ergot.
Guaiac.
Iron Chloride, and other salts of iron.
Manganese Dioxide.
Myrrh.
Pennyroyal.
Potassium Permanganate.
Pulsatilla: tincture.
Quinine.
Rue.
Savine.
Strychnine.
Tansy.

Errhines (*Sternutatories*).

Cubebs.
Sanguinarine.
Saponin.
Veratrine.
White Hellebore.

Escharotics (*Caustics*).

Acid, Acetic, Glacial.
Acid, Arsenous.
Acid, Carbolic.
Acid, Carbolic, Iodized.
Acid, Chromic.
Acid, Dichloracetic.
Acid, Lactic.
Acid, Nitric.
Acid, Trichloracetic.
Alum, Burnt.
Copper Sulphate.
Iodine.
Mercury Bichloride.
Potassa.
Silver Nitrate.
Soda.
Sodium Ethylate.
Zinc Chloride.
Zinc Sulphate.

Expectorants.

Acid, Benzoic.
Ammoniac.

Ammonium Carbonate.
Ammonium Chloride.
Ammonium Salicylate.
Antimony and Potassium Tartrate.
Antimony salts in general.
Apocodeine Hydrochlorate.
Apomorphine Hydrochlorate.
Balsam Peru.
Balsam Tolu.
Benzoates.
Cetrarin.
Emetine, in small doses.
Glycyrrhizin, Ammoniated.
Grindelia.
Ipecac.
Lobelia.
Oil Pinus Sylvestris.
Oil Santal.
Oil Turpentine.
PilocarpineHydrochlor.
Potassium Iodide.
Pyridine.
Sanguinarine.
Saponin.
Senegin.
Squill.
Tar.
Terebene.
Terpene Hydrate.
Terpinol.

Galactagogues.

Acid, Lactic.
Castor Oil: topically.
Extract Malt.
Galega.
Jaborandi.
PilocarpineHydrochlor.
Potassium Chlorate.

Gastric Tonics (Stomachics).

Alkalies: before meals.
Aromatics.
Berberine Carbonate.
Bismuth salts.
Bitters.
Carminatives.
Cetrarin.
Chamomilla Compound
Hydrastis.
Ichthalbin.
Nux Vomica.
Orexine Tannate.
Quassin.
Seng.
Strychnine.

Germicides.—See Antiseptics and Disinfectants.

Hematinics.—See also, Tonics.

Acid, Arsenous; and arsenical compounds.

Carnogen.
Cetrarin.
Ext. Bone-marrow.
Gaduol.
Globon.
Hemo-gallol.
Hemol.
Hemoglobin.
Ichthalbin.
Iron compounds.
Levico Water.
Manganese compounds.
Pepto-mangan.

Hemostatics.—See Styptics and Hemostatics.

Hepatic Depressants.

LESSENING BILE :
Alcohol.
Lead Acetate.
Purgatives: many of them.
Morphine.
Opium.
Quinine.

LESSENING UREA :
Colchicum.
Morphine.
Opium.
Quinine.

LESSENING GLYCOGEN:
Arsenic.
Antimony.
Codeine.
Morphine.
Opium.
Phosphorus.

Hepatic Stimulants.

Acid, Benzoic.
Acid, Nitric.
Acid, Nitrohydrochlor.
Aloes.
Ammonium Chloride.
Amyl Nitrite.
Antimony.
Arsenic.
Baptisin.
Benzoates.
Calomel.
Colocynth.
Euonymin.
Hydrastine Hydrochlorate.
Ipecac.
Iron.
Mercury Bichloride.
Podophyllin.
Potassium and Sodium Tartrate.
Resin Jalap.
Sanguinarine.
Sodium Bicarbonate.
Sodium Phosphate.
Sodium Pyrophosphate.
Sodium Salicylate.
Sodium Sulphate.

Hypnotics(Soporifics).

Amylene Hydrate.
Bromidia.
Cannabine Tannate.
Chloral Hydrate.
Chloral-Ammonia.
Chloralose.
Chloralamide.
Chloralimide.
Duboisine Sulphate.
Hyoscine Hydrobrom.
Hyoscyamine.
Morphine.
Narceine.
Paraldehyde.
Sulfonal.
Tetronal.
Trional.
Urethane.

Intestinal Astringents—See Astringents.

Irritants.

RUBEFACIENTS:
Acetone.
Ammonia.
Arnica.
Burgundy Pitch.
Canada Pitch.
Capsicum.
Chloroform.
Iodine.
Melissa Spirit
Menthol.
Mustard.
Oil Turpentine.
Oleoresin Capsicum.
Spirit Ants.
Volatile Oils.

PUSTULANTS :
Antimony and Potassium Tartrate.
Oil Croton.
Silver Nitrate.

VESICANTS :
Acid, Acetic, Glacial.
Cantharidin.
Chrysarobin.
Euphorbium.
Mezereon.
Oil Mustard.

Laxatives.—See Cathartics.

Motor Depressants.

Acid, Hydrocyanic.
Aconite.
Amyl Nitrite.
Amyl Valerianate.
Apomorphine Hydrochlorate.
Bromalin.
Bromides.
Bromoform.

Chloral Hydrate.
Chloroform (l a r g e doses).
Coniine Hydrobromate.
Curare.
Gelsemium.
Gold Bromide.
Lobelia.
Muscarine.
Nitrites.
Nitroglycerin.
Physostigmine.
Quinine: large doses.
Sparteine Sulphate.
Veratrum Viride.

Motor Excitants.

Alcohol.
Atropine.
Belladonna.
Brucine.
Camphor.
Chloroform.
Convallarin.
Ignatia.
Nux Vomica.
Nicotine.
Picrotoxin.
Pilocarpine Hydrochlorate.
Pyridine.
Rhus Toxicodendron.
Strychnine.

Mydriatics.

Atropine.
Cocaine.
Daturine.
Duboisine Sulphate.
Gelseminine.
Homatropine Hydrobromate.
Hyoscine Hydrobromate.
Hyoscyamine.
Muscarine.
Mydrine.
Scopolamine Hydrobromate.

Myotics.

Arecoline Hydrobromate.
Eserine (Physostigmine).
Morphine.
Opium.
Muscarine Nitrate: internally.
Pilocarpine Hydrochlorate.

Narcotics.—*See also, Hypnotics.*

Chloroform.
Chloral Hydrate.
Conium.
Hyoscyamine.
Hypnotics.
Morphine.
Narceine.
Narcotine.
Opium.
Rhus Toxicodendron.
Stramonium.

Nervines.—*See Antispasmodics, A n odynes, Sedatives, Anesthetics, Motor Depressants, Motor Stimulants, Narcotics.*

Nutrients.— *See Hematinics and Tonics.*

Oxytocics (*Ecbolics*).

Acid, Salicylic.
Cimicifugin.
Cornutine.
Cotton-Root Bark.
Ergot.
Hydrastine.
Hydrastinine Hydrochlorate.
Pilocarpine Hydrochlorate.
Potassium Permanganate.
Quinine.
Rue.
Savine.
Sodium Borate.
Stypticin.

Parasiticides. — *S e e Antiseptics and Disinfectants.*

Ptyalagogues. — *See Sialogogues.*

Purgatives.— *See Cathartics.*

Pustulants.—*See Irritants.*

Refrigerants.

Acid, Citric.
Acid, Phosphoric, Dilute.
Acid, Tartaric.
Ammonium Acetate.
Magnesium Citrate.
Magnesium Sulphate.
Potassium Bitartrate.
Potassium Citrate.
Potassium Nitrate.
Potassium Tartrate.
Sodium Nitrate.
Sodium Tartrate.

Resolvents (*Discutients*).

Acid, Perosmic.
Arsenic.
Cadmium Iodide.
Gaduol.
Ichthalbin: internally.
Ichthyol: topically.
Iodides.
Iodine.
Iodipin.
Iodole.
Iodo-hemol.
Levico Water.
Mercurials.
Thiosinamine.

Respiratory Depressants.

Acid, Hydrocyanic.
Aconite.
Chloral.
Chloroform.
Conium.
Gelsemium.
Muscarine.
Nicotine.
Opium.
Physostigma.
Quinine.
Veratrum Viride.

Respiratory Stimulants.

Aspidosperma (Quebracho).
Aspidospermine.
Atropine.
Caffeine.
Cocaine.
Duboisine Sulphate.
Strychnine.

Restoratives. — *See Hematinics, Tonics.*

Rubefacients. — *See Irritants.*

Sedatives, Cardiac (or Vascular).— *See Cardiac Sedatives.*

Sedatives (Nerve).— *See also, Depressants.*

Acetanilid.
Acid, Hydrobromic.
Acid, Hydrocyanic.
Acid, Valerianic.
Allyl Tribromide.
Amylene Hydrate.
Amyl Nitrite.
Anemonin.
Antipyrine.
Antispasmin.
Bromalin.
Bromides.
Bromidia.
Bromipin.
Bromo-hemol.
Bromoform.
Butyl-Chloral.
Caesium and Ammonium Bromide.
Camphor.
Camphor, Monobrom.
Cannabine Tannate.
Celerina.
Chloral Hydrate.
Chloroform.
Cocaine.
Codeine.
Conium.
Duboisine Sulphate.
Eserine.
Ether.
Ethyl Bromide.
Ethylene Bromide.
Gallobromol.
Hyoscine Hydrobrom.

Hyoscyamine.
Hyoscyamus.
Lactucarium.
Lobelia.
Morphine.
Narceine.
Neurosine.
Paraldehyde.
Peronin.
Scopolamine Hydrobromate.
Solanin.
Stramonium : tincture.
Sulfonal.
Urethane.
Valerian, and Valerianates.
Validol.

Sialagogues (*Ptyalogogues*).

Acids and Alkalies.
Antimony compounds.
Capsicum.
Chloroform.
Eserine.
Ginger.
Iodine compounds.
Mercurials.
Mezereon.
Muscarine.
Mustard.
Pellitory.
Pilocarpine Hydrochlor.
Pyrethrum.

Soporifics.—*See Hypnotics.*

Spinal Stimulants.
—*See also, Motor Excitants.*

Alcohol.
Atropine.
Camphor : small doses.
Ignatia.
Nux Vomica.
Picrotoxin.
Strychnine.

Sternutatories.—*See Errhines.*

Stimulants, Bronchial.—*See Expectorants.*

Stimulants, Various.
—*See Gastric, Hepatic, Renal, Spinal, Vascular, etc.*

Stomachics.—*See Gastric Tonics.*

Styptics and Hemostatics.

Acid, Gallic.
Acid, Tannic.
Acid, Trichloracetic.
Alum.
Antipyrine.
Copper Sulphate.
Creolin.
Ferropyrine.
Hamamelis.
Hydrastinine Hydrochlorate.
Iron Subsulphate.
Iron Sulphate.
Iron Terchloride.
Lead Acetate.
Manganese Sulphate.
Oil Turpentine.
Silver Nitrate.
Stypticin.

Sudorifics.—*See Diaphoretics.*

Teniafuges.—*See Anthelmintics.*

Tonics, Cardiac.—*See Cardiac Stimulants.*

Tonics, General.—*See also, Hematinics.*

VEGETABLE TONICS :

Absinthin.
Baptisin.
Bitters.
Bebeerine.
Berberine Carbonate.
Cinchona alkaloids and salts.
Cod-Liver Oil.
Columbin.
Eucalyptus.
Gaduol.
Hydrastis.
Hydroleine.
Quassin.
Salicin.

MINERAL TONICS :

Acids, Mineral.
Acid, Arsenous; and its salts.
Acid, Hypophosphorous.
Acid, Lactic.
Bismuth salts.
Calcium Glycerinophosphate.
Cerium salts.
Copper salts : small doses.
Gold salts.
Glycerinophosphates.
Hemo-gallol.
Hemol.
Hypophosphites.
Ichthalbin.
Iron compounds.
Levico Water.
Manganese compounds.
Phosphorus.

Tonics, Nerve. — *See Nervousness, Neurasthenia, Neuritis, Opium Habit, in Part II.*

Vaso-Constrictors.

Ergot and its preparations.
Hydrastinine Hydrochlorate.
Hydrastine Hydrochlor.
Stypticin.

Vaso-Dilators.

Amyl Nitrite.
Ether.
Erythrol Tetranitrate.
Nicroglycerin.
Potassium Nitrite.
Sodium Nitrite.
Spirit Nitrous Ether.

Vascular Sedatives and Vascular Stimulants. — *See Cardiac Sedatives, and Cardiac Stimulants.*

Vermicides.—*See Anthelmintics.*

Vesicants.—*See Irritants.*

When in immediate need

of Drugs or Chemicals not at hand, any pharmacist is in a position to use our EMERGENCY DEPARTMENT, which is in operation every day in the year, Sundays and Holidays included, until 9 p. m. — Hurry orders reaching us after regular business hours will receive prompt attention, — *provided :*

1—that they come by WIRE ;

2—that they call for MERCK'S *chemicals or drugs* (no other brands being in stock with us) ;

3—and that the quantity and nature of the goods admit of their being sent through the MAILS.

As it is impossible for us to ascertain in each instance the identity of a Physician who might wish to make use of this department, we must insist (for the proper protection of the Profession against the unauthorized purchase of poisons, etc.; as well as in due recognition, by us, of the established usage in the traffic with medicines and drugs) that every such order be transmitted through an established Pharmacist ; and pharmacists, when telegraphing orders to us, should always mention their jobber to whom the article is to be charged.

We trust that this Department will prove of value in cases of emergency and immediate need.

MERCK & CO., New York.